D1562790

GAS AND LIQUID CHROMATOGRAPHY IN ANALYTICAL CHEMISTRY

GAS AND LIQUID CHROMATOGRAPHY IN ANALYTICAL CHEMISTRY

Roger M. Smith
Reader in Analytical Chemistry,
Department of Chemistry,
Loughborough University of Technology,
Loughborough, Leics., LE11 3TU, UK

JOHN WILEY & SONS
Chichester · New York · Brisbane · Toronto · Singapore

Library of Congress Cataloging-in-Publication Data:

Smith, Roger M.
Gas and liquid chromatography in analytical chemistry / Roger M. Smith.
p. cm.
Includes bibliographies and index.
ISBN 0 471 90980 7
1. Gas chromatography. 2. High performance liquid chromatography.
I. Title.
QD79.C45S65 1988
543'.0894—dc19 87-35095
CIP

British Library Cataloguing in Publication Data:

Smith, Roger M.
Gas and liquid chromatography in analytical chemistry.
1. Gas chromatography 2. Liquid chromatography
I. Title
543'.0896 QD79.C45

ISBN 0 471 90980 7

Typeset by Photo·graphics, Honiton, Devon
Printed in Great Britain by St Edmundsbury Press, Bury St Edmunds, Suffolk.

CONTENTS

PREFACE

Since the first report of chromatography by Tswett in 1906 and the pioneering work of Martin and Synge in 1941, chromatographic techniques have taken a dominant place in the methods available to the analytical chemist. They have become an indispensable component of many chemical industries for the assessment of the purity and identity of raw materials and products, of many aspects of health care and of the quality of the environment. However, along with many other analytical methods, these important instrumental techniques often constitute only a small part of the general education of many undergraduates in chemistry, but they will frequently become a major and essential part of their work in industry or research. Although there is already an extensive research literature, both journals and detailed and advanced monographs, covering all aspects of the applications and methods of chromatography, there are few modern texts for the newcomer that can act as an introduction to this field. The present book has come out of the author's involvement over the last 16 years in lecturing on chromatography as part of instrumental analysis to undergraduate and postgraduate students, and in particular as the organiser of analytical chemistry short courses, originally on high-performance liquid chromatography and gas–liquid chromatography, and now also on supercritical fluid chromatography, given annually at Loughborough University of Technology for analytical chemists from industry and government service. The book is aimed at technical staff in industry, senior undergraduates, postgraduate research students and new graduates, who find a need to use chromatographic methods in their work but have limited practical experience or background in the field and wish to go further than simply operating an instrument. The emphasis has been largely on the more traditional role of analytical chromatography in chemistry, with only references to its wide and growing application in biochemistry and biotechnology and as a preparative technique.

As well as introducing the major techniques of gas–liquid chromatography, thin-layer chromatography and high-performance liquid chromatography, the book also provides a framework that the reader can use to explore further topics by the inclusion of comprehensive bibliographies. The aim throughout has been to provide a practical viewpoint for the practising analyst. Many concepts that have aided the advance of the method but are no longer used and much of the theoretical background have been deliberately omitted or referred to only for reference. Because in each laboratory the samples will differ, no attempt has been made to provide a compilation of applications, but the Appendices list the main literature sources to provide a starting point for method determination.

The whole area of chromatography is still rapidly developing, and the author is conscious that in some areas this has been a selective view, but each week more than the equivalent of this book is added to the research literature of chromatography.

Loughborough 1988 Roger M. Smith

CHAPTER 1

INTRODUCTION

1.1 ANALYTICAL CHROMATOGRAPHY

The two principal problems facing an analytical chemist are the accurate and reproducible determination of very small quantities of an analyte and the need to be able to carry out determinations when the analyte of interest is present as a minor component in a complex matrix of potentially interfering substances. These factors have led to two general approaches. Firstly, the use of highly specific analytical methods that will only respond to the analyte of interest, ignoring all other components; and secondly, the employment of powerful separation techniques in which the components of the mixture are physically separated, so that each component can be determined individually using a relatively non-specific detector.

The first of these approaches is typified by the atomic absorption spectroscopic analysis of a specific element in a complex mixture, such as a metal alloy or crude ore sample. The selectivity of this combination of the hollow-cathode lamp and monochromator ensures that only the element of interest will respond. However, although this approach is applicable to the determination of different elements in an inorganic sample, the vast number of often closely related organic compounds contain only a limited range of elements. In this case, the individual components in a sample can only be discriminated by using a separation method before the measurement step.

The separation of a mixture of compounds can be carried out by many techniques, with differing degrees of efficiency and resolving power. These include simple filtration or crystallisation methods in which one compound is partly or largely separated as a solid from the rest of a liquid mixture.

For volatile compounds, greater separation can be achieved by distillation of the sample to give a series of fractions with different vapour pressures. Other useful techniques are based on the partitioning of a sample between two immiscible phases. In its simplest form, the phases are held in a separating funnel, the analyte being distributed largely into one phase by a suitable selection of the two phases.

If the two immiscible phases in a partitioning system are not held statically but move relative to one another through a tube or column, the analytes will be carried with the moving phase through the column. However, they will be partly retained depending on their interaction with the second static phase, and even small differences in the distribution coefficients will cause the components to be separated (Figure 1.1). These three characteristic features, namely two immiscible phases, the relative movement of the phases and the distribution of the analytes between the phases, are present in all chromatographic separation methods. By altering the nature of the phases or changing the experimental conditions, such as the temperature, the degree of separation and its selectivity can be altered. Once the components have been separated, they can be detected with either a universal or specific detector to give a chromatogram (Figure 1.2).

Although both phases could move through the column in opposite directions, this is experimentally difficult to achieve, and usually one phase (the stationary phase) is held static and the second phase (the mobile phase) travels past the stationary phase under gravity, pressure, or by capillary action. The stationary phase can be either a solid, or a liquid spread as a thin film on the walls of the column or over an inert support. The mobile phase can be either a gas, a liquid, or a supercritical fluid and is often termed the "eluent" if the analytes are carried (or eluted) out of the column system before detection. This basic concept has developed into a wide range of analytical chromatographic methods broadly based into two main groups depending on whether the mobile phase is a gas or a liquid (Table 1.1).

The distribution and retention of an analyte in a chromatographic separation may result from adsorption onto a solid stationary phase and/or a partition process between gas and/or liquid phases. The extent of the interaction of an analyte with the stationary phase is characteristic of the individual compound. Thus a compound will be retained to the same extent whether it is injected on its own as a pure compound or as a component of a complex mixture. Retention properties can therefore be used as a method for the identification of the components of a mixture by comparison with authentic standards. For most quantitative and analytical separations, an instrumental method of detection is used to increase reproducibility and sensitivity.

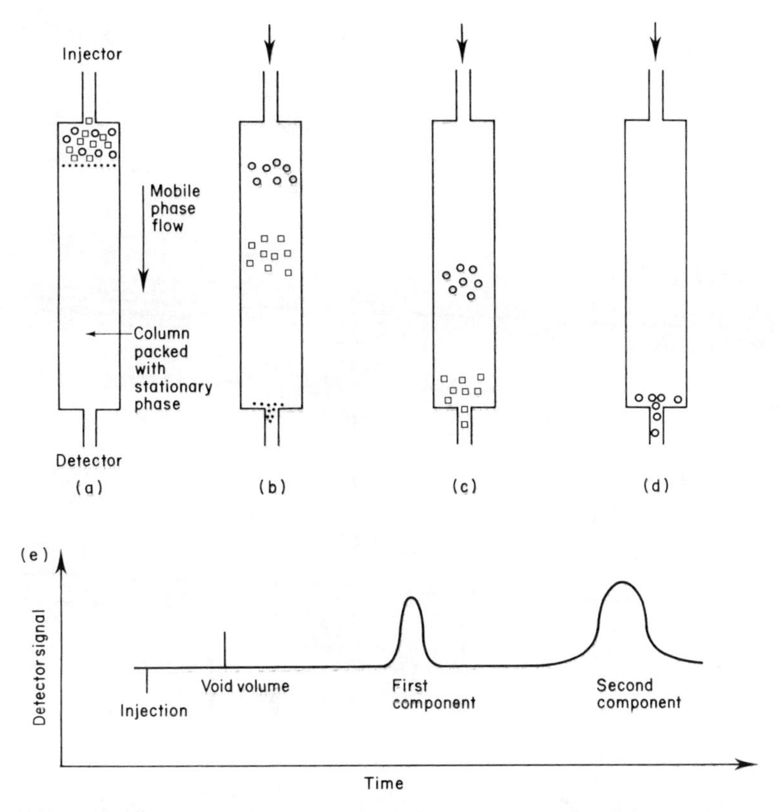

Figure 1.1 Chromatographic separation on a column. (a) Introduction of the sample. (b) Elution of unretained components at column void volume. (c) Elution of more weakly retained compound. (d) Elution of more strongly retained component. (e) Chromatogram recorded by detector at end of the column.

By using separation systems of different sizes the chromatographic technique can be used for preparative separations (kilograms down to grams) or to carry out sensitive analytical separations (down to 10^{-13} g). The large-scale methods have found widespread application in organic and industrial chemistry for the separation and purification of the products of synthetic reactions or the isolation of natural products from crude plant or fungal sources.

This book will concentrate on the uses of chromatographic methods, in particular gas–liquid chromatography (GLC), thin-layer chromatography (TLC) and high-performance liquid chromatography (HPLC), as analytical techniques for the identification and quantification of analytes. These

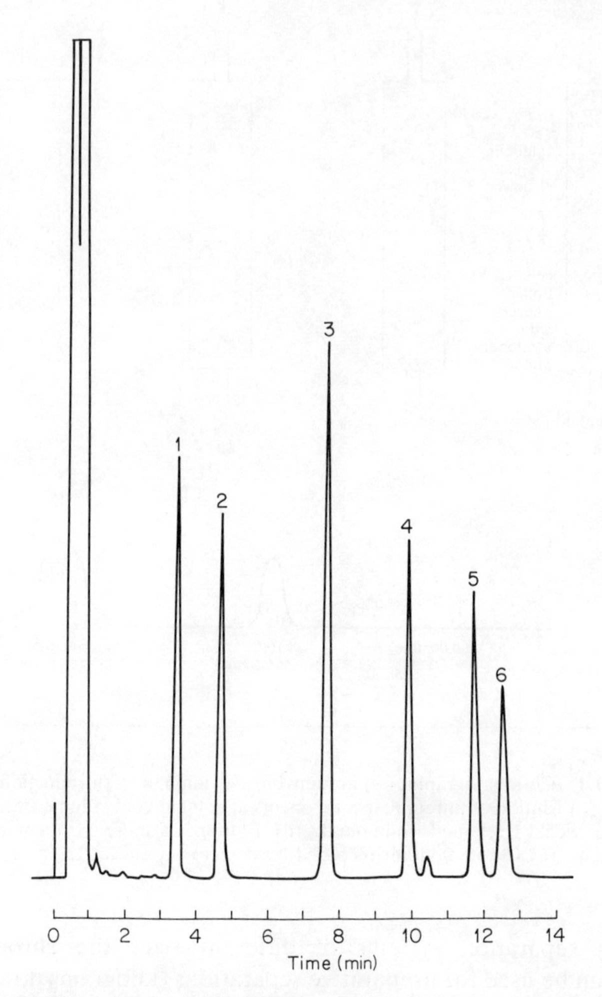

Figure 1.2 Chromatogram of a mixture of aromatic amines separated by gas chromatography using a column 1.5 m × 4 mm packed with 6% Carbowax 20M on KOH-treated 80–100 mesh Diatomite C. Temperature programmed from 100 to 190°C at 8°C min^{-1}. Nitrogen carrier gas at 40 ml min^{-1} with flame ionisation detection. Peaks: 1, n-hexylamine; 2, n-heptylamine; 3, *N*,*N*-dimethylaniline; 4, *N*-ethylaniline; 5, *o*-ethylaniline; 6, *p*-ethylaniline. (P.J. Ridgeon, *Gas chromatography separations*, 3rd ed., Pye Unicam, Cambridge, 1976, p. 15. Reproduced by permission of Philips Analytical.)

Table 1.1 Methods of chromatography.

Mobile phase	Stationary phase	Chromatographic method	Abbreviation
Gas	Liquid	Gas–liquid chromatography	GLC
	Bonded Liquid	Gas–liquid chromatography	GLC
	Solid	Gas–solid chromatography	GSC
Liquid	Liquid	Liquid–liquid chromatography	LLC
		Droplet counter-current chromatography	DCC
	Bonded liquid	High-performance liquid chromatography (reversed- phase)	HPLC
	Solid	High-performance liquid chromatography (normal- phase)	HPLC
		Thin-layer chromatography	TLC
Supercritical fluid	Bonded liquid	Supercritical fluid chromatography	SFC

methods have found widespread application in the food and pharmaceutical industries, and in health care, forensic science, environmental monitoring and other fields important in modern life. The term "chromatography" has also come to include a number of related separation techniques using similar equipment but based on different physical concepts. These include size exclusion chromatography (SEC) based on the molecular size differences of the analytes and ion-exchange chromatography (IEC) in which the separation depends on the interactions between ionised groups on the analytes and the stationary phase; these techniques will be briefly included as appropriate.

1.2 ORIGINS OF CHROMATOGRAPHY

Although there are claims that separations based on methods similar to chromatography were known in the Middle Ages, these were largely empirical without any deliberate intention to use the technique for chemical analysis. An early recorded separation was carried out about the end of the 19th century as part of an investigation by D.T. Day in the USA into the origin of different crude oils and their formation [1, 2]. He suggested that the differences were a result of the movement of the oil through the rocks of the Earth's crust. He simulated the effects on a small scale in the laboratory by passing an oil sample through a column of Fuller's earth. He showed that the early fractions from the column differed in their composition from later fractions. Although this appears to have been a crude form of chromatographic separation, he did not exploit his observations or provide a explanation of the phemomenon.

The honour of being the father of chromatography goes to a Russian botanist, Micheal Tswett (or Tsvett), who was working on the separation of plant pigments [3–6]. In 1906, he reported a separation method in which extracts of plants were placed onto the top of a short column of a finely powdered material, which was then washed through with a solvent to give different bands of the chlorophylls (Figure 1.3). He found that different column materials gave different separations and that the sharpness of his bands depended on the fineness of the material packed into the column. He coined the word "chromatography" from the Greek χρωμα (colour) γραφια (writing) to describe the new technique, even though he was aware that it could also be used for the separation of colourless compounds.

However, his results suggested that there were more components in the plant extracts than could be isolated by traditional chemical techniques and it was felt that his method must be causing degradation of the samples. Although his separation was later shown to be correct, the method was not adopted in other laboratories. No further work was carried out until 1931 when Kuhn, Winterstein and Lederer virtually reinvented the technique for the analysis of the carotenoids and xanthophylls [7, 8]. The method was again not widely applied although it retained some interest. A significant analytical development took place in 1941 when Brockmann published a method for the standardisation of the chromatographic activity of alumina as a stationary phase [9]. By first totally drying the alumina and then specifically deactivating it by the addition of controlled amounts of water, a reproducible retention activity could be obtained. This could be tested with standard dyestuffs. This process enabled reproducible results to be obtained in different laboratories, thereby satisfying an essential prerequisite for any analytical method to gain wide acceptance.

Up to this time, all the methods had employed a solid stationary phase and a liquid mobile phase. An important advance was made in 1941 by Martin and Synge, two biochemists working at the Wool Research Institute in Leeds, who were trying to improve the separation of amino acids. They first examined the use of a continuous liquid–liquid partitioning system in which the two solvents moved in opposite directions through a tube [10]. They rapidly realised that it was the relative rather than the absolute movement that was needed. In a paper that represents the foundation of modern liquid chromatography, they reported the concept of partition chromatography using a stationary liquid and a mobile liquid [11]. They described the processes that were taking place and derived a relationship

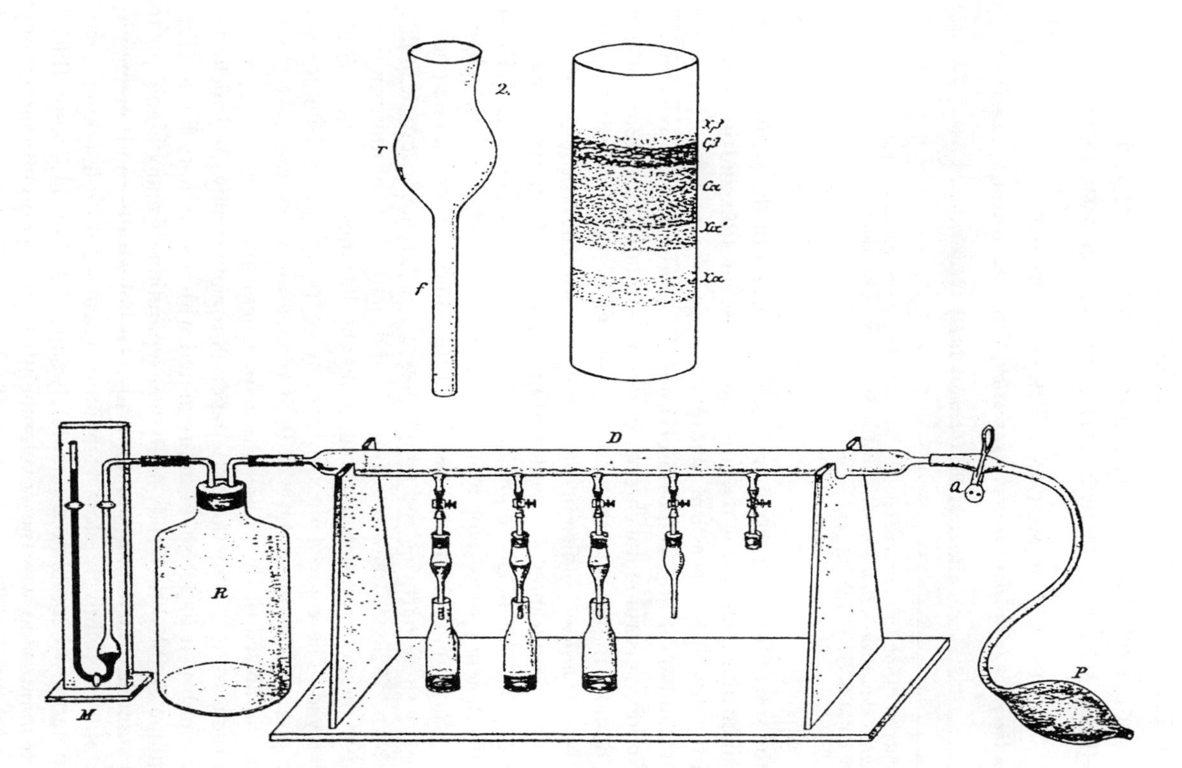

Figure 1.3 The original chromatographic apparatus constructed by M. Tswett and reported in 1906. Insert: typical separation of plant chlorophylls. (Reproduced with permission from *J. Chem. Educ.*, 1967, **44**, 238.)

151. A NEW FORM OF CHROMATOGRAM EMPLOYING TWO LIQUID PHASES

1. A THEORY OF CHROMATOGRAPHY

2. APPLICATION TO THE MICRO-DETERMINATION OF THE HIGHER MONOAMINO-ACIDS IN PROTEINS

By A. J. P. MARTIN and R. L. M. SYNGE

From the Wool Industries Research Association, Torridon, Headingley, Leeds

(*Received 19 November 1941*)

Figure 1.4 Benchmark paper by Martin and Synge, which established the foundations of modern chromatography in both theory and practice. (Reprinted by permission from *Biochem. J.*, 1941, **35,** 1358. Copyright © 1941 The Biochemical Society, London.)

between the separation and the distribution coefficient of the samples. They also explained the concept of the efficiency of separation and in doing so placed chromatography on a firm theoretical basis (Figure 1.4). This paper included a number of important observations and provides the benchmark paper for modern instrumental chromatography. Although they were working with two liquids, they noted that "the mobile phase need not be a liquid but may be a vapour", thus predicting their own later development of gas–liquid chromatography. In studies to increase the separation efficiency, they showed that the height equivalent to a theoretical plate (HETP), a measure of efficiency, was dependent on the linear flow rate of the eluent through the column and on the diffusivity of the sample in the mobile phase, and that "very small particles and a high pressure difference" were desirable to obtain high efficiency. However, they found that a lack of uniformity of particle sizes and flow through the column meant that these conditions could not be achieved. These problems led them away from systems using a solid stationary phase. The importance of their pioneering work was recognised in 1952 by the award of the Nobel Prize for Chemistry.

Martin then went on to develop paper chromatography in which the stationary phase is a paper sheet impregnated with a stationary liquid [12]. This technique enabled the ready analytical separation of many highly polar biological samples including amino acids, carbohydrates and pigments. However, it is a relatively slow technique, each separation taking some hours, and has now been almost completely superseded by TLC and HPLC.

The next major step was the development of gas–liquid chromatography in 1952 by Martin and James, while at the National Institute for Medical Research at Mill Hill in London. They demonstrated the separation of

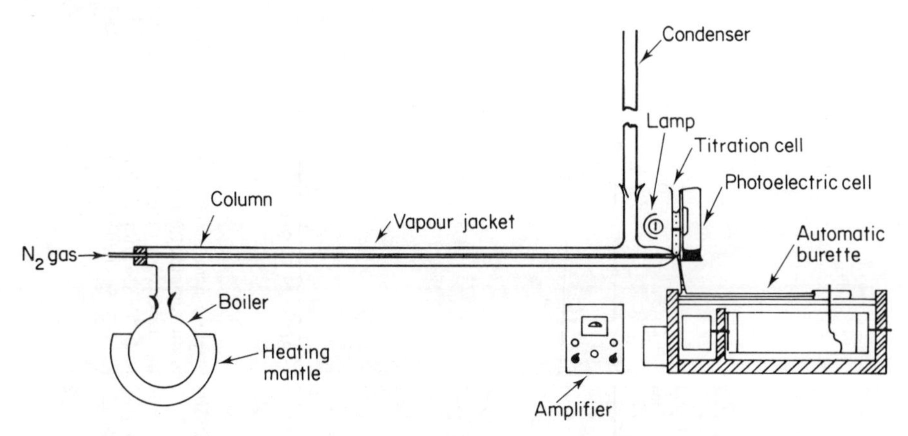

Figure 1.5 The first gas–liquid chromatograph (GLC) instrument built by James and Martin, showing the titration cell at the end of the column. Boiling solvent provided a constant-temperature jacket and the acids (or bases) were detected by titration with base (or acid) using an automatic burette. (Reproduced by permission of *Manufacturing Chemist*.)

volatile carboxylic acids and amines using a gas as a mobile phase. The stationary phase was a liquid film on an inert support, which was packed into a tube heated by a boiling solvent. The separated components were quantified using a titrimetric detector (Figure 1.5) [13, 14]. By modern standards the chromatographic system was very simple but the separation (Figure 1.6) was dramatically better than fractional distillation. These first reports were rapidly followed by an explosion of work in many other laboratories and, within three years, gas–liquid chromatographic equipment was commercially available [15] and the rapid development of modern-day gas–liquid chromatography as a widely used routine analytical method was under way.

The first step was the adoption of the thermal conductivity detector, which was already known in the relatively inefficient method of gas–solid chromatography. Although this chromatographic technique had preceded GLC [16], it has always been very limited in its application and even now is not widely used. A major step was the invention of the flame ionisation detector (FID) almost simultaneously in 1958 by Harley, Nel and Pretorius in South Africa [17] and by McWilliam and Dewar in Australia [18, 19]. This detector increased the sensitivity of GLC by more than a thousandfold and enabled the use of more efficient columns and smaller samples. This was followed by the introduction by Lovelock of the argon ionisation [20] and electron affinity detectors [21]. The latter is the forerunner of the

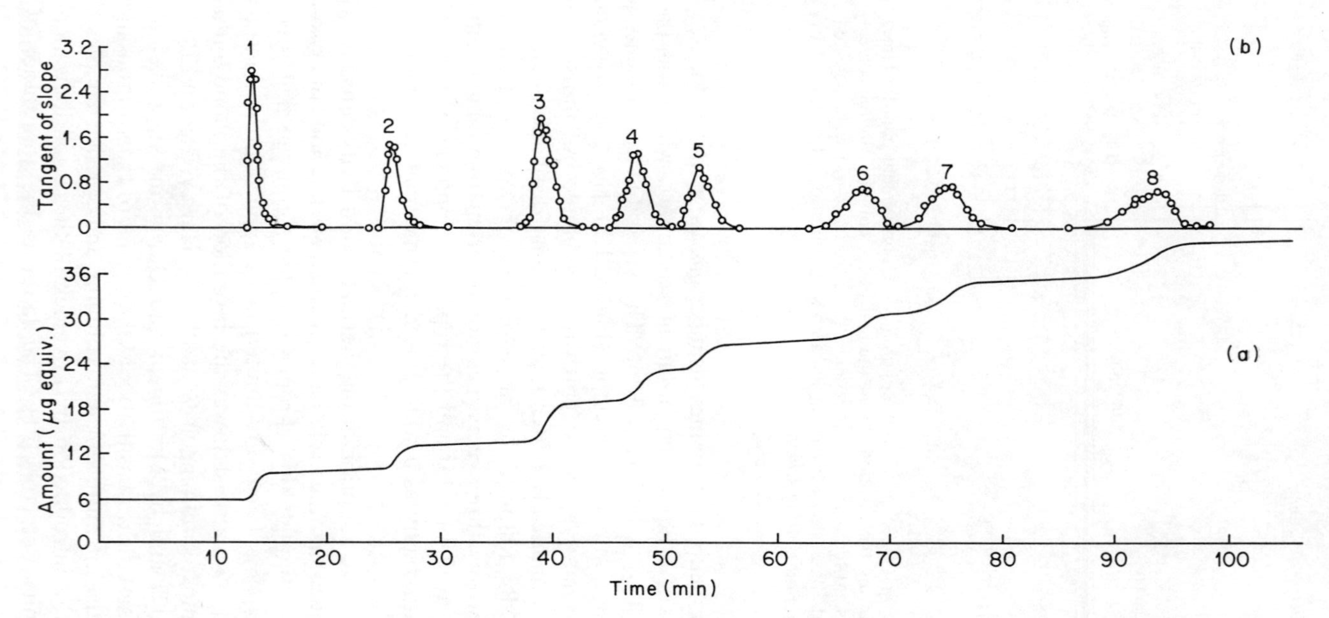

Figure 1.6 Separation of C_1–C_5 carboxylic acids on first gas–liquid chromatograph. Column 10% stearic acid in DC 550 silicone oil on Kieselguhr (Celite 545) 11 ft × 4 mm i.d. Carrier gas nitrogen at flow rate of 18.2 ml min^{-1}. Column temperature 137°C. Detection by automatic titrimetry with 0.038 M sodium hydroxide, indicator phenol red. (a) Experimental titration curve from detector. (b) Differential of experimental curve. Peaks: 1, acetic acid; 2, propionic acid; 3, isobutyric acid; 4, n-butyric acid; 5, α,α-dimethylpropionic acid; 6, isovaleric acid; 7, α-methylbutyric acid; 8, n-valeric acid. (Reprinted by permission from *Biochem. J.*, 1952, **50,** 679. Copyright © 1952 The Biochemical Society, London.)

modern electron capture detector, one of the most widely used selective detectors.

The potential efficiency and separation power of the column was dramatically altered by the proposal by Golay for open-tubular or capillary columns [22, 23] a concept that, although initially of great interest and application in the petroleum industry [24, 25], has only recently realised its full potential with the introduction of first glass and now fused silica open-tubular columns (Figure 1.7) [26, 27]. A period of first rapid and then steadier development followed, leading to the present-day GLC instrument with solid-state amplifiers, full microprocessor control of temperature and flow rates and bonded-phase fused silica columns, but still based essentially on the concepts of the first GLC instruments.

The development of methods in which the mobile phase is a liquid, leading to the modern high-performance liquid chromatograph, was much less rapid. Rather than one major step as in GLC with the work of Martin and James, there were a series of stages each marking a small advance. The primary practical difference in liquid chromatography is that diffusion rates in a liquid are much smaller than in a gas. For maximum efficiency the particle size of the stationary phase must therefore be correspondingly smaller. In 1956, Moore and Stein [28] reported the development of a dedicated amino-acid analyser, which was based on an ion-exchange separation but contained many of the features of a liquid chromatograph. It used a programmed oven, solvent gradient elution, a post-column reaction detector and photometric detection. The next stage was taken in 1956 when E. Stahl standardised thin-layer chromatography [29]. In this method the separation is carried out on a thin layer of small particles of a stationary absorbent, which have been coated onto a flat plate.

It was not until the late 1960s and early 1970s that significant progress took place in liquid column chromatography with theoretical and practical papers from a number of laboratories. A major problem was that the large stationary-phase silica particles used in gravity-flow preparative column chromatography gave only low efficiency separations. However, the theoretically desirable much smaller particles (3–10 μm) were not available. To circumvent this problem, Kirkland and others developed relatively large pellicular beads (37–44 μm) [30, 31]. These had a thin porous surface layer (2 μm deep) over an impervious solid core. They therefore behaved in a similar way to a 2 μm particle and could be used as a solid stationary phase or could be coated to give a liquid stationary phase. Although this approach proved useful and the columns could be readily packed to give efficient analytical separations (Figure 1.8), the solid core was wasted space and this type of material is rarely used today. However, it prompted the development

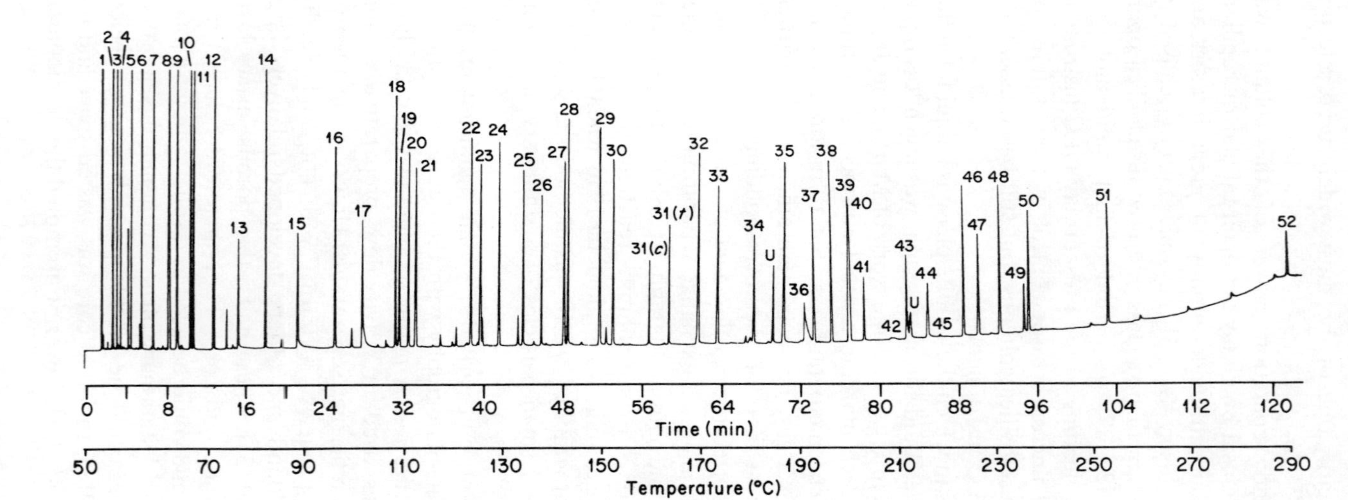

Figure 1.7 Separation of fragrance components on a bonded Supelcowax column 30 m × 0.25 mm i.d. Film thickness 0.25 μm. Column temperature 2 min at 50°C then programmed at 2°C min^{-1} to 280°C. Helium carrier gas and flame ionisation detector. Peaks: 1, hexane; 2, ethyl acetate; 3, ethyl alcohol; 4, ethyl propionate; 5, α-pinene; 6, camphene; 7, β-pinene; 8, δ-3-carene; 9, myrcene; 10, limonene; 11, 1,8-cineole; 12, γ-terpinene; 13, n-octanol; 14, 6-methyl-5-hepten-2-one; 15, n-nonanal; 16, L-menthone; 17, n-decanal; 18, linalool; 19, linalyl acetate; 20, laevo-bornyl acetate; 21, β-caryophyllene; 22, lavandulol; 23, α-terpineol; 24, benzyl acetate; 25, citronellol; 26, nerol; 27, α-ionone; 28, geraniol; 29, phenyl ethyl alcohol; 30, *cis*-jasmone; 31, nerolidol (*cis* and *trans*); 32, cedrol; 33, cedryl acetate; 34, heliotropine; U, unknown; 35, cinnamic alcohol; 36, methyl jasmonate; 37,hexadecanolide; 38, tonalid; 39, oxalide; 40, coumarin; 41, ethyl stearate; 42, musk xylol; 43, vanillin; U, unknown; 44, phytol; 45, musk ambrette; 46, phenyl ethyl benzoate; 47, musk T; 48, phenyl ethyl phenyl acetate; 49, musk ketone; 50, ethyl behenate; 51, benzyl isoeugenol; 52, cinnamyl cinnamate. (Reprinted with permission of Supelco, Inc., Bellefonte, PA 16823.)

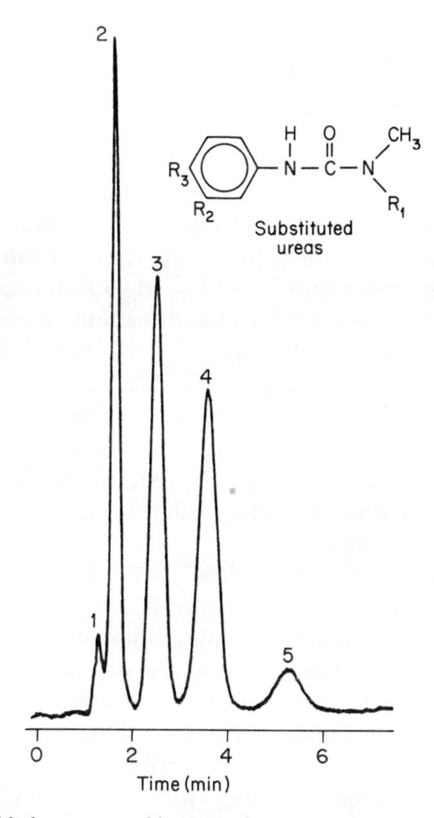

Figure 1.8 Liquid chromatographic separation of substituted urea herbicides on a liquid-coated pellicular bead column. Column 500 mm × 2.1 mm i.d. packed with 1.0% β,β′-oxypropionitrile on 37–44 μm controlled surface porosity (CSP) support. Mobile phase dibutyl ether at 1.14 ml min^{-1}. Peaks: 1, impurity; 2, Linuron (R_1 = OCH_3; R_2, R_3 = Cl); 3, Diuron (R_1 = CH_3; R_2, R_3 = Cl); 4, Monuron (R_1 = CH_3; R_2 = H; R_3 = Cl); 5, Fenuron (R_1 = CH_3; R_2, R_3 = H). (J. J. Kirkland, *J. Chromatogr. Sci.*, 1969, **7,** 7. Reproduced from the *Journal of Chromatographic Science* by permission of Preston Publications. A Division of Preston Industries Inc.)

of pumps and detectors [32], which were then available to take advantage of the next major development.

Rather than coating silica particles with a liquid film of the stationary phase, which could be washed off by the mobile phase, it was shown that a long-chain hydrocarbon chemically bonded as a stable "fur" to the surface of the silica beads would behave as a stable stationary liquid phase [33]. This led to a much more stable chromatographic system. Initially, long thin columns (1 mm i.d. × 500–1000 mm) were used, with consequent problems of high back-pressures (up to 6000 psi), and led to the term "high-pressure liquid chromatography" (HPLC) being used to describe the technique.

By the late 1970s following a better understanding of the theory of liquid chromatography [34, 35], it was recognised that relatively short columns packed with fine microparticulate (10 or 5 μm) stationary-phase materials could be used to give similar or better efficiencies but with faster separations and lower back-pressures (Figure 1.9) [36]. High-performance liquid chromatography (also abbreviated to HPLC) or simply liquid chromatography rapidly became the accepted description. Since then HPLC has expanded enormously and it is now the largest sales area for scientific equipment. In recent years there has been a renewed interest in long narrow columns to give comparable efficiencies to capillary columns in GLC and over a million theoretical plates have been achieved [37], but the penalties of the prolonged time (over 1000 min) required for the separation and limited sample capacity have limited their acceptance.

For the analytical chemist, the years since 1940 have been a period of growth and excitement in which great advances in separation methods were announced and then applied worldwide within weeks. At a more stately pace, all the chromatographic methods are still developing and in recent years have been influenced, as with many other areas of analytical chemistry, by the impact of microcomputers. Together GLC and HPLC now account for nearly half of all chemical analyses carried in the world. This has led to an extensive and growing literature, with over eight journals currently devoted solely to this one technique (see Appendix 1). It is estimated that about 150 000 research papers on GLC, TLC and HPLC have been published between the introduction of the techniques and 1986, and new papers continue to appear at a high rate.

1.3 SELECTION OF A CHROMATOGAPHIC METHOD

One of the problems facing the newcomer to chromatography is the selection of the most suitable chromatographic method for a particular sample. This is primarily dependent on the properties and structure of the analyte but the nature of the matrix of the sample, the reasons for the analysis and the availability of suitable equipment are also important. The answer will come, partly from an understanding of the capabilities and limitations of the different analytical techniques and partly from experience gained over a number of years of practical experience.

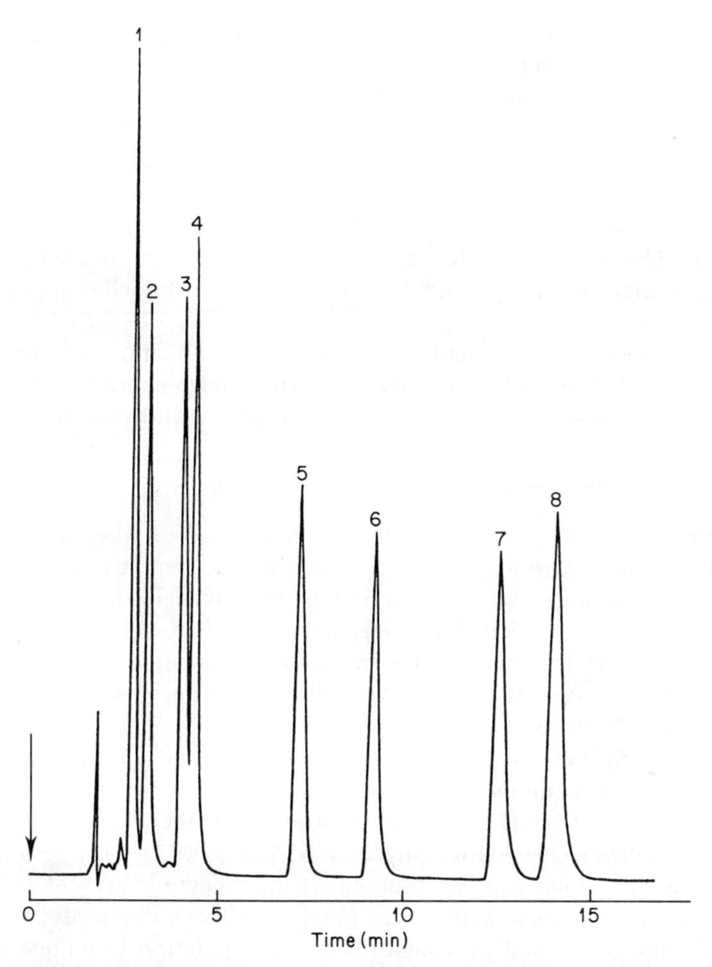

Figure 1.9 HPLC separation of mixture of aromatic amines and alcohols on 5 μm silica column. Column LiChrosorb Si 60, 5 μm, 250 mm × 3.2 mm i.d. Mobile phase isooctane–dichloromethane–isopropanol 90:9:1 at 1.5 ml min^{-1}. Ultraviolet spectroscopic detection at 254 nm. Peaks: 1, 2,6-diethylaniline; 2,2-methyl-6-ethylaniline; 3, *o*-isopropylaniline; 4, *o*-ethylaniline; 5, 2-phenyl-2-propanol; 6, α-methylbenzyl alcohol; 7, benzyl alcohol; 8, cinnamyl alcohol. (Reproduced by permission of E. Merck, Darmstadt)

a. Nature of analyte

For gas chromatographic analysis the analyte must be capable of entering the gas phase and therefore this technique, as either gas–liquid or gas–solid chromatography is restricted to gases and to solids and liquids that are volatile and thermally stable within the normal operating range of the instrument (up to 300°C).

Liquid chromatography has fewer restrictions and can be used to analyse any liquid or any solid that will dissolve in the mobile phase. However, some samples, particularly those without ultraviolet chromophores, may be difficult to detect at low concentrations. Thus for a wide range of organic compounds either gas or liquid chromatographic techniques can potentially be used. Unsuitable samples can also be chemically treated to form derivatives that are volatile (for GLC) or have a strong chromophore (for HPLC).

Insoluble solids, such as rigid polymers, are virtually impossible to analyse directly, but if degraded to smaller fragments under controlled conditions or by chemical reaction they may be able to give useful information.

b. Matrix of the sample

For many samples the chromatographic technique can be used both to separate the analyte from a complex matrix and to determine its concentration and identity. Complex and time-consuming preanalytical work-up procedures may therefore be avoided. However, it is important that residues of the samples do not remain on the chromatographic system as contaminants. A prefractionation step is therefore often still required but this may be filtration or a simple solvent extraction.

If the sample contains both volatile and involatile constituents, liquid chromatographic methods may require less sample preparation than for gas chromatography. If it is desirable to observe all the constituents of the sample, thin-layer chromatography may be necessary, as any residues left on the sampling point can be observed. If the analyte is present as a trace component in a complex matrix, then highly sensitive and selective detection methods may be needed or a high-resolution separation technique such as open-tubular gas chromatography may be used.

c. Reasons for analysis

One of the most widely used analyses is the rough qualitative examination of reaction products from organic syntheses. In this case TLC is ideal, as accuracy and sensitivity are not prime requirements. For process or quality control, clinical studies or environmental monitoring, greater selectivity and good quantitative accuracy are needed. These can normally only be achieved by an instrumental method, such as GLC or HPLC. These methods are also easier to automate when large numbers of similar samples are being examined.

The ability of chromatographic techniques to separate the components of a sample can also be used preparatively to give pure constituents on either an analytical or industrial scale. A single TLC plate can be used for up to 20 mg of a sample and yield sufficient pure components for most structural techniques including infrared, ultraviolet, nuclear magnetic resonance and mass spectroscopy. Larger samples can be separated using multiple runs. GLC and HPLC can also be used by scaling up the dimensions of the column. In the industrial plant, the separation of kilograms of sample at a time has been achieved with a throughout of tonnes per year.

BIBLIOGRAPHY

History of chromatography

L.S. Ettre and A. Zlatkis (Eds.), *75 years of chromatography—a historical dialogue*, J. Chromatogr. Library, Vol. 17, Elsevier, Amsterdam, 1979.
L.S. Ettre, "The development of chromatography", *Anal. Chem.*, 1971, **43** (December), 20A–31A.
J.N. Done, G.J. Kennedy and J.H. Knox, "Revolution in liquid chromatography", *Nature*, 1972, **237**, 77–81.
L.S. Ettre and C. Horvath, "Foundations of modern liquid chromatography", *Anal. Chem.*, 1975, **47**, 422A–446A.
L.S. Ettre, "Evolution of liquid chromatography: a historical overview", *High-Perform. Liquid Chromatogr.*, 1980, **1**, 1–74.

REFERENCES

1. D.T. Day, "Report of paper 'Experiments on the diffusion of crude petroleum through Fuller's earth' Presented at a meeting of the Geological Society of Washington", *Science*, 1903, **17**, 1007–1008.
2. H. Weil and T.I. Williams, "History of chromatography", *Nature*, 1950, **166**, 1000–1002.
3. H.H. Strain and J. Sherma, "Micheal Tswett's contributions to sixty years of chromatography", *J. Chem. Educ.*, 1967, **44**, 235–237.
4. M. Tswett, "Adsorption analysis and chromatographic methods. Application to the chemistry of the chlorophylls", *Ber. Deutsch. Botan. Ges.*, 1906, **24**, 384. (Translation and comments H.H. Strain and J. Sherma, *J. Chem. Educ.*, 1967, **44**, 238–242).
5. K. Sakodynsky, "M.S. Tswett—his life", *J. Chromatogr.*, 1970, **49**, 2–17.
6. K. Sakodynskii, "The life and scientific works of Micheal Tswett", *J. Chromatogr.*, 1972, **73**, 303–360.
7. R. Kuhn and E. Lederer, "Zerlegung des Carotins in seine Komponenten", *Berichte*, 1931, **64**, 1349–1357.
8. R. Kuhn, A. Winterstein and E. Lederer, "Zur Kenntis der Xanthophylle", *Z. Physiol. Chem*, 1931, **197**, 141–160.

9. H. Brockmann and H. Schodder, "Aluminiumoxyd mit abgestuftern Adsorptionsvermögen zur chromatographischen Adsorption (Aluminium oxide of graduated adsorption capacity for chromatographic adsorption)", *Berichte*, 1941, **74**, 73–78.
10. A.J.P. Martin and R.L.M. Synge, "Separation of the higher monoamino-acids by counter-current liquid–liquid extraction: the amino-acid composition of wool", *Biochem. J.*, 1941, **35**, 91–121.
11. A.J.P. Martin and R.L.M. Synge, "A new form of chromatogram employing two liquid phases. 1. A theory of chromatography. 2. Application to the micro-determination of the higher monoamino-acids in proteins", *Biochem. J.*, 1941, **35**, 1358–1368.
12. R. Consden, A.H. Gordon and A.J.P. Martin, "Qualitative analysis of proteins: a partition chromatographic method using paper", *Biochem. J.*, 1944, **38**, 224–232.
13. A.T. James and A.J.P. Martin, "Gas–liquid partition chromatography: the separation and microestimation of volatile fatty acids from formic acid to dodecanoic acid", *Biochem. J.*, 1952, **50**, 679–690.
14. A.T. James and A.J.P. Martin, "Gas–liquid partition chromatography. A technique for the analysis of volatile materials", *Analyst*, 1952, **77**, 915–932.
15. "Vapour phase chromatography" and "Gas chromatograph", *Rev. Sci. Instrum.*, 1955, **26**, 990.
16. L.S. Ettre, "The development of gas adsorption chromatography", *Am. Lab.*, 1972, October, 10–16.
17. J. Harley, W. Nel and V. Pretorius, "Flame ionization detector for gas chromatography", *Nature*, 1958, **181**, 177–178.
18. I.G. McWilliam and R.A. Dewar, "Flame ionization detector for gas chromatography", *Nature*, 1958, **181**, 760.
19. I.G. McWilliam, "The origin of the flame ionization detector", *Chromatographia*, 1983, **17**, 241–243.
20. J.E. Lovelock, "A sensitive detector for gas chromatography", *J. Chromatogr.*, 1958, **1**, 35–46.
21. J.E. Lovelock and S.R. Lipsky, "Electron affinity spectroscopy—a new method for the identification of functional groups in chemical compounds separated by gas chromatography", *J. Am. Chem. Soc.*, 1960, **82**, 431–433.
22. M.J.E. Golay, "Theory and practice of gas–liquid partition chromatography with coated capillaries", in V.J. Coates, H.J. Noebels and I.S. Fagerson (Eds.), *Gas chromatography (1957)*, Academic Press, New York, 1958, pp. 1–14.
23. M.J.E. Golay, "Vapor phase chromatography and telegrapher's equation", *Anal. Chem.*, 1957, **29**, 928–932.
24. R.P.W. Scott, "The construction of high-efficiency columns for the separation of hydrocarbons", in D.H. Desty (Ed.), *Gas chromatography 1958*, Butterworths, London, 1958, pp. 189–199.
25. D.H. Desty, A. Goldup and B.H.F. Whyman, "The potentialities of coated capillary columns for gas chromatography in the petroleum industry", *J. Inst. Petroleum*, 1959, **45**, 287–298.
26. K. Grob, "Twenty years of glass capillary columns. An empirical model for their preparation and properties", *J. High Res. Chromatogr., Chromatogr. Commun.*, 1979, **2**, 599–604.

27. L.S. Ettre, “Open-tubular columns: evolution, present status, and future”, *Anal. Chem.*, 1985, **57**, 1419A–1438A.
28. S. Moore and W.H. Stein, “Procedures for the chromatographic determination of amino acids on four percent crosslinked sulphonated polystyrene resins”, *J. Biol. Chem*, 1954, **211**, 893–906.
29. E. Stahl (Ed.), *Thin-layer chromatography*, 2nd ed., Allen and Unwin, London, 1969.
30. J.J. Kirkland, “High-speed liquid chromatography with controlled surface porosity supports”, *J. Chromatogr. Sci.*, 1969, **7**, 7–12.
31. J.J. Kirkland, “Columns for modern analytical liquid chromatography”, *Anal. Chem.*, 1971, **43** (October), 36A–48A.
32. H. Felton, “Performance of components of a high pressure liquid chromatography system”, *J. Chromatogr. Sci.*, 1969, **7**, 13–16.
33. J.J. Kirkland, “High speed liquid-partition chromatography with chemically bonded organic stationary phases”, *J. Chromatogr. Sci.*, 1971, **9**, 206–214.
34. L.R. Snyder, “Column efficiencies in liquid adsorption chromatography: past, present and future”, *J. Chromatogr. Sci.*, 1969, **7**, 352–360.
35. J.H. Knox and M. Saleen, “Kinetic conditions for optimum speed and resolution in column chromatography”, *J. Chromatogr. Sci.*, 1969, **7**, 614–622.
36. J.J. Kirkland, “High-performance liquid chromatography with porous silica microspheres”, *J. Chromatogr. Sci.*, 1972, **10**, 593–599.
37. H.G. Menet, P.C. Gareil and R.H. Rosset, “Experimental achievement of one million theoretical plates with microbore liquid chromatographic columns”, *Anal. Chem.*, 1984, **56**, 1770–1773.

CHAPTER 2

BASIC CONCEPTS OF CHROMATOGRAPHY

2.1 PRINCIPLES AND DEFINITIONS

a. Basic concepts

Ever since the start of chromatography, attempts have been made to devise equations that would describe in detail the processes occurring during a chromatographic separation and to confirm the validity of these models by their comparison with experimental results. The basic theory of chromatography was rapidly established and has been described by Giddings and by Snyder (see Bibliography). However, many of these studies and the theoretical models and conclusions are of academic rather than practical interest to the chromatographer. Indeed, many of the details are still the subject of debate and continuing study. However, the basic concepts and theories underlying all chromatographic techniques are effectively the same, whether the mobile phase is a gas, liquid, or supercritical fluid or if the stationary phase is a liquid or solid. They can therefore be described using a common set of basic equations.

This discussion will therefore be confined to a general view of the theory of chromatography and will concentrate on those aspects which both directly affect the separation process and can be altered or manipulated by the operator to improve the separation process. However, this will often lead to a simplification of the current view, but if required more rigorous details are given in the References and Bibliography.

In practice, the limitations of the practical design of the components of the chromatograph and its operation with real samples often prevent the operator from achieving the advantages proposed by theory, but an awareness of the directions indicated by the physicochemical studies can frequently lead to an improvement in the separation.

In setting out to establish and optimise a chromatographic system, there are four principal factors that the operator must consider. Firstly, one must consider the resolution of the separation, which is expressed as the ability to separate the compound of interest from the other components of the sample. Secondly, the sensitivity for the analyte of interest needs to be optimised. This is often linked with the selectivity of the detector and the presence of interferents. Thirdly, in order for the analyst to have confidence in the results of an analysis, the technique must be reliable and the results reproducible. Finally, the time taken to carry out the analysis is an important practical factor. It is often necessary to choose a compromise between the maximum separation, which may take a considerable time, and the efficient economic use of the relatively expensive time of the operator and capital depreciation of the instrument. Of these factors, the sensitivity and selectivity of detection are primarily a function of the choice of the detector but to some extent sensitivity is dependent on the sharpness of the chromatographic peak. The reproducibility of the analysis is largely dependent on good instrument design, manufacturing technology and operator experience. The remaining factors of chromatographic resolution and speed of analysis are interrelated and can often be significantly improved by the operator, based on an understanding of the separation process.

If the peaks for two components of a sample are partially overlapping, then their accurate quantification can be difficult and their identification will be uncertain because of the possible presence of unresolved minor components (Figure 2.1a). The separation of the components can be improved either by increasing the difference in their retention times (Figure 2.1b) or by retaining the same relative retentions but making the chromatography more efficient so that the peaks are narrower (Figure 2.1c). In order to be able to select the best method to use, we must understand the factors that cause sample retention and peak broadening. These will be examined in terms of elution chromatography, but the same concepts could also be applied to planar separations such as TLC.

Many of the terms, abbreviations and definitions used in chromatography are confusing, because, as the technique has developed over the last 80 years, the individual modes of chromatography have often been studied in isolation. Although in recent years standard definitions and symbols have been proposed by IUPAC [1] and ASTM [2, 3], these often differ and are contrary to current common usage and the styles adopted in the principal chromatography journals. These problems of nomenclature have been

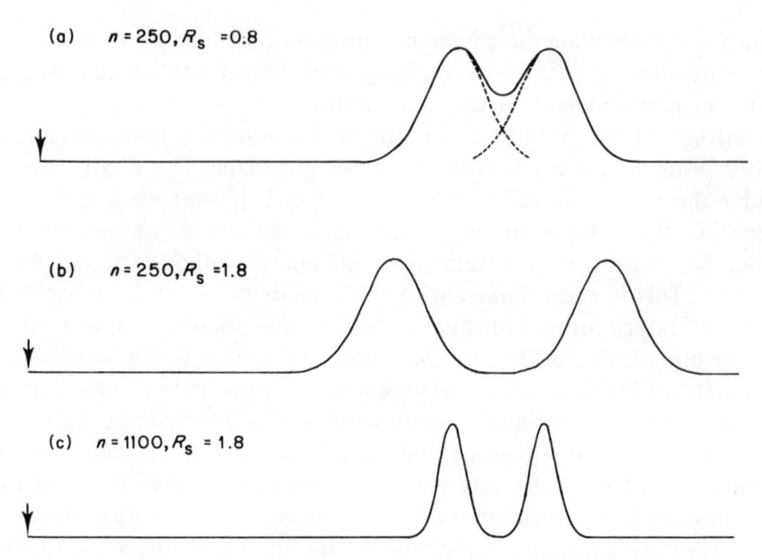

Figure 2.1 (a) Chromatogram with two unresolved peaks. (b) Effect of increasing the separation of the peaks but with the same efficiency as in (a). (c) Effect of increasing the efficiency of the column but with the same separation as in (a).

discussed in detail by Ettre [4–6]. Generally in this book the IUPAC definitions will be adopted with two important exceptions, which will follow current usage in the chromatography literature: k' will be used for capacity factor (rather than k) and t_0 (rather than t_M) for the column void volume. A full listing of the terms and symbols is given in Appendix 3 and the cases when confusion may occur will be noted in this chapter.

b. Chromatographic retention

The retention time (t_R) of a compound in a chromatographic separation is the time after injection for the peak maximum to be eluted (Figure 2.2). In theory the value of significance is the volume of the mobile phase required to elute the analyte, compared to the volume of the column. In practice this would be difficult to measure and therefore time is used, which is much easier. The column volume is defined as the volume accessible to the eluent or the total volume less the volume of the packing. It is usually referred to as the column void volume (t_0 but IUPAC uses t_M for the mobile phase time) or sometimes as the column dead volume. It is often determined as the time taken for an unretained sample to travel through the column. The difference between the analyte retention time and the column void volume

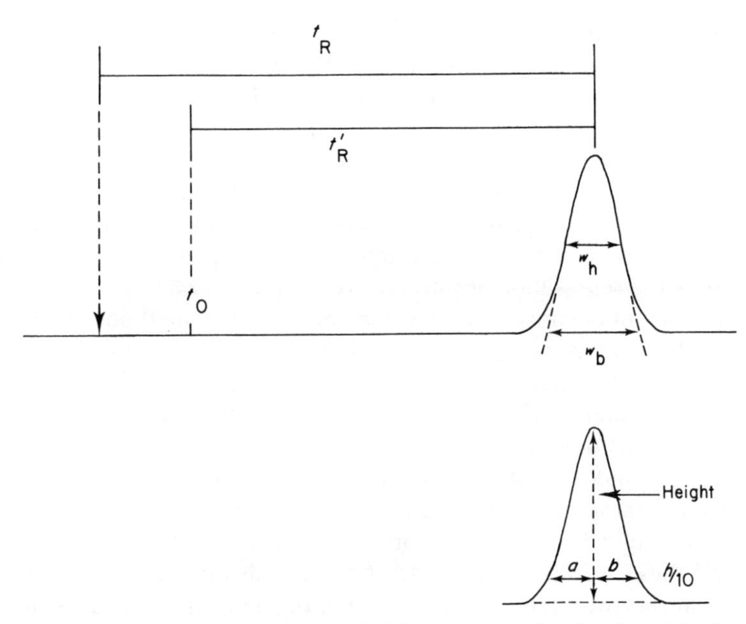

Figure 2.2 A typical chromatogram. Variables: t_R, retention time from injection; t_0, column void volume; t'_R, adjusted retention time; w_b, width of peak at base; w_h, width of peak at half-height; a and b width of front and back of the peak at 10% of the peak height.

is the adjusted retention time (t'_R), which is the time an analyte is retained on the column compared to an unretained compound (Equation 2.1)

$$t'_R = t_R - t_0 \tag{2.1}$$

and is therefore directly related to the interaction of the analyte with the stationary phase. The capacity factor (k') is an alternative method for reporting the retention of an analyte, which is routinely used in HPLC but rarely in GLC (Equation 2.2).

$$k' = \frac{t_R - t_0}{t_0} \tag{2.2}$$

The capacity factor compares the adjusted retention time of the analyte to the retention time of the mobile phase. This is effectively the ratio of the time the analyte molecules spend in the stationary phase (where they are stationary) to their time in the mobile phase (where they are moving down the column). This ratio should be independent of the flow rate of the mobile phase or the physical dimensions of the column. It should therefore be a useful method for comparing retentions on columns of different dimensions or with different pumps or instruments. However, in practice the accurate determination of the column void volume of an HPLC column can be very difficult. Different "unretained' marker compounds or other techniques give different values [7, 8] and as yet chromatographers have not been able to agree on a standard method or precise definition (see also section 13.3). In GLC the front edge of the solvent peak is often used to determine the void volume and, except for the separation of gas samples or low molecular-weight analytes, represents an unretained sample.

The fundamental concept in all chromatographic methods is that the sample is distributed between the two phases as it passes through the column. It is differences in these interactions for different analytes that cause different retentions and produce the separation. The distribution of an analyte by partitioning between two phases can only occur between two immiscible gas and/or liquid phases and such methods are referred to as partition chromatography. With liquid–solid or gas–solid separations the analyte is retained by adsorption onto the surface of the stationary solid phase, and these methods can be described as adsorption chromatography. However, in the two methods the retention of the analyte behaves in much the same way and the proportioning of the sample between the phases is best described by the use of the term "distribution constant". In reality the situation is often a mixture of the two modes as even with liquid phases, part of the retention can often be attributed to interactions with the underlying solid support material. The use in HPLC of bonded stationary phases, which are liquids chemically bonded to a solid support, or the use in GLC of cross-linked liquid phases and solid polymer columns, would further confuse any attempt to divide methods dogmatically into adsorption and partition chromatographies.

The analyte, when it is in the column, will be distributed between the two phases according to the amount of each phase and its distribution constant (K) (Equation 2.3), which

$$K = \frac{\text{concentration in unit volume of the stationary phase}}{\text{concentration in unit volume of the mobile phase}} \tag{2.3}$$

which is a characteristic physical property of the analyte and depends only on the structure of the analyte, the nature of the two phases and the temperature.

The proportion of the analyte in each phase will therefore be the product of the distribution constant and the phase ratio (α) = (V_S/V_M) (Equation 2.4)

$$k' = K\frac{V_S}{V_M} = \frac{t'_R}{t_0} \tag{2.4}$$

in which V_S and V_M are the volumes of the stationary and mobile phases in the column, respectively, and depend only on the physical dimensions of the column. The phase ratio is therefore identical for all the components of a sample in a particular separation. The ratio of the analyte in the two phases has already been defined as the capacity factor (k'). As capacity factors are related to retention times (Equation 2.2), there is therefore a direct relationship between distribution constant and the retention time of an analyte (Equation 2.4). The retention of an analyte is dependent only on the distribution constant and the phase ratio and should be independent of the presence of small amounts of any other components of the sample. This conclusion has the important consequence that the retention of an analyte will be the same whether it is injected as a pure compound or as a component of a mixture. Chromatographic retention times can therefore be used for quantitative identification.

For two compounds to be separated on a particular chromatographic system, they must have different distribution constants. Conversely, any compounds with the same distribution constants will have the same retentions and will not be separated. If a compound has a very large distribution constant, it will be retained on the column for a long time, or if very small the analyte will elute with the unretained sample. Usually the conditions of stationary phase and mobile phase in HPLC, or temperature in GLC, are adjusted so that an analyte of interest has a capacity factor between 1 and 10. In GLC, higher temperatures will make a sample more volatile, favouring the gas phase and decreasing the distribution constant.

Thus if the chromatographer wishes to improve the separation of two compounds with very similar retentions either the mobile phase, the stationary phase, or the temperature of the system must be altered. This should change the distribution constants of the two compounds. Unfortunately, in practice it is often difficult to predict the effects of a change and the separation of the two peaks may improve or may become worse. Although some guidance can be obtained from the structure of the analytes and a knowledge of the properties of different stationary phases, usually the only method is to make the change experimentally and examine the results.

In GLC almost all separations are carried out using either nitrogen, hydrogen, or helium as the mobile phase. Their partition properties are effectively similar so that only the stationary phase and temperature can be changed. Unless the two compounds have very dissimilar structures and hence differ markedly in their interactions with the stationary phase (for example Figure 2.3), changing the temperature for similar compounds usually has relatively little effect because the distribution of both will be affected in the same way.

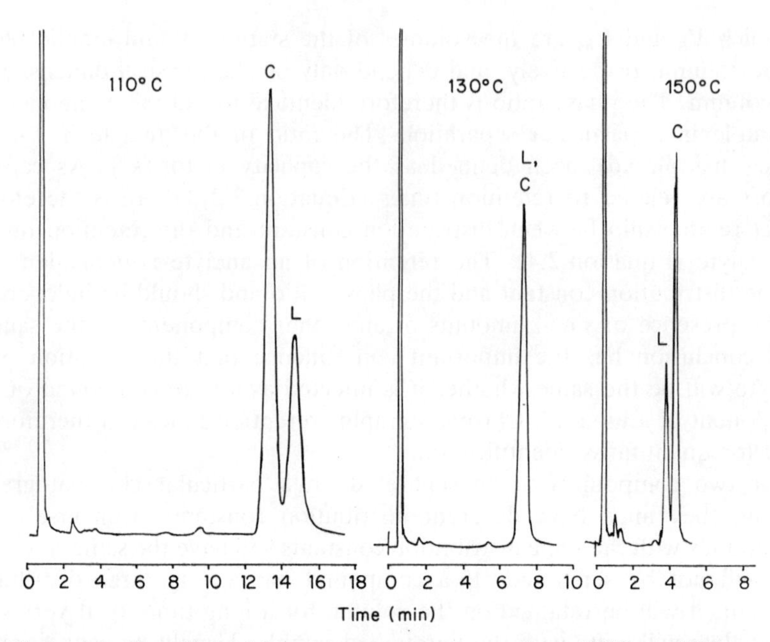

Figure 2.3 The effect of increasing the temperature from 110°C to 130°C to 150°C on the GLC separation of camphor (C) and linalool (L) on a 10% Carbowax 20M column. At higher temperatures the retention of both compounds decreases. Because of their different interactions with the column they first coelute at 130°C and then the peak order is reversed at 150°C.

The greater versatility of HPLC is possible because both the stationary and mobile phases can be altered but temperature changes are very restricted. Mobile-phase changes, such as a change from methanol to tetrahydrofuran, can cause quite marked changes in elution order (Figure 2.4) but the effects are not easy to predict.

In their early work Martin and James showed that if the distribution constant was independent of the quantity of the analytes, then a chromatographic peak should be Gaussian in shape (Chapter 1). However, particularly in gas–solid chromatography or if there is a large difference between the polarity of the analytes and the stationary phase, in practice the distribution isotherm (or ratio between the two phases) may not be constant but will change with the size of the sample. Because the concentration across the analyte peak changes from low at the front and tailing edges to a maximum in the centre, these different parts of the peak will therefore have different distributions and will move through the column at different speeds. This causes the peak shape to become distorted. Typically, as the sample size

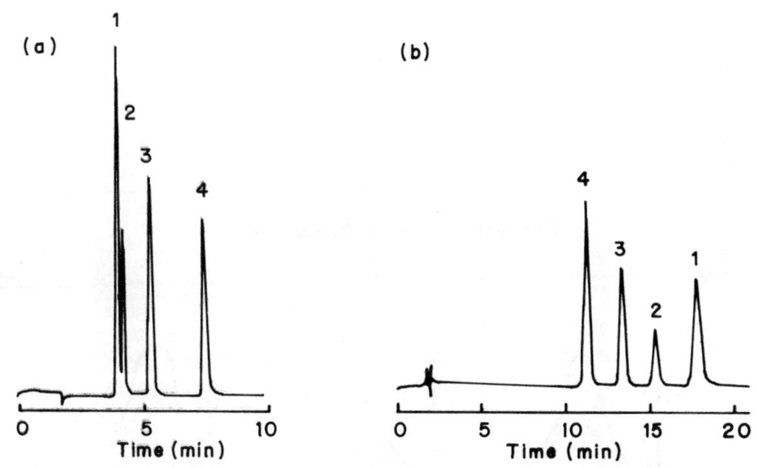

Figure 2.4 Chromatograms illustrating the difference in selectivity caused in HPLC by changing the mobile phase from (a) methanol–water 50:50 to (b) tetrahydrofuran–water 25:75. Analytes: 1, *p*-nitrophenol; 2, *p*-dinitrobenzene; 3, nitrobenzene; 4, methyl benzoate. Column ODS bonded to Hypersil 5 μm, 150 mm × 4.6 mm i.d. Ultraviolet detection at 254 nm. (Reproduced by permission of Elsevier Science Publishers from N. Tanaka, H. Goodell and B. L. Karger, *J. Chromatogr.*, 1978, **158,** 223.)

increases, the distribution into the mobile phase decreases which results in tailing peaks (Figure 2.5).

The extent of tailing of a peak is measured as the ratio of the widths of the front and back of the peak at 10% of the peak height (tailing = b/a, Figure 2.2). More accurate representations can be obtained by measuring the statistical moments of the peak but these are little used in routine assays [9].

c. *Separation efficiency*

Ideally, an analyte placed on a column as a sharp band should spread out as little as possible during the separation, as any broadening of the peaks can cause overlap and a loss of resolution. The efficiency of a column is a measure of the broadening of the sample peak as it passes through the column [10]. This is expressed as the number of theoretical plates on the column (n), but in many papers and texts N is used. The expression "theoretical plate" came originally from distillation theory but the connection is best ignored. The efficiency can be determined experimentally from a chromatogram as the square of the ratio of the retention of the analyte divided by the peak broadening (Equation 2.5).

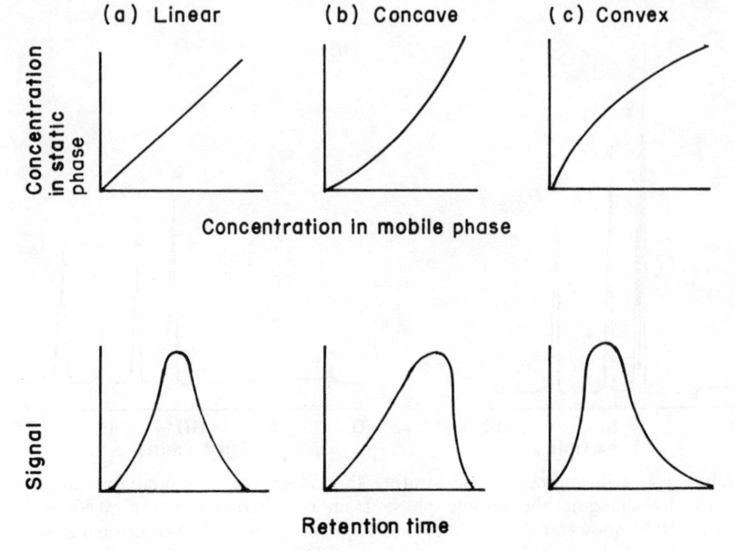

Figure 2.5 The influence on the peak shape of analytes with non-linear adsorption isotherms. (a) Linear isotherm giving a Gaussian peak; (b) and (c) non-linear isotherms.

$$n = \left(\frac{t_R}{\sigma}\right)^2 = 16\left(\frac{t_R}{w_b}\right)^2 = 5.54\left(\frac{t_R}{w_h}\right)^2 \qquad (2.5)$$

The peak broadening is defined as the standard deviation of the retention times of the individual molecules (σ). More practically this value can be related to the width of the peak at its base (w_b) or at half-height (w_h) (Figure 2.2). Assuming that the peak shape is a Gaussian curve, the width at the base is 4σ and the width at half-height is 2.35 σ, leading respectively to the factors 16 and 5.54 in Equation 2.5.

It is often difficult to accurately measure the width of the base of a peak because it requires an estimation of the extrapolation of the sides of the curve at their steepest slope. However, it is much easier to measure the width of the peak at its half-height (w_h) by measurement between the recorder traces, and this method is the most frequently used. The efficiency of a peak is dimensionless and should be the same for all peaks in a separation. Later peaks in a chromatogram run under constant elution conditions are therefore progressively broader than earlier peaks. In practice, very rapidly eluted peaks usually give poorer efficiencies, because they are affected more significantly by extra-column band-broadening factors, whose influence is virtually constant for all peaks. Very narrow peaks are also more prone to errors in the measurement of the peak width. Because of the difficulties in measuring narrow peak widths, small differences in efficiencies should not be regarded as significant.

Typical values of the efficiency for packed GLC columns would be n = 500–2000, and for open-tubular columns 30 000–100 000. In HPLC the efficiency is often expressed as plates/metre even if shorter columns are being used and typical values would be 30 000–60 000 for 5 μm particles and 10 000–20 000 for 10 μm particles. Thus the frequently used 25 cm analytical column could have 7000–12 000 plates.

An alternative expression, which is often used in open-tubular GLC, is the effective peak efficiency N (Equation 2.6).

$$N = 16\left(\frac{t'_R}{w_b}\right)^2 = 5.54\left(\frac{t'_R}{w_h}\right)^2 \quad (2.6)$$

If N has been used for the efficiency then N_{eff} will be used for the effective efficiency. These values are based on the adjusted retention time (t'_R) instead of the retention time (t_R). Because of the long column void volume of open-tubular columns, the normal efficiency calculation will often give unrealistically high values for rapidly eluted peaks with capacity factors less than 1. For peaks with longer retention times the values of n and N converge.

In order to be able to compare columns of different lengths, the efficiencies can be expressed in a different format as the *height equivalent to a theoretical plate* (HETP) h (Equation 2.7).

$$h = L/n \quad (2.7)$$

This value is often alternatively given the symbol H. The HETP is the distance along the length of the column equivalent to one theoretical plate of efficiency and is independent of the column length L (typically in millimetres).

The effective plate height can be used to calculate the analogous *height equivalent to an effective theoretical plate* (HEEPT) H.

Another method that has been used to express efficiency in capillary and other high-resolution systems is the Separation or Trennzahl number (TZ), which is derived from the retention times and peak widths of adjacent homologous n-alkanes (x and x+1) (Equation 2.8).

$$TZ = \frac{t_{R(x+1)} - t_{R(x)}}{w_{h(x+1)} + w_{h(x)}} - 1 \quad (2.8)$$

In practical terms, it is the number of analyte peaks that could be placed with a resolution of 1.0 between the two n-alkane peaks. The larger the value the more components of a mixture could theoretically be resolved.

d. Resolution

The resolution (R_s) of two components in a chromatogram is determined from the difference in their retention times and the widths of the peaks (Equation 2.9, Figure 2.6a)

$$R_S = \frac{2(t_{R2} - t_{R1})}{(w_{b1} + w_{b2})} \tag{2.9}$$

A resolution of $R_s = 1.0$ represents an overlap of 2% (or 98% separation) and a resolution of 1.25 represents 99.4% or almost complete separation (Figure 2.6b). Resolutions greater than 1.0 are usually sufficient for accurate quantification. Peaks with a resolution of 0.7 (Figure 2.6c) overlap severely and accurate quantification will be difficult, particularly if the peaks have different heights. With poorer resolutions, no valley may be present between the peaks and a minor component may be lost as a shoulder under a major component (Figure 2.6d).

The equations for efficiency (Equation 2.5) and resolution (Equation 2.9) for two peaks with similar retentions can be combined to illustrate their relationship (Equation 2.10).

$$R_s = \frac{\sqrt{n}}{4}\left(\frac{\alpha - 1}{\alpha}\right)\left(\frac{k'_2}{1 + k'_2}\right) \tag{2.10}$$

In this equation α is the ratio of the capacity factors of the two peaks (k'_2/k'_1). An important conclusion from this equation is that the resolution of a separation depends on the square root of the efficiency. However, efficiency is directly proportional to column length. To increase the resolution of a separation twofold, it is therefore necessary to increase the length of the column fourfold, which may cause undue resistance to flow.

2.2 OPTIMISATION OF SEPARATION EFFICIENCY

As seen in the previous section the resolution of two peaks may be improved by changing the separation conditions, if this causes an increase in the relative retentions of the analytes. Alternatively, if the band spreading of the peaks can be reduced, then greater resolution can be achieved with the same separation. A detailed understanding of this section is not needed for much practical chromatography but the chromatographer should be aware of the qualitative implications and trends, as they should influence the design of a separation system and suggest directions for improvements. Newcomers to practical chromatography may find it best to scan this section only briefly

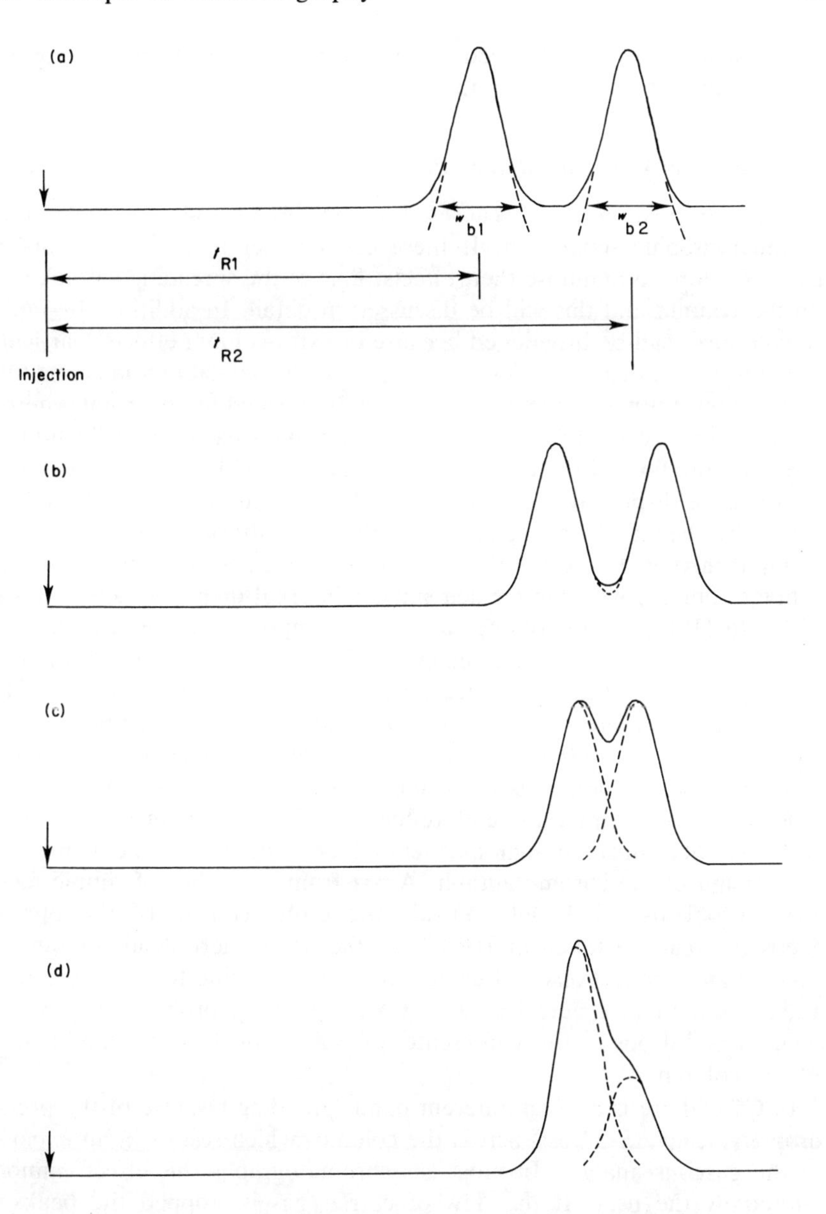

Figure 2.6 (a) Measurement of the resolution of two peaks in a chromatogram. (b) Chromatogram with a resolution of 1.25 showing baseline separation. (c) Chromatogram with a resolution of 0.7 showing serious overlapping and incomplete separation of the two components. (d) Chromatogram with a resolution of 0.6 showing the loss of a minor component under the edge of a major peak.

and return for a more detailed examination after gaining more experience and reading later sections on the practice of chromatography.

a. *Factors affecting band spreading*

There are a number of causes for the broadening of peaks during chromatographic separation. If these can be identified then the operator can take steps to minimise their effects. Part of the spreading will take place on the column and this will be discussed in detail. In addition the injected sample peak can be broadened because of extra-column effects, particularly the presence of dead volumes in the injector, the detector, connecting tubing and column fittings. A dead volume may be formed at any point where the tubing/connection is broadened so that stagnant areas occur, which are not swept by the flow of the eluent. The sample will eddy into these points and thus will be dispersed over a greater volume of the eluent. In GLC, these effects can be largely ignored, because the high diffusion rates in a gas result in rapid mixing. However, the effect may be significant with narrow open-tubular columns with their much smaller internal diameters and small flow rates. In HPLC, dead volume can be very important because of the lower diffusion rates, which reduce mixing, and even small additional volumes of 10–20 μl can significantly reduce efficiencies. The smaller the scale of the instrument the more significant the problem, and with microbore columns (internal diameter (i.d.) less than 2 mm) even the detector flow cell may cause problems unless it is designed for use with low flow rates.

Basically, the limitation and reduction of dead volumes are a matter of good mechanical design and should be inherent in the manufacture and design of the chromotagraph. Apart from a number of simple factors, these problems are largely outside the direct control of the operator. Steps that can be taken in HPLC are the use of zero-dead-volume couplings in any connections and using narrow-bore tubing to link the injection valve, column and detector. A serious operating problem occurs if the column bed drops as this can create quite a major dead volume at the top of the column.

In GLC there is also an inherent band spreading because of the pressure drop (typically 15–30 psi) across the column, which results in an expansion of the gaseous analyte. In most gas chromatographs this effect cannot be altered by the user. If the flow of carrier gas is stopped the peaks will continue to expand and, on restarting the flow, efficiency will have been lost. The volumes of liquid mobile phases are virtually unchanged on compression so, despite the higher pressures used to pump the eluent through the column, this factor can be ignored in HPLC.

Each of the different sources of band spreading (column, tubing, injector, detector, etc.) can be regarded as independent so that the overall band spreading can be expressed in terms of the HETP as the sum of the individual variances for the total system (Equation 2.11).

$$h = \Sigma\sigma^2/L \tag{2.11}$$

However, most band spreading occurs on the column as a consequence of the kinetics of the separation process and this topic has been the subject of much interest and research. A number of detailed studies are listed in the Bibliography. Some of the factors that affect the spreading can be selected or changed by the operator. A knowledge of the reasons for their effects can assist the user to optimise a separation and explain many commonly adopted practices.

One of the most widely used models for the causes of band spreading on the column is based on the work of a Dutch group under van Deemter, who in 1956 [11] derived an equation with three main components and related these to the average mobile-phase flow rate u (Equation 2.12) [11].

$$h = A + \frac{B}{u} + Cu \tag{2.12}$$

Three principal areas were identified as contributing to the band broadening; eddy diffusion (A term), molecular diffusion in the mobile phase (B term), and mass transfer effects in the stationary and mobile phases (C term). The equation does not agree totally with practical observation but it can give a useful qualitative representation of many of the chromatographic processes. Recently some of the conclusions and terms have been questioned by Hawkes [12]. Other researchers have suggested modifications to the details of the equation and alternative expressions have also been proposed [13].

We can examine the details of the contributions from the three terms in the van Deemter equation as separate sources of variance. In each case the aim of the chromatographer is to reduce the value of the term and hence of the overall variance. The exact values of the individual terms will not be discussed but it will be useful to discuss the trends and the effect on each of the terms of altering the practical aspects of the separation system. The discussion is based on the "random walk' model of Giddings in which the movements of the individual analyte molecules are studied rather than regarding the analyte as a whole [13]. The same factors can be applied to all the different methods of chromatography but the values and relative importance of the terms will differ because of differences in the physical properties of the mobile phases, in particular the diffusion rates and viscosities (Table 2.1).

Table 2.1 Typical order-of-magnitude values of the physical properties of mobile phases used in chromatography.

Mobile phase	Density ($g\ cm^{-3}$)	Diffusion rate ($cm^2\ s^{-1}$)	Viscosity ($g\ cm^{-1}\ s^{-1}$)
Gas	10^{-3}	10^{-1}	10^{-4}
Supercritical fluid	0.3	10^{-3} to 10^{-4}	10^{-4} to 10^{-3}
Liquid	1	$<10^{-5}$	10^{-2}

(Reproduced with permission from U. van Wasen, I. Swaid and G. M. Schneider, *Angew. Chem., Int. Ed.*, 1980, **19**, 575–587.)

i. Eddy diffusion

As analyte molecules travel through the column, they can follow different pathways around the particles of the stationary phase, some shorter and others longer. These variations in the distance travelled or "multipaths' therefore cause the bands to spread out (*A* term, Equation 2.13, Figure 2.7)

$$\sigma^2 = 2\ L\lambda d_p \qquad (2.13)$$

The spreading depends on the particle size and a geometrical packing factor (λ), whose value increases with decreasing particle size (d_p) but is in the region of 1–2. Thus smaller particles will give more efficient columns as long as they can be uniformly packed. In GLC, dry packing the stationary phase with particles of diameter 0.125–0.150 μm (100–120 mesh) usually represents the optimum. Smaller particles are harder to pack uniformly and cause a higher pressure drop across the column. For open-tubular columns this term is equal to zero as the liquid phase is coated only on the walls of the column. The expression also suggests that in HPLC smaller particles are desirable; however, compared to other causes of band spreading in liquid chromatography, the influence is minor. The literature on the value and overall significance of this term has been recently reviewed [14] and Hawkes has proposed that it could be omitted altogether [12].

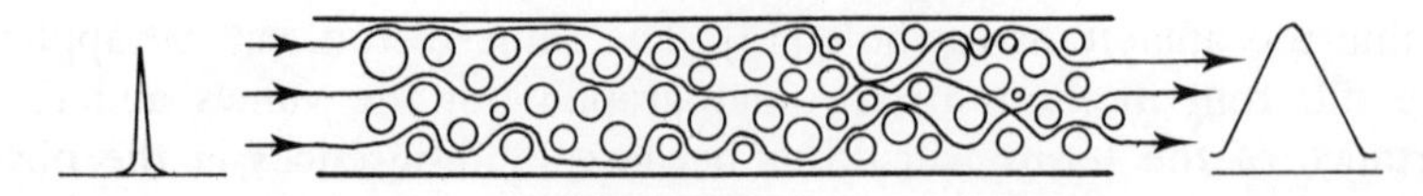

Figure 2.7 Band spreading due to eddy diffusion around the particles in a packed column causing differences in the path lengths through the column.

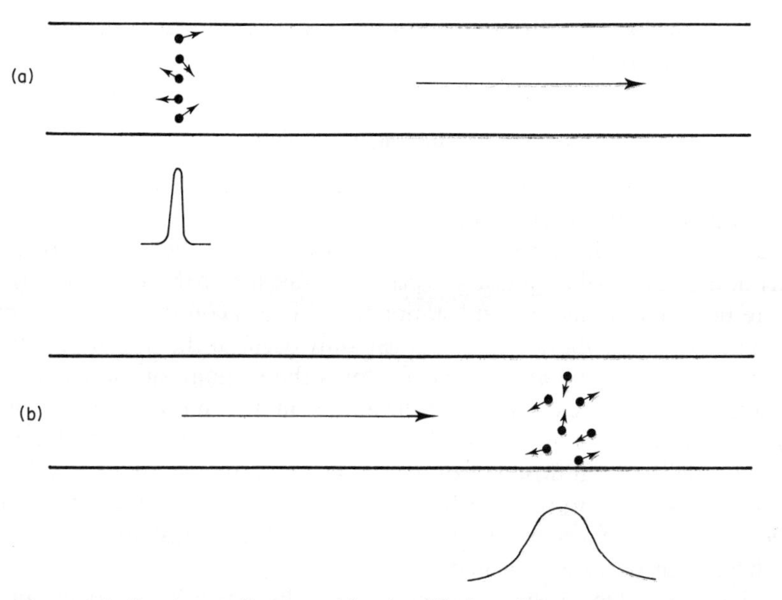

Figure 2.8 Band spreading caused by the diffusion of the analyte molecules in the mobile phase while passing through the column. (a) Initial band. (b) Diffusion of the band with time.

ii. Molecular diffusion

The molecules of any analyte dissolved in a fluid will diffuse in all directions with time. Part of the diffusion will be along the axis of the column and will therefore result in an axial spreading of the peak (Figure 2.8). The extent of the spreading is directly proportional to the coefficient of diffusion of the analyte in the mobile phase (D_M) and the total time the sample is in the mobile phase (L/u)(B term, Equation 2.14).

$$\sigma^2 = \frac{2D_M L\gamma}{u} \tag{2.14}$$

Here γ is a geometrical or tortuosity factor dependent on the nature of the packing. For packed columns it has a value of about 0.6–0.8 or for open-tubular columns a value of 1.0. The diffusion rate (D_M) is dependent on the temperature and pressure of the mobile phase and therefore spreading is reduced by low temperatures and higher column pressures. In GLC, this spreading will also be reduced if a higher-molecular-weight carrier gas is used, as their diffusion rates are lower. This term is quite significant in GLC and can cause undesirable spreading at low flow rates. However, because of the much lower diffusion coefficients in liquids, it can usually be disregarded in HPLC.

iii. Resistance to mass transfer

This is the most important factor for both GLC and HPLC and has been widely studied both experimentally and theoretically. It is often divided into two components, the resistance to mass transfer in the stationary phase ($C_S u$) and in the mobile phase ($C_M u$). The transfer of the molecules of the analyte between the mobile and stationary phases is continually taking place to maintain the distribution ratio but can only occur at the interface between the two phases. As the analyte passes down the column, at the leading and tailing edges of a peak the concentrations in the mobile phase will be changing. However, both phases have a finite thickness into which analyte molecules will have diffused. Consequently, because the analyte molecules have to diffuse back to the interface to respond to the changes in the mobile phase, there will be a time lag before the concentration distribution between the phases can be re-established.

At the front edge of the peak the mobile phase will be relatively rich in the analyte, whereas the stationary phase will be deficient in analyte. If the diffusion to the interface is slow, the mobile-phase concentration will overrun the corresponding concentration in the stationary phase and, as the equilibration belatedly takes place, the peak will be broadened. The extent of the broadening will depend on the diffusion rates of the analyte in the two phases. Because the effect depends on the time taken for the analyte to reach the interface, it will be time-dependent, and the broadening will be worsened as the flow rate of the mobile phase increases. At the tail of the peak, the opposite effect occurs and the stationary phase will be relatively rich in the analyte and any delay in equilibration will stretch the tail of the peak (Figure 2.9).

The effect due to a liquid stationary phase will depend on the thickness of the film (d_f), the diffusion coefficient of the analyte in the stationary phase (D_S) and a geometrical factor (q), whose value depends on the nature of the packing (for open-tubular columns $q = 2/3$) (C_S term, Equation 2.15).

$$\sigma^2 = \frac{Lqk' d_f^2 u}{(1+k')^2 D_S} \qquad (2.15)$$

In practice, these factors suggest that for increased efficiency the stationary phase should be present as a thin film. Thus for a column in which the liquid phase is spread over the surface of an inert support material, the surface area of the support should be as large as possible. This effect means

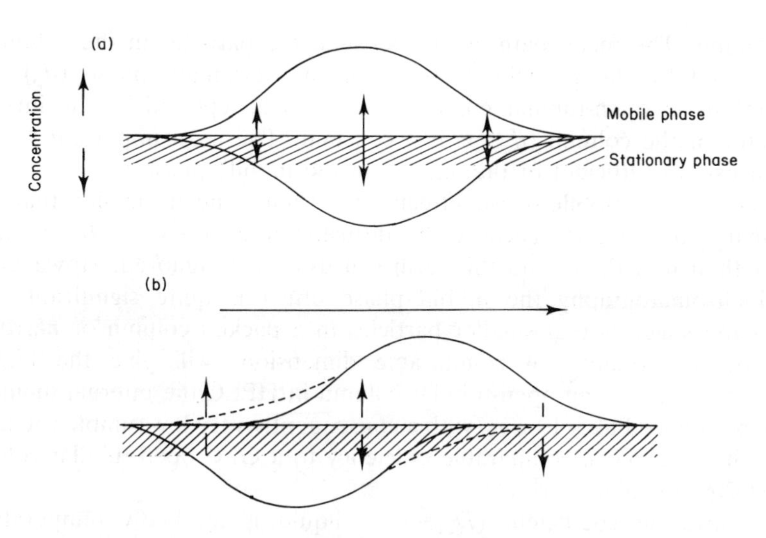

Figure 2.9 The concentration profile of the analyte at the interface of a two-phase partition system. (a) As a static distribution. (b) The changes as the mobile phase moves past the stationary phase and diffusion-controlled processes in the stationary liquid phase and the mobile phase cause band broadening.

that if increasing the proportion of stationary phase is used as a method to increase the retention of an analyte, it will also result in a thicker film and hence some loss of efficiency. This effect is particularly observed in open-tubular columns, as increased film thickness is often also used to increase sample capacity. However, because low-viscosity liquids may form globules on the surface of an open-tubular column, more viscous stationary phases with poorer diffusion characteristics may have to be used to obtain a uniform film thickness. Thus a thin film of a polymerised or crosslinked phase is frequently employed. A less viscous phase may be preferable on a support-coated packed column, where the rougher support surface will prevent agglomeration. With solid stationary phases, the exchange of the analyte on to and off the surface is so rapid that this term can be largely ignored.

The second part of the term for the resistance to mass transfer represents the effects of diffusion in the mobile phase (C_M term, Equation 2.16).

$$\sigma^2 = \frac{Lwf(k')d_p{}^2u}{D_M} \tag{2.16}$$

The affinity of the analyte for the mobile phase is expressed by a function $f(k')$ of the capacity factor, which has only been precisely described for open-tubular columns but implies that increased retention can cause band broadening. The mean path length between the particles in the column is represented by the particle diameter of the stationary phase (d_p). For separations in open-tubular columns this term is replaced by the internal diameter of the column (d_c). w is a constant. The most important term is the diffusion coefficient of the analyte in the mobile phase (D_M).

In GLC, the mobile-phase effects are usually much smaller than the stationary-phase effects because the diffusion rate in a gas (D_M) is much higher than in a liquid and this term can usually be ignored. However, in liquid chromatography the mobile-phase effect is quite significant. The expression suggests that smaller particles in a packed column or narrower open-tubular columns of comparable dimensions will give the highest efficiency. Thus for an open-tubular column in HPLC the internal diameter has to be reduced to 5–10 μm (with a consequently very low sample capacity) before it can achieve comparable efficiency to a GLC open-tubular column of 100–200 μm diameter.

The diffusion coefficient (D_M) in a liquid is markedly temperature-dependent and a reasonable approximation can be calculated using the Wilke–Chang equation (Equation 2.17)

$$D_M = \frac{7.4 \times 10^{-12} T (\psi N_{\text{eluent}})^{0.5}}{\eta V_{\text{solute}}^{0.6}} \tag{2.17}$$

where T is temperature (K), ψ is eluent association factor (non-polar 1.0, methanol 1.9, water 2.6), M_{eluent} is eluent molecular weight, η is eluent viscosity (mPa) and V_{solute} is solute molecular volume (cm^3 mol^{-1}). Thus the diffusion coefficient depends on the size of the analyte molecule, and in practice increasing temperature in HPLC is more likely to be used to increase the efficiency of the separation of larger molecules, such as steroids or peptides, rather than simple aromatic compounds.

Both the expressions for mass transfer effects are simplifications, as in reality the mobile phase in a packed column is also held as a stagnant phase in the pores of the packing material and in pools of different depths caused by irregularities in the surface of the packing material. The layer of a coated stationary phase is also variable in depth. These aspects have been the subject of considerable interest to theoretical and physical chemists. The main conclusion of practical benefit is that the efficiency is increased by uniformity of pore size.

iv. Overall band-spreading equation

The three factors representing band spreading on the column can be brought together by combining Equations 2.13 to 2.16. This gives an equation in the same form as the van Deemter equation (Equation 2.12) [11, 12]. As has been seen, exact numerical values cannot be given for some of the expressions but a graphical representation can illustrate the contributions of the different factors (Figure 2.10). In this plot of HETP against flow rate, the *A* term is independent of the flow rate. At low flow rates the *B* term dominates, whereas as the flow rate increases above a minimum optimum HETP value, the *C* term becomes more important. Thus for any chromatographic system there should be an optimum flow rate for maximum efficiency. At higher flow rates, the efficiency decreases because of mass transfer effects but the time required to carry out an analysis decreases. Often this loss in efficiency has to be balanced against the economic advantage that a faster separation represents in reduced staff time and greater equipment usage.

In GLC carrier gas-flow rates in the region of 15–20 ml min^{-1} for a 2–3 mm i.d. packed column would be below the optimum flow. These are not very different from the typical operating flow rates of 30–40 ml min^{-1} so that for maximum efficiency the optimum value should be determined. The flow rates used for separations on open-tubular GLC columns are usually much above the optimum value in order to reduce the retention times

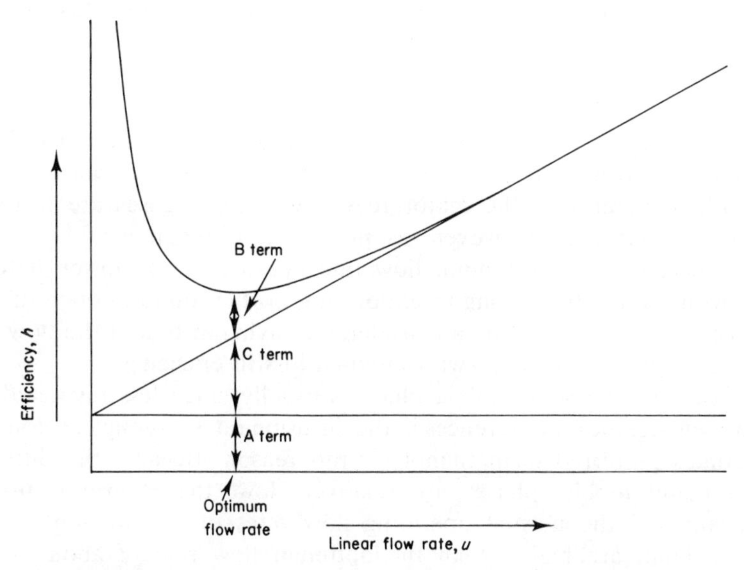

Figure 2.10 Relationship between eluent flow rate and the efficiency of a chromatographic system based on the van Deemter curve. The three terms are: *A*, contribution from eddy diffusion; *B*, contribution from molecular diffusion; *C*, contribution from mass transfer effects.

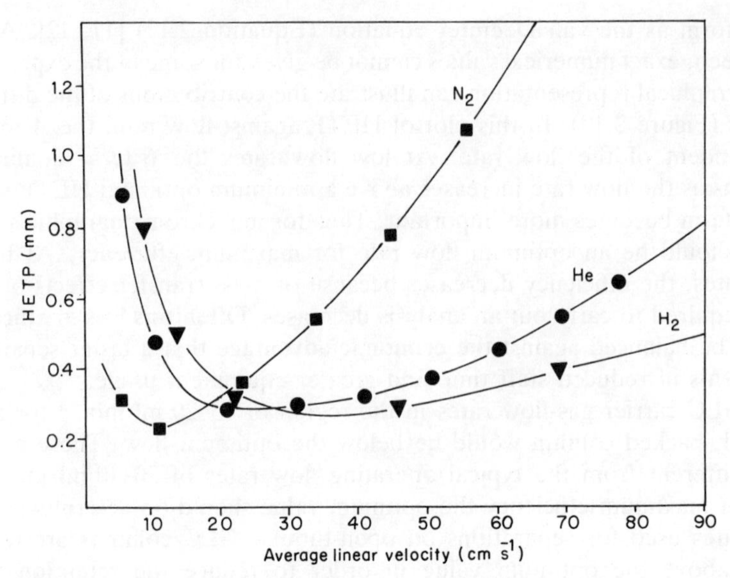

Figure 2.11 Experimental van Deemter curves for open-tubular columns in gas–liquid chromatography showing the effect of the different carrier gases, nitrogen, hydrogen and helium. Column glass WCOT, 25 m × 0.25 mm, OV-101; C_{17} at 175°C; k = 4.94. (Reproduced by permission of Hewlett Packard.)

(Figure 2.11). In this region the *C* term is important and the much flatter curve obtained with hydrogen or helium as the carrier gas, because of their higher diffusion rates, is the major reason why these gases are favoured compared to nitrogen. However, as predicted, nitrogen would give the highest efficiency at its optimum flow rate because of the lower *B* term. Clearly with the relatively long retention times often found in open-tubular chromatography, there will be a considerable saving in time if the flow rate can be substantially increased with minimal loss in efficiency.

In HPLC the choice of mobile phase is usually much less restricted but there are few significant differences in the diffusion rates although acetonitrile is sometimes preferred to methanol for this reason. Because the diffusion rates in liquid mobile phases are relatively low, the *B* term is usually insignificant and the normal operating flow rates of 1–2 ml min^{-1} for a 4.6 mm column are higher than the optimum flow rate of about 0.5 ml min^{-1} (0.05 cm s^{-1}) (Figure 2.12). Significant improvements in the efficiency, particularly for higher-molecular-weight analytes, can usually be made by increasing the temperature because of the effect on the diffusion rates.

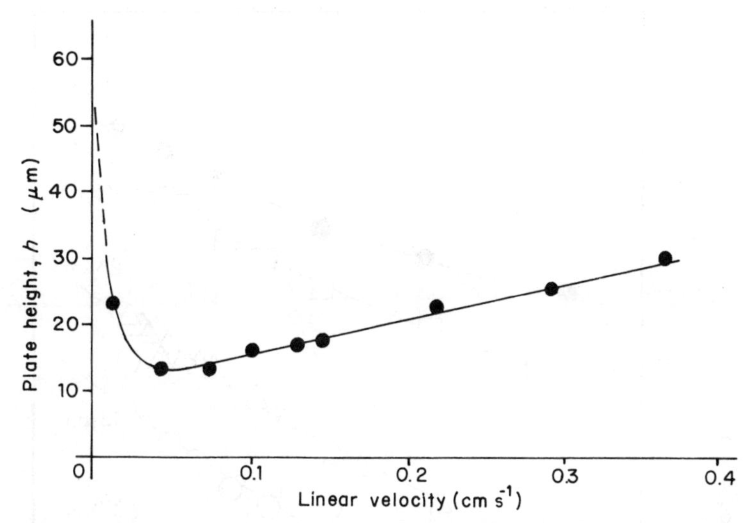

Figure 2.12 Experimental van Deemter curve in HPLC for the separation of anisole (k' = 3.8) on a CP-Spher C-18 column (250 mm × 4.6 mm i.d.) showing the much-reduced contribution from molecular diffusion in liquid chromatography. Mobile phase methanol–water 60:40. (Reproduced by permission of Chrompack.)

v. *Conclusions from van Deemter equation*

Consideration of the factors affecting efficiency in chromatographic systems suggests that a number of approaches can be adopted to maximise resolution. These can also be demonstrated experimentally in both GLC and HPLC. The columns should be uniformly packed with small particles, although if these are too small high-back pressures may limit this approach. It may also become difficult to grade the particles uniformly or to pack them evenly into the column. The effect of different particle diameters in HPLC can be clearly demonstrated (Figure 2.13). The mobile-phase flow rate should be optimised and there are penalties for using too low a flow rate in both time and efficiency. In contrast, by exceeding the optimum flow rate, a time advantage may be gained, which can compensate for lost efficiency.

In GLC, a high-molecular-weight carrier gas has advantages for packed columns but lower molecular-weight gases are preferred for open-tubular columns. In HPLC, temperature can be used to increase efficiency but the most important factor is the particle size of the stationary phase. Open-tubular columns in both techniques have the advantage of avoiding the disturbance to the flow due to the packing material, but for high efficiency in HPLC the internal column diameter should be similar to the optimum

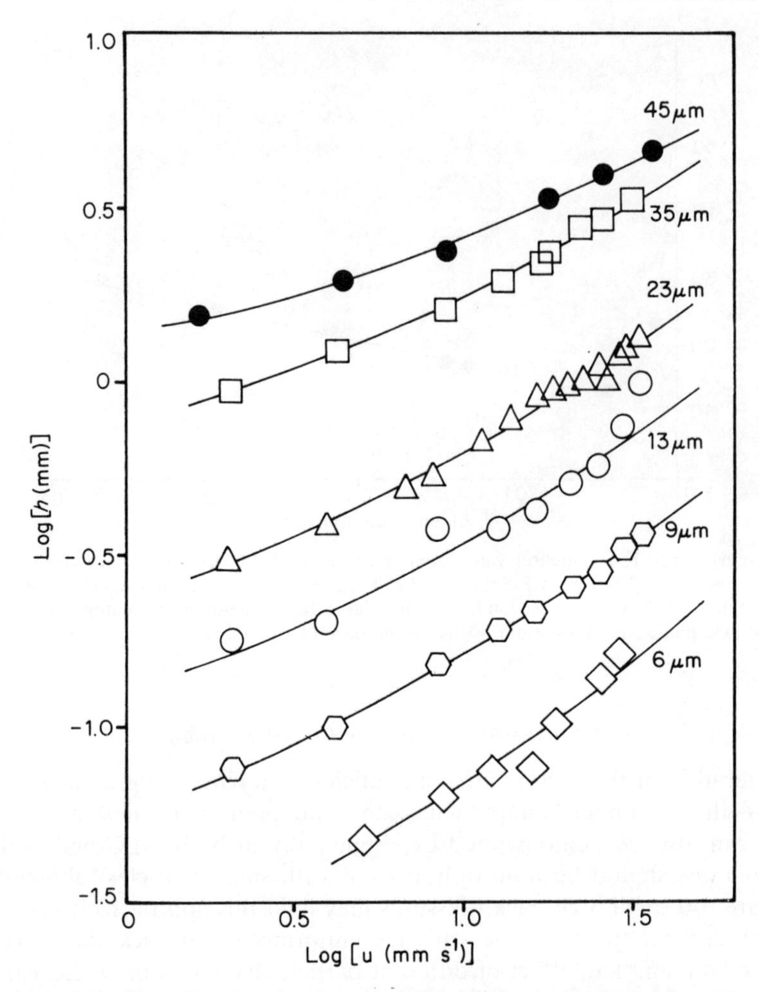

Figure 2.13 Plot of log h against log u (u is the mobile-phase velocity) for silica gel columns of various particle sizes, demonstrating the increase in efficiency that can be obtained with smaller particle sizes. (Reproduced from the *Journal of Chromatographic Science* by permission of Preston Publications, A Division of Preston Industries, Inc.)

Figure 2.14 (opposite) Comparison of gas–liquid chromatographic (GLC) and supercritical fluid chromatographic (SFC) separation of polynuclear aromatic hydrocarbons from coal tar. (a) GLC carried out on an SE-54 open-tubular column 20 m × 0.3 mm i.d., film thickness 0.25 μm. Temperature programmed from 40°C to 265°C at 4°C min^{-1} after 4 min isothermal. Carrier gas hydrogen; flame ionisation detector. (b) SFC separation on an SE-54 capillary column 34 m × 50 μm, i.d., 0.25 μm film thickness. Carbon dioxide as the mobile phase, density programmed from 0.225 to 0.70 g ml^{-1} at 0.005 g ml^{-1} min^{-1} after 15 min isoconfertic (constant density) period. Flame ionisation detection. (Reprinted with permission from J. C. Feldstec and M. L. Lee, *Anal. Chem.*, 1984, **56**, 619A–628A. Copyright (1984) American Chemical Society.)

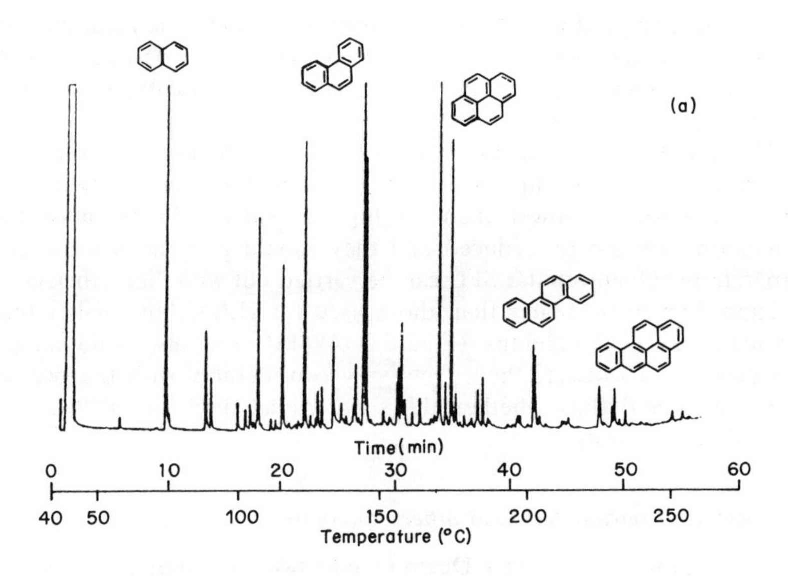
(a)
Time(min)
0
10
20
30
40
50
60
40
50
100
150
200
250
Temperature (°C)

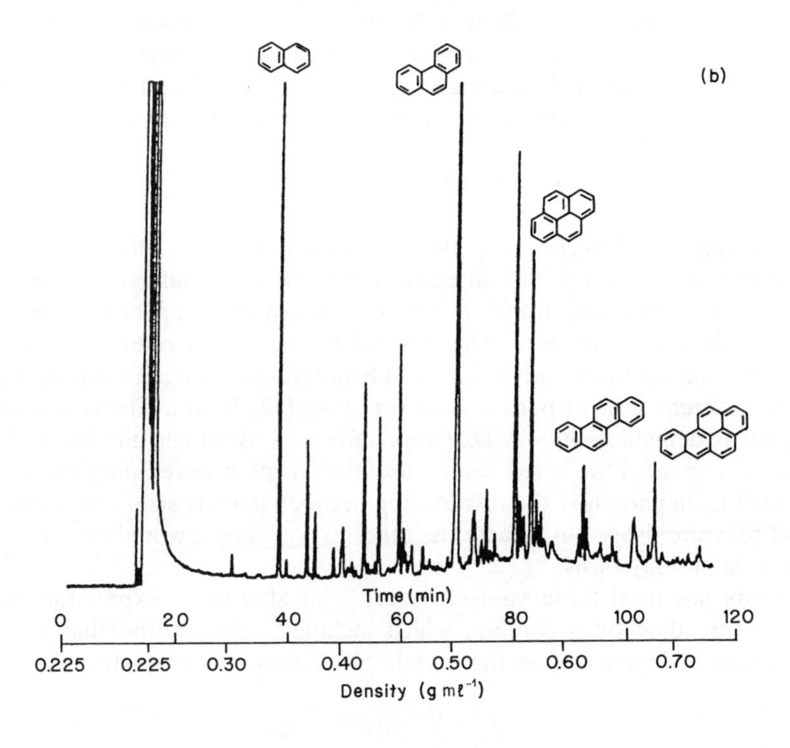
(b)
Time (min)
0
20
40
60
80
100
120
0.225
0.225
0.30
0.40
0.50
0.60
0.70
Density (g mℓ⁻¹)

particle size (typically 3–5 μm) for a packed column. The resulting columns are so narrow that they currently present unacceptable experimental problems of limited sample capacity and poor analyte detectability, which preclude their routine application.

The van Deemter equation can also be used to illustrate the potential advantages of using supercritical fluids as mobile phases. Because their diffusion rates are lower than for liquids (Table 2.1) the mass transfer limitations are greatly reduced and they should give much more efficient separations. Open-tubular SFC can be carried out with high efficiency using column diameters greater than those used for HPLC but smaller than the comparable GLC columns (Figure 2.14) [15] (see also Bibliography in Chapter 16). However, the effects vary considerably with the pressure of the system and thus whether the supercritical fluid has a "gas" like or "liquid"-like density.

b. Reduced parameters and other equations

One problem with the van Deemter equation is that it assumes that the three terms are independent, whereas it can be shown that there is a relationship between eddy diffusion and resistance to mass transfer in the mobile phase. This led Giddings to propose a coupled equation [13], which gave a better correlation with experimental results (Equation 2.18)

$$h = (B/u) + C_S u + \frac{1}{1/A + 1/(C_M u)} \tag{2.18}$$

With open-tubular columns, the equation reduces to the van Deemter equation as the term $A = 0$. In many physicochemical studies of chromatography, it is particularly useful to be able to express band broadening using dimensionless parameters. This has led to the use of reduced parameters and has the additional advantage that comparisons between columns packed with different sizes of particles are very easy [12, 16–18]. The reduced plate height is defined as $h_r = h/D_p$ (sometimes h is used) and the reduced flow rate as $\nu = ud_p/D_M$. It has been a useful concept in developing the theories of HPLC, in particular the relationship between particle size, time of analysis and pressure drop, but as with the other expressions it would not be needed on a day-to-day basis.

Knox has used these values to derive an alternative expression for the band spreading on a column, which includes a recognition that the eddy diffusion is dependent on the mobile-phase flow rate (Equation 2.19).

$$h_r = \frac{B}{\nu} + A\nu^{0.33} + C\nu \tag{2.19}$$

Typical numerical values for HPLC for many of the equations discussed in this chapter have been given in a recent review by Meyer [19].

BIBLIOGRAPHY

J.C. Giddings, *Dynamics of chromatography,* Part 1, *Principles and theory*, Chromatographic Science Series, Vol. 1, Marcel Dekker, New York, 1965 (reissued 1986).

J.Å. Jönsson (Ed.), *Chromatographic theory and basic principles*, Chromatographic Science Series, Vol. 38, Marcel Dekker, New York, 1987.

B.L. Karger, L.R. Snyder and C. Horvath, *Introduction to separation science*, John Wiley, New York, 1973.

J.A. Perry, *Introduction to analytical gas chromatography: history, principles, and practice*, Chromatographic Science Series, Vol. 14, Marcel Dekker, New York, 1981, Chs. 6 and 7.

A.S. Said, *Theory and mathematics of chromatography*, Alfred Hüthig, Heidelberg, 1981.

P. Sewell and B. Clark, *Chromatographic separations*, John Wiley, Chichester, 1987.

L.R. Snyder, *Principles of adsorption chromatography. The separation of nonionic organic compounds*, Marcel Dekker, New York, 1968.

REFERENCES

1. H.M.N.H. Irving, H. Freiser and T.S. West, *Compendium of analytical nomenclature. Definitive rules 1977*, Pergamon, Oxford, 1978 (IUPAC Orange Book).
2. *Standard practice for liquid chromatography terms and relationships,* ASTM E 682, American Society for Testing and Materials, Philadelphia, 1979.
3. *Standard recommended practice for gas chromatography terms and relationships,* ASTM E 355, American Society for Testing and Materials, Philadelphia, 1977.
4. L.S. Ettre, "The nomenclature of chromatography. I. Gas chromatography", *J. Chromatogr., Chromatogr. Rev.*, 1979, **165**, 235–256.
5. L.S. Ettre, "The nomenclature of chromatography. II. Liquid chromatography", *J. Chromatogr., Chromatogr. Rev.*, 1981, **220**, 29–63.
6. L.S. Ettre, "The nomenclature of chromatography. III. General rules for future revisions", *J. Chromatogr., Chromatogr. Rev.*, 1981, **220**, 65–69.
7. R.J. Smith, C.S. Nieass and M.S. Wainwright, "A review of methods for the determination of hold-up volume in modern liquid chromatography", *J. Liquid Chromatogr.*, 1986, **9**, 1387–1430.
8. G.E. Berendson, P.J. Schoenmakers, L. de Galen, G. Vigh, Z. Varga-Puchony and J. Inczédy, "On the determination of the hold-up time in reversed phase liquid chromatography", *J. Liquid Chromatogr.*, 1980, **3**, 1669–1686.
9. J.J. Kirkland, W.W. Yaw, H.J. Stoklosa and C.H. Dilks, "Sampling and extra-column effects in high-performance liquid chromatography; influence of peak skew on plate count calculations", *J. Chromatogr. Sci.*, 1977, **15**, 303–316.
10. B.A. Bidlingmeyer and F.V. Warren, "Column efficiency measurement", *Anal. Chem.*, 1984, **56**, 1583A–1596A.

11. J.J. van Deemter, F.J. Zuiderweg and A. Klinkenberg, "Longitudinal diffusion and resistance to mass transfer as causes of nonideality in chromatography", *Chem. Eng. Sci.*, 1956, **5**, 271–289.
12. S.J. Hawkes, "Modernization of the van Deemter equation for chromatographic zone dispersion", *J. Chem. Educ.*, 1983, **60**, 393–398.
13. E. Grushka, L.R. Snyder and J.H. Knox, "Advances in band spreading theories", *J. Chromatogr. Sci.*, 1975, **13**, 25–37.
14. C.L. De Ligny, "The contribution of eddy diffusion and of the macroscopic mobile phase velocity profile to plate height in chromatography. A literature investigation", *J. Chromatogr., Chromatogr. Rev.*, 1970, **49**, 393–401.
15. M. Novotny, "Capillary separation methods: a key to high efficiency and improved detection capabilities", *Analyst*, 1985, **109**, 199–206.
16. G.R. Laird, J. Jurand and J.H. Knox, "New columns for old in liquid chromatography", *Proc. Soc. Anal. Chem.*, 1974, **11**, 310–317.
17. J.H. Knox and H.P. Scott, "*B* and *C* terms in the van Deemter equation for liquid chromatography", *J. Chromatogr.*, 1983, **282**, 297–313.
18. J.H. Knox, "Practical aspects of liquid chromatography theory", *J. Chromatogr. Sci.*, 1977, **15**, 352–364.
19. V.R. Meyer, "High-performance liquid chromatographic theory for the practitioner", *J. Chromatogr., Chromatogr. Rev.*, 1985, **334**, 197–209.

CHAPTER 3

GAS–LIQUID CHROMATOGRAPHY: INSTRUMENTATION

3.1 INTRODUCTION TO GAS–LIQUID CHROMATOGRAPHY

Gas–liquid chromatography (GLC) has found widespread application in many areas of chemical analysis and a wide range of compounds can be examined. The separation is carried out using a gaseous mobile phase, which transports the analyte through a column containing the liquid stationary phase. The retention of the analyte depends on the degree of its interaction with the liquid phase and its volatility. The column can be heated, typically within the range 50–300°C, to increase the volatility of the sample and reduce retention times. The limitations on the suitability of compounds for analysis by GLC are that the analyte must have a significant vapour pressure and be thermally stable at some point within this temperature range. GLC can therefore be applied to all gases, most non-ionised small and medium-sized (up to about C_{25}) organic molecules (either liquid or solids) and many organometallic compounds, but it cannot be used for organic or inorganic salts or for macromolecules (either synthetic or biological polymers). In some cases, chemical modification to form a more volatile or more stable derivative can convert an unsuitable analyte into a suitable form for analysis. If involatile components are present in a sample, care must be taken that they are not deposited at the start of the column where they can intefere with subsequent analyses.

As an analytical technique, GLC is very simple to operate and is capable of high resolution, selectivity and sensitivity. For many years it dominated separation methods until the relatively recent development of HPLC as an instrumental technique. The methods and instrumentation have changed little over the years and the main recent advances have been in the use of microelectronics for instrumental control and data collection and a great increase in the use of open-tubular columns compared to packed columns. Almost all separations in gas chromatography use a liquid stationary phase. The corresponding technique with a solid stationary phase, gas–solid chromatography (GSC), is relatively unimportant except in a few specialised areas of gas analysis. This book will therefore concentrate almost entirely on GLC but the instrumentation and detection methods of the two techniques are virtually identical.

GLC has been the subject of a large number of journal articles and monographs, and a number of general texts are listed in the Bibliography. Books on more specialised topics will be listed in subsequent chapters and compilations of the applications of GLC will be found in Appendix 1.

3.2 INSTRUMENTATION FOR GAS–LIQUID CHROMATOGRAPHY

The basic design of a gas chromatograph has remained virtually unaltered since the earliest commercial instruments were produced in the 1960s. The essential instrumental components are a regulated flow of carrier gas, an injection port to introduce the sample into the carrier gas, a temperature-controlled oven containing the stationary phase in an analytical column (see Chapter 4), and a detector and recording device (see Chapter 5) to monitor the components of the sample as they are eluted from the column (Figure 3.1). The separation can be carried out by using a packed column in which the stationary phase is a thin liquid film coated onto an inert support material. Alternatively, open-tubular or capillary columns can be used in which the stationary phase is spread as a thin film on the walls of a narrow tube.

Over the years gas chromatography has been shown to be a generally reliable and reproducible technique with few mechanical or instrumental problems. However, as with any sophisticated technique, the equipment has to be well maintained and operated following reasonable precautions. Some hints on practical operation and problems are discussed in Appendix 2.

a. *The carrier gas*

The selectivity of any chromatographic system can be altered by changing either the liquid or stationary phases as this will change the distribution constants. However, in GLC the choice of a carrier gas is limited by practical constraints and is not normally a factor that can be used to improve the separation of two compounds. The carrier gas should be non-reactive so that it does not alter the analyte, non-toxic and non-flammable, so that it

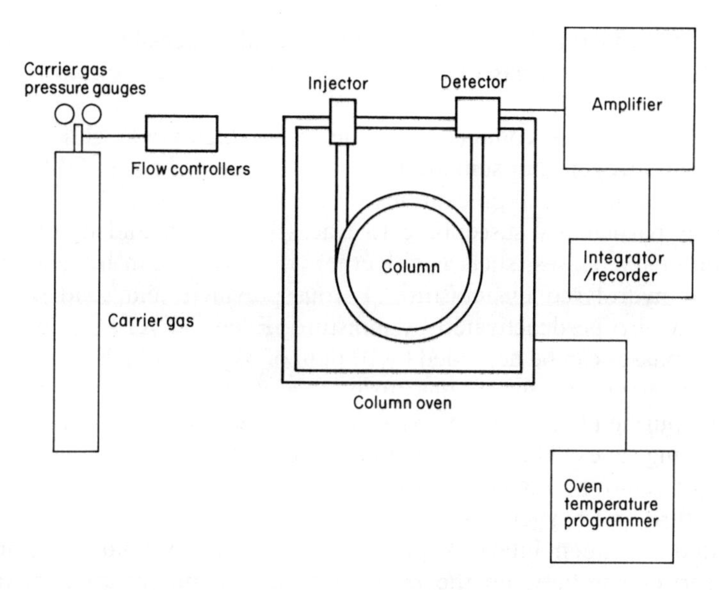

Figure 3.1 Schematic diagram of gas–liquid chromatograph.

can be vented at the end of the instrument without risk to the operator, and cheap, in order to reduce running costs as it is discarded after use. The theoretical background of band spreading suggested that gases with a higher molecular weight and hence a low diffusivity will give the highest optimum efficiency. The only gas that fully satisfies these requirements is oxygen-free nitrogen. An alternative is helium, but it is more expensive and has a low molecular weight. It is much cheaper in the USA than in Europe and is correspondingly more popular there. Argon could be possible but it is much more expensive and offers few advantages. Hydrogen has also been widely used although it is highly flammable and mixtures with air can be explosive. Its use therefore requires precautions, and a leak sensor in the column oven is desirable. Hydrogen may also react with the analyte to give hydrogenated artefacts, which give additional peaks in the chromatogram [1]. Hydrogen and helium are preferred for separations on open-tubular columns. Although the efficiency at the optimum flow rate is poorer than with nitrogen, the van Deemter curves are flatter (see Figure 2.11). These gases can therefore be used at higher flow rates without significant loss in efficiency and thus give shorter retention times. They are also employed when a thermal conductivity detector is being used as their high thermal conductivities enhance the sensitivity for most organic compounds (Section 5.2.a). The choice of carrier gas is therefore largely determined by the column or detector required for the separation. Occasionally specialised gas mixtures are used as the carrier gas to enhance sensitivity with the electron capture detector (Section 5.3.a).

For most purposes commercial cylinders of high-grade gases, such as oxygen-free nitrogen, are suitable for use as the carrier gas in GLC, but for some applications special precautions should be taken. The electron capture detector is particularly susceptible to traces of oxygen and moisture, and some stationary phases, such as polyethers or polyols, can be degraded by oxygen or hydrolysed by moisture. Porous polymers and solid stationary phases may also be deactivated by moisture. Even the very stable siloxane stationary phases can be degraded by 10 ppm of oxygen at high temperatures. In these cases and whenever the purity of a carrier gas is suspect, the gas should be purified before use by passing it through a series of gas adsorption and/or drying tubes. These are commercially available and contain activated carbon (to remove organic impurities), Drierite, or molecular sieve traps (for moisture and oxygen). The purification system should be checked and reactivated at frequent intervals. Care should be taken to avoid contamination of the carrier gas between the traps and the column by using metal gas lines. Nylon tubing, although widely used, is porous and moisture or oxygen may permeate into the system. Hydrogen for flame detectors (see Chapter 5) is normally used directly from the cylinder. Gas purification systems may be needed on compressed air supplies for detectors, especially if they are provided from a compressor as traces of oil may be present.

The carrier gas is usually controlled in two stages. Firstly the outlet pressure of the cylinder is fixed by a diaphragm pressure valve to about 30–70 psi (2.1–4.8 bar). The flow rate through the column is then regulated with a mass flow controller valve on the chromatograph. This should ensure a constant flow rate irrespective of changes in the back-pressure of the column. For maximum reproducibility the flow controllers should be thermostated to control the gas density. Typical flow rates for 3 mm i.d. packed columns would be 30–40 ml min^{-1}. In older chromatographs the valves were usually set manually and were therefore difficult to reproduce, but some recent instruments have digitally controlled or servo-operated flow controllers. The flow rates though open-tubular columns are much lower (2–4 ml min^{-1}) and in many instruments a finely adjusted pressure controller is used instead of a mass flow controller, because the latter cannot compensate for the changes in the split ratio during the injection sequence (see later).

The flow rate of carrier gas through the column is usually measured at the detector end of the column by means of a bubble-flow meter. The time taken for the carrier gas to carry a soap bubble up a calibrated tube of known volume is measured with a stop watch. A rapid check that the carrier gas is being eluted from the column is always a useful first step in cases of operating problems with a chromatograph.

b. Sample injection

The aim of the injection is to introduce the sample as a sharp band into the carrier gas stream. The samples for GLC can be either liquids, solids, or gases. Solids and in most cases liquids are usually injected using a syringe as dilute solutions in a volatile solvent. The solvent is unretained and passes straight through the column to give an overloaded solvent peak. The front edge of this peak is often used as a marker of the the column void volume (see Figure 1.2). Except for samples of volatile solvents, neat liquids are rarely injected because they volatilise too slowly or are too viscous to be rapidly expelled from the syringe needle.

The injection takes place through a septum, which is made of rubber or elastomer, and seals as the needle is withdrawn. This septum can be a source of operational problems as repeated injections will eventually cut into the seal and allow the carrier gas to leak. In addition, traces of the sample may be deposited or adsorbed onto the inner surface of the septum. If the column temperature is then raised in a programmed separation, these compounds may be released to give spurious peaks on the chromatogram. Septa can also contain plasticisers and volatile monomer components, which can be released in a similar manner. A separation should therefore be carried out with a solvent blank. If necessary, a high-temperature grade of septum, with a lower bleed rate, should be used, but these are often harder and more difficult to pierce. In some instruments, part of the carrier gas is bled to waste from just below the septum to reduce these problems.

Gases can be injected using a gas-tight syringe (50–1000 μl) but, because the volume of a gas is so dependent on its pressure, reproducibility can be a problem. Higher precision can be achieved by using a six-port sampling valve. A section of tubing of known volume is filled with the sample at atmospheric pressure and this is then switched into the carrier gas flow in a similar manner to the injection of a liquid sample in HPLC (see Figure 10.10). Because the pressures involved in GLC are much lower than in HPLC, the mechanical requirements of seals and components in the gas injection valve are not so demanding and the valve can be much simpler and cheaper.

Although solids are invariably injected as dilute solutions, they can be qualitatively injected using a specially designed syringe needle with a cavity in the tip, but slow volatilisation can cause problems.

i. Packed column injection

The normal technique with packed columns is to inject between 0.1 and 10 μl of a dilute solution or between 0.01 and 0.2 μl of a neat liquid using a 1, 5 or 10 μl syringe either into a space at the top of the column or into the bed of the stationary phase. Larger samples of pure liquids would usually overload the detector or column. Most instruments have a heated injection zone, which is set at 25–40°C higher than the column temperature to ensure rapid volatilisation of the sample. The length of the syringe needle should be selected so that when the needle is fully inserted, the injection will occur in the centre of the heated zone.

Injection onto the bed of the column is believed to give more reproducible peaks and higher efficiency. However, part of the stationary phase will be in the heated injector zone and the liquid phase may be degraded if its maximum recommended temperature is exceeded (see Chapter 4).

ii. Open-tubular column injection

Open-tubular columns have only a thin film of the liquid phase on the inner wall of the column and have a much lower sample capacity than packed columns. To avoid overloading the stationary phase, a number of specialised techniques have been used for sample introduction. Because these columns can separate large numbers of components in a single sample, they are often used for complex mixtures containing components with a wide range of volatilities, such as essential oils or petroleum samples. Any injection technique must therefore deliver all these components equally to the column without discrimination. Three main techniques have been used, namely split, splitless and on-column injection, and each has its own advantages and disadvantages. These methods have been the subject of intensive research and debate over the last 5–10 years and more detailed discussions can be found in the books and references listed in the Bibliography.

Wide-bore open-tubular columns with internal diameters of 750, 630, or 550 μm or narrower columns with particularly thick layers of the stationary phase will have a greater sample capacity (but lower efficiency) and the same injection techniques as for packed columns can be used.

Split injection

This is the simplest method and is ideal if only a limited number of components are present or very accurate quantitative results are not required. The injection port is fitted with two valves, one acting as a septum purge and allowing a small flow of carrier gas from just below the septum, and the second taking carrier gas from the bottom of the injection port near the column inlet (Figure 3.2a). The upper septum bleed valve is largely designed to remove any compounds being lost from the septum and is usually set to a very low flow rate (2–5 ml min^{-1}). In the split mode, the lower valve is used to adjust the ratio of carrier gas going to waste in the atmosphere compared to flow onto the column. Typically ratios of 10:1 to 500:1 are used so that only a small proportion of the sample is transferred onto the column. The analytes should be present as 0.01–10% of the sample solution and normally small sample volumes of 0.1–0.5 μl are injected.

Although this method of injection is easy to set up and use, it can discriminate against high-boiling components. This has been attributed to the pressure pulse and change in the vapour viscosity produced by the vaporisation of the solvent, which directs a higher proportion of the more volatile components onto the column (Figure 3.3). However, these discrimination effects are reproducible and in quantitative studies can be compensated for by the use of standard samples. For samples containing analytes with a limited range of volatility, few discrimination effects are observed. However, because much of the sample is vented to waste, the split injection technique does not enable the maximum sensitivity to be obtained from the sample, which is particularly important when limited amounts of the sample are available.

Splitless injection

This technique is particularly suitable for samples containing trace compounds in low concentrations, when it is necessary to transfer as much of the sample as possible onto the column to increase the sensitivity. It uses the same split/splitless injection port as for split injection (Figure 3.2b) but with a medium or wide open-tubular column (0.3 mm i.d. or greater). In this case the lower valve is closed at the time of injection and only a small flow-septum purge flow is used. A relatively large volume (0.5–5 μl) of the sample in a volatile solvent can be injected. The analytes should be present in low concentrations (no more than 50 ppm) to prevent overloading the column. The injection takes place into a heated zone over a period of up to 20 s, so that all the sample is transferred to the column. After a predetermined period (such as 40–60 s) the lower valve is opened for 20–60 s to purge the injection port and prevent residual sample from causing tailing of the peaks.

This long injection period would normally cause band broadening, so the vaporised sample has to be concentrated as a sharp band on the top of the

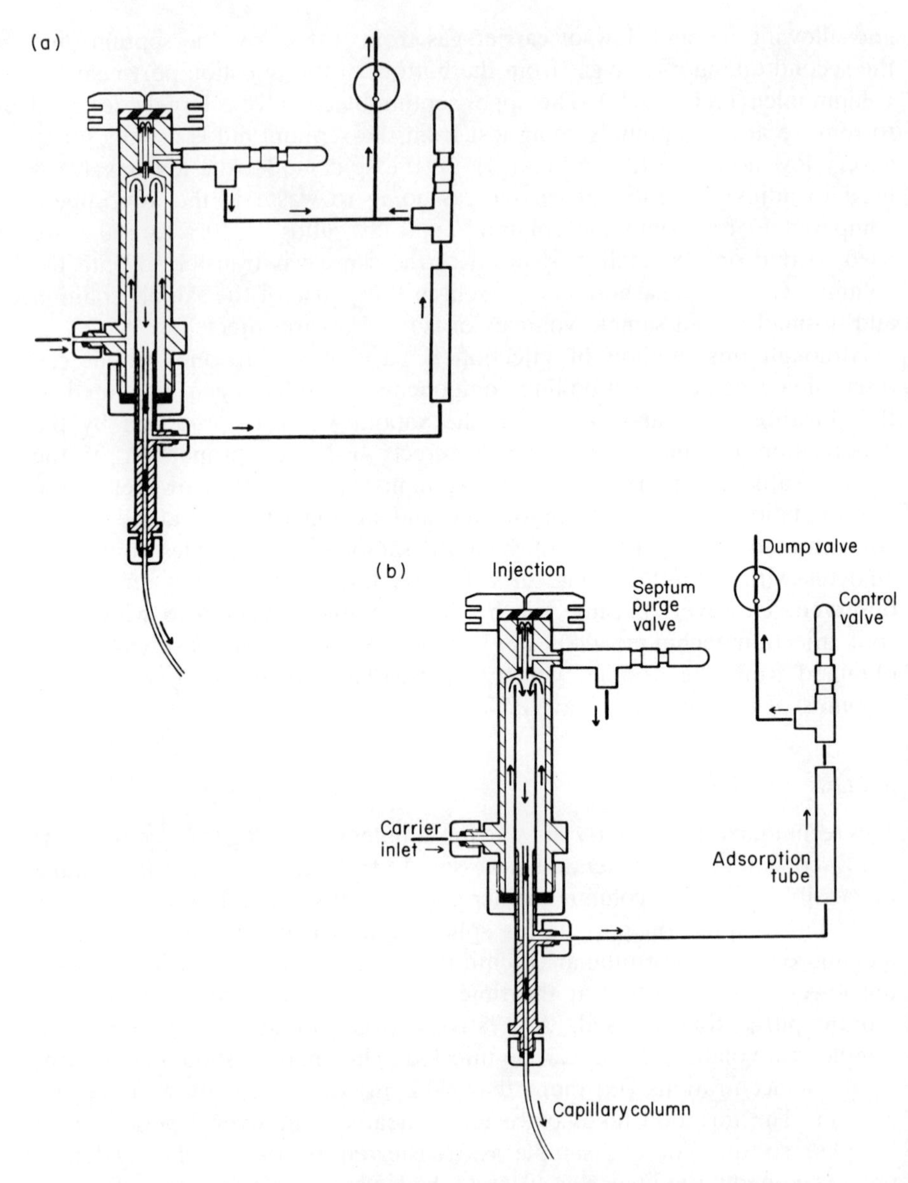

Figure 3.2 Injector for open-tubular columns. (a) Split injection mode: both the septum and bottom purge valves are open so that only a small proportion of the sample is carried onto the column. (b) Grob-type splitless mode: during injection the bottom valve is closed, and after a fixed time (e.g. 40 s) the dump valve is opened to purge any residue from the injection port. (Reproduced by permission of Scientific Glass Engineering.)

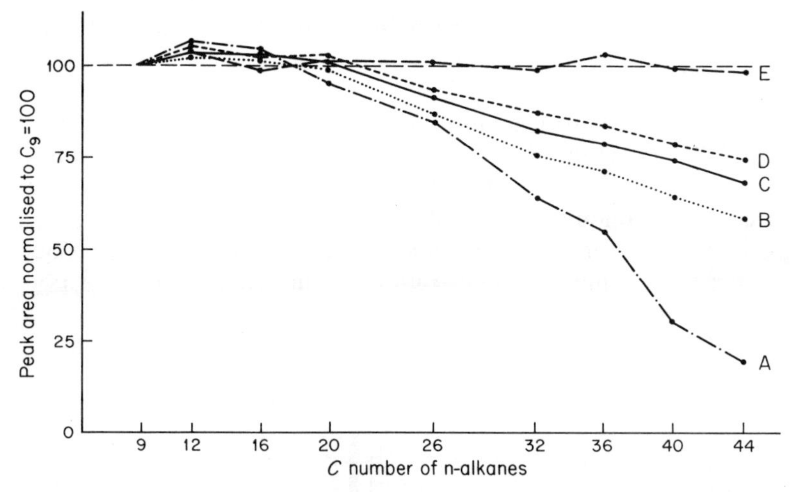

Figure 3.3 Examples of discrimination effects on injection of a homologous series of alkanes using different techniques. Curves A to D show the loss of the higher homologues with split injection using different syringe handling methods. Curve A, sample in syringe needle during injection. Curves B–D, sample in barrel during injection, with either: B, immediate injection through a cold needle; C, using the solvent wash method (Appendix 2); D, injection after allowing the needle to heat up. Curve E, cold on-column injection method, avoiding any analyte losses. (Reproduced by permission of Dr. Alfred Hüthig Verlag from *J. High Res. Chromatog., Chromatogr. Commun.*, **2**, 20 (1979).)

column. This can be achieved either by cooling the column (cold-trapping) or using the solvent effect.

In the cold-trapping method the column is held at least 100–150°C below the boiling points of the analytes of interest. They are trapped at the top of the column and the solvent is unretained. The column temperature is then programmed to release the analytes and enable separation to take place. This method will therefore not work if volatile analytes are present in the sample, as they would not be retained. Because the changes in the purging can cause large differences in the flow rate through the injector, these methods preclude the use of a mass flow controller and the column flow rate is controlled by applying a constant carrier gas pressure.

In the solvent effect or "Grob' method a solvent with a relatively high boiling point, such as octane (b.p. 126°C), is used. On injection, the volatilised solvent condenses at the start of the cool column to form a thick layer on the wall of the column, which increases the sample capacity. This assists in the trapping of the analytes, holding them in the first few coils of the column. For this method to operate the initial column temperature must be 10–30°C lower than the boiling point of the solvent. Once the injector has been purged, the column temperature is increased and separation takes place.

A number of companies now also offer a programmed temperature vaporiser (PTV) system or "cold split/splitless injector', in which the temperature of a split/splitless injector port can be cooled during the injection of the sample as a liquid. The injector is then rapidly heated to transfer the sample components to a cool column before programming begins. Because this avoids a high injector temperature it can reduce decomposition and sample changes.

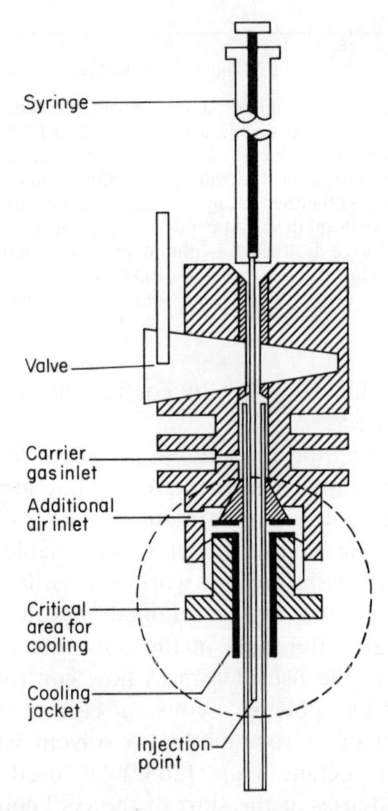

Figure 3.4 Injector for direct cold on-column injection onto a capillary column showing rotating value and insertion of needle into the bore of the column. (Reproduced by permission of Dr Alfred Hüthig Verlag from *J. High Res. Chromatogr., Chromatogr. Commun.*, **2**, 368 (1979).)

On-column injection

Unlike the previous two methods in which the sample is first volatilised and then passed to the column, with on-column injection the sample is placed directly into the column in the liquid phase, so that sample decomposition or loss is reduced. This method is also suitable for dilute samples, 0.01 to 300 ppm. The injection port contains a valve (Figure 3.4) rather than the normal septum. The injection is carried out using a very fine quartz capillary needle, which, for a column with an internal diameter of 0.25 mm, would have an external diameter of 0.17 mm. The needle is closely fitting in the valve and blocks the escape of sample during the injection step. Sometimes a short length of uncoated tubing (two to three coils) is placed at the start of the column as a "retention gap', which results in increased efficiency, because on evaporation the analytes are focused onto the start of the coated column.

The top of the column is cooled with a external cooled air flow to 10–30°C below the boiling point of the solvent. The sample is rapidly injected to form a liquid film on the wall of the column. The temperature of the column is then raised to separate the solvent and analytes. The advantages of this method are that the injection conditions are very mild and less decompositon of the sample occurs than with the other techniques. All the sample is transferred to the column so that there can be no discrimination or needle volatilisation effects. This method is therefore particularly suitable for samples with a wide range of components of different volatilities (Figure 3.3E).

Although handling the fine needle and the injection valve requires greater operator skill, in recent years the valve operation has been automated and this can now be regarded as a routine method.

iii. Automatic injection

If multiple samples are being analysed, an automatic injection system can be used, which mimics the manual injection process but can be programmed to take samples from a carousel or rack of samples. Each of the different injection techniques can be used and the injection system together with the column oven temperature are usually controlled by an integrator or separate microcomputer. As well as enabling a higher sample throughput to be achieved by unattended overnight operation, automatic injection usually increases the reproducibility of the results. It should eliminate much of the variation caused by the injection technique of the individual operator. For this reason, some manufacturers now offer one-shot automatic injectors, which can handle a single sample and, by controlling every stage of the injection sequence, can increase reproducibility.

c. *Column oven*

The temperature of the column is controlled by the oven. In early gas chromatographs, the oven had a high thermal mass to ensure temperature stability. Now it is considered more important to maintain a uniform temperature throughout the oven and column. Modern instruments therefore use a circulating air bath, in which air is driven by a fan over a heater and around the oven to ensure rapid mixing. This is particularly important for open-tubular columns because of their low thermal mass. Even small irregularities in temperature at different points on the column can cause loss of efficiency or produce spikes on the peaks, the so-called Christmas tree effect [2]. The low thermal mass of the oven also enables the operator to change the temperature rapidly and reproducibly during an analysis. The temperature in the oven is controlled by thermocouples and feedback proportional heaters to within ±0.1°C of the set values and the temperature can be programmed to increase at up to 30–40°C min^{-1}.

For simple samples, separations are usually carried out under isothermal conditions of constant temperature. This method is also preferred when accurate retention measurements are needed for the identification of analytes, as it is easy to control and reproduce. However, many samples contain components with a wide range of volatility. If these are analysed isothermally at a high temperature, the first components will be rapidly eluted and will probably be unresolved (Figure 3.5). However, if a lower column temperature is used, the overall retention time of the separation will become uneconomically long and the later peaks will be so broad that they may be lost as baseline drift. For these samples it is very useful to program the oven temperature to increase at specified points during the analysis (Figure 3.6). This enables the volatile compounds to be resolved at a low temperature and the elution of the later less-volatile components to be speeded up as the temperature increases. The programmer on most instruments can include a series of timed events including initial isothermal periods, pauses at intermediate fixed temperatures, different rates of temperature rise, a final isothermal period and automated cooling-down of the column to the initial temperature at the end of the run, ready for a new cycle (Figure 3.7). To carry out these operations, microcomputers have now largely taken over from mechanical timers.

At the end of the run, the oven has to be cooled below the initial temperature setting to avoid any temperature hysteresis caused by the

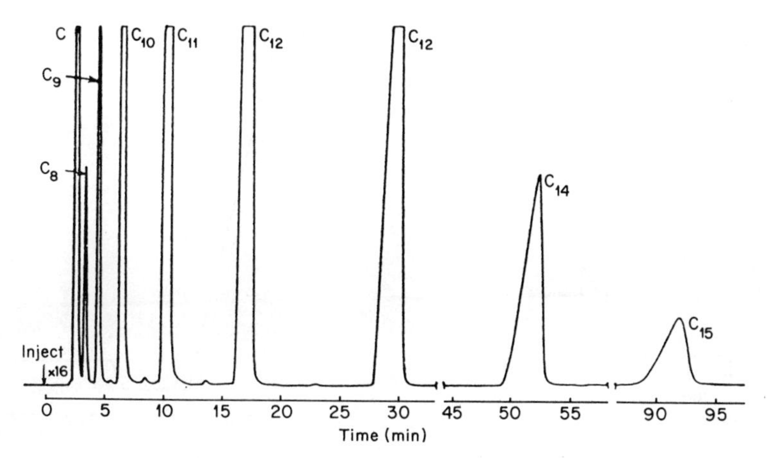

Figure 3.5 Isothermal separation of homologous series of alkanes at 150°C. Column 3% Apiezon L on 100–120 mesh VarAport 30. Carrier gas helium at 10 ml min^{-1}. (Reproduced by permission of Varian Associates from McNair and Bonelli, *Basic gas chromatography*, p. 190.)

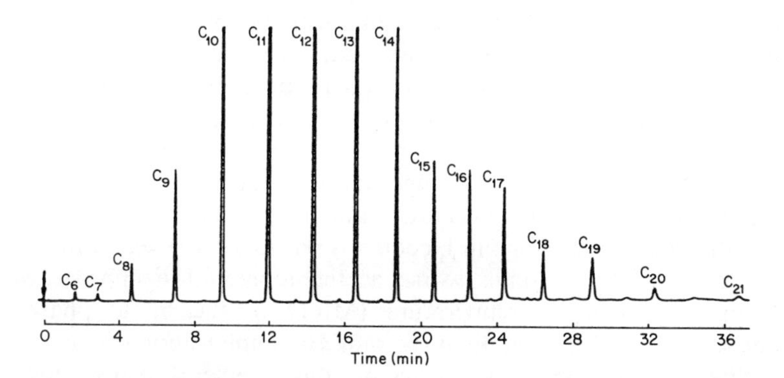

Figure 3.6 Sample and column as Figure 3.5 but temperature-programmed separation showing the greater range of hydrocarbons, which can be resolved in a shorter time. (Reproduced by permission of Varian Associates from Mc Nair and Bonelli, *Basic gas chromatography*, p. 190.)

thermal mass of the oven. It must then be equilibrated at the starting conditions before the next injection can take place. As a result the cycle time of a programmed separation can take much longer than an isothermal separation. The advantage of the better separation from the programmed separation must then be balanced against the extra time required, and in many cases isothermal separation at a lower temperature can achieve a

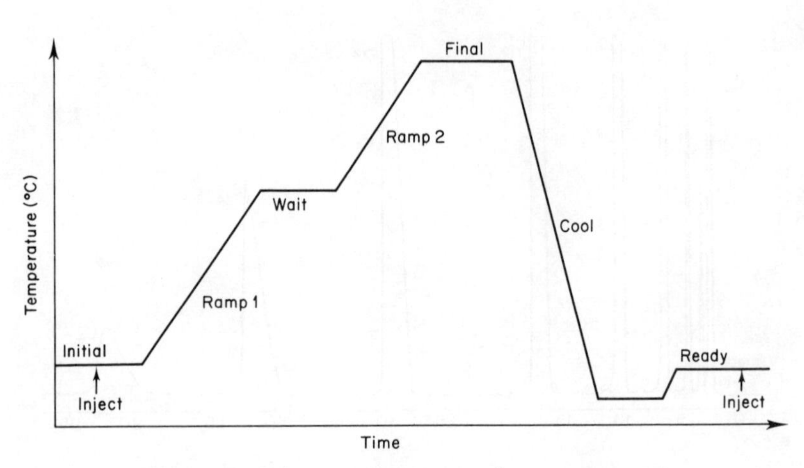

Figure 3.7 Column oven programmed temperature cycle.

comparable result. Temperature programming has the additional advantage that the increasing temperature also sharpens the peak shapes of the later components. The increase in peak width with retention time found with isothermal separation can thus be avoided, giving increased peak heights and sensitivity. Programmed operation is particularly useful for samples, such as crude oils, which contain a very wide range of materials.

Hovever, temperature programming can cause a drift in the baseline output. This is because of small changes in the background signal as residues are eluted from the column or because of the thermal or chemical degradation of the stationary phase. Column bleeding is particularly severe with packed columns and usually becomes worse as the temperature approaches the maximum recommended temperature (MRT) of the liquid phase (see Chapter 4). This problem can limit the range of amplification of the detector signal that can be used. To overcome this problem, many older gas chromatographs were designed to use two columns and two detectors in parallel. The columns should contain identical packing material and the signal from the reference column detector is subtracted from that from the analytical column detector to compensate for the bleeding. An alternative technique is to store the background signal from a single analytical column in a microprocessor memory and carry out an electronic subtraction on subsequent runs. In practice the dual-column instruments are often operated using two different columns and separate detectors to avoid the need for repeated change columns.

If possible the operating conditions that cause column bleeding should be avoided, as they are indication of column degradation or contamination.

Both may result in a detrimental change in the composition and selectivity of the stationary phase and may cause changes in the reproducibility of the chromatographic separation. These problems rarely arise with open-tubular columns because the stationary phases are generally more stable and only single-column instruments are used.

BIBLIOGRAPHY

General texts on gas–liquid chromatography

F.L. Bayer, "Gas chromatographic equipment", *J. Chromatogr. Sci.*, 1986, **24**, 549–568.

A. Braithwaite and F.J. Smith, *Chromatographic methods*, Chapman and Hall, London, 1985.

R.L. Grob (Ed.), *Modern practice of gas chromatography*, 2nd ed., John Wiley, New York, 1985.

E. Heftmann (Ed.), *Chromatography. Fundamentals and applications of chromatographic and electrophoretic methods*, Part A, *Fundamentals and techniques*, Part B, *Applications*, J. Chromatogr. Library, Vol. 22, Elsevier, Amsterdam, 1983.

W. Jennings, *Analytical gas chromatography*, Academic Press, New York, 1987.

H.M. McNair and E.J. Bonelli, *Basic gas chromatography*, Varian, Palo Alto, 1969.

J.A. Perry, *Introduction to analytical gas chromatography: history, principles, and practice*, Chromatographic Science Series, Vol. 14, Marcel Dekker, New York, 1981.

C.F. Poole and S.A. Schuette, *Contemporary practice of chromatography*, Elsevier, Amsterdam, 1984.

J. Willett, *Gas chromatography*, John Wiley, Chichester, 1987.

A. Zlatkis and V. Pretorius (Eds.), *Preparative gas chromatography*, John Wiley, New York, 1971.

Injection techniques in gas–liquid chromatography

K. Grob, *Classical split and splitless injection in capillary GC*, Alfred Hüthig, Heidelberg, 1986.

K. Grob, *On-column injection in capillary gas chromatography*, Alfred Hüthig, Heidelberg, 1987.

P. Sandra (Ed.), *Sample introduction in capillary gas chromatography*, Vols. 1 and 2, Alfred Hüthig, Heidelberg, Vol. 1 1985, Vol. 2, 1986.

REFERENCES

1. P.A.P. Liddle and A. Bossard, "A case of artefact formation when using hydrogen as carrier gas in capillary gas chromatography", *J. High Res. Chromatogr., Chromatogr. Commun.*, 1985, **7**, 646–647.
2. F. Munari and S. Trestianu, "Thermal peak splitting in capillary gas chromatography", *J. Chromatogr.*, 1983, **279**, 457–472.

CHAPTER 4

GAS–LIQUID CHROMATOGRAPHY: COLUMNS AND STATIONARY PHASES

4.1 GAS CHROMATOGRAPHY COLUMNS

The column is the heart of the chromatographic system and determines the efficiency and selectivity that can be achieved in the separation. If the components of a mixture are not resolved on the column, then even sophisticated microprocessor control of the instrumentation or data interpretation will not enable accurate quantification or confident identification. This is one reason why, despite the microprocessor revolution in instrumentation, many GLC separations are virtually unchanged since the days of the valve amplifier.

However, this is also the area of GLC that has changed most significantly in the last few years. Up to the early 1980s almost all separations were carried out using packed columns in which the liquid stationary phase was coated on an inert support packed in a metal or glass tube. For many chromatographers this was the instinctive choice of method for routine analyses. Few separations were carried out on open-tubular columns in which the stationary phase was coated on the wall of a metal capillary tube. These columns gave much higher efficiencies but were relatively expensive to purchase and more complicated to use.

In recent years, first glass and then the much more robust and widely applicable fused silica open-tubular columns have been introduced. The stationary phase is now usually a crosslinked polymer, which routinely gives high efficiency and reproducibility. In addition, there has been a greater understanding of the techniques required for sample injection, enabling the full potential of these columns to be readily realised (Section 3.2.b). Although open-tubular columns are likely to play an increasingly important role in the future, packed columns are still useful for less complex separations because of their cheapness, simplicity of preparation and ease of use, with a great robustness to operator error and injection technique.

a. *Packed columns*

There are three components to most packed columns, the column tubing, the support material and the liquid stationary phase. Many of the stationary phases are also used in open-tubular columns and are discussed later (Section 4.2).

i. Column tubing

The tubing for the column needs to be chemically inert to prevent sample decomposition, thermally stable and capable of being formed into a coil so that a length of 1–5 m can fit inside a reasonably sized oven. The usual materials that have been used are copper, stainless steel, or glass. At one point there was interest in nickel but the results were inconsistent. Glass-lined tubing has been proposed to give the inertness of glass with the mechanical strength of metal. Typical column sizes are 1–3 m long with a 1/8 or 1/4 inch outside diameter and 2–3 mm inside diameter. The mixture of metric and imperial units is a consequence of the use of standard pipe fittings for connections and fittings. Narrower columns are sometimes claimed to be more efficient but without apparent theoretical justification. However, it can be more difficult to pack the stationary phase uniformly into the wider columns. The diameter of the coil has no effect as long as it is at least ten times the diameter of the tubing. With tighter coils, the path length of the gas flow on the inside and outside of the coil may be significantly different. The tubing is coiled before it is packed.

Glass columns are generally very popular, usually 1/4 inch o.d., as the packing can be examined visually for the build-up of contamination from involatile residues, decomposition of the stationary phase and for settling of the packing material. The inner surface of a glass column is usually silylated to reduce its interaction activity with polar samples (see section 4.4). In early work, it was feared that metal surfaces caused decomposition of sensitive samples or dehydration reactions, but most of these effects are now attributed to problems in the injection port.

Each different make and sometimes each model of gas chromatograph frequently requires a column with different dimensions and shape, so

Figure 4.1 Electron micrograph of Gas Chrom G, a diatomaceous earth support material, showing individual skeletons.

columns are often not interchangeable between instruments, although some manufacturers are better than others.

ii. Support material

By spreading the liquid stationary phase over an inert support, a thin film with a large surface area can be presented to the gas phase. The support material therefore needs to be inert and unreactive towards the analytes and the liquid phase, to have a large surface area to give the thinnest film for a particular loading, and to be mechanically strong to withstand the packing procedure without fragmentation. The particles should have a uniform particle and pore size so that they can be packed regularly and thus give the optimum column efficiency.

Most support materials in use are based on diatomaceous earths (kieselguhrs), which are obtained from geological deposits of the skeletons of diatoms (single-cell algae) (Figure 4.1) and consist primarily of silica with minor metallic impurities. In order to increase the particle size, the diatomaceous earths are first calcinated, either alone to give red firebrick, or with a sodium carbonate flux to give a grey or white filter aid. These materials are then broken up and graded to give the chromatographic

Table 4.1 Support materials for GLC.

Material	Colour	Packed density (g ml^{-1})	Surface area (m^2 g^{-1})	Maximum liquid loading (%)
Chromosorb P	Pink	0.47	4.0	30
Chromosorb W	White	0.24	1.0	15
Chromosorb G	Grey	0.58	0.5	5
Chromosorb A	(Preparative)	0.48	2.7	25
Chromosorb T (temperature limit 250°C)	(Teflon)	0.42	7.5	5

supports. One of the most widely used series of support materials is the Chromosorbs, and similar or equivalent materials are also available from other sources (Table 4.1).

Calcinated firebrick yields Chromosorb P (pink), which is a hard support with a large surface area. It is mainly used for the separation of hydrocarbons and moderate-polarity samples and can accept a high liquid-phase loading. However, polar samples may suffer interactions with the surface unless it is further treated. The flux-calcinated materials are used to form either the friable support Chromosorb W (white) or the hard support Chromosorb G (grey, oyster-white), both of which are suitable for polar samples. Chromosorb W therefore needs careful handling to avoid the formation of fines but can accept a higher loading than Chromosorb G, which has a lower surface area (Table 4.1). The related material Chromosorb A is designed for preparative-scale separations and can accept a high surface loading. It is mechanically stronger and this reduces the formation of fines on packing.

All the support materials can be obtained in a range of particle sizes (Table 4.2). For analytical columns either 80–100 or 100–120 mesh supports are usually used and for preparative columns 40–60 or 60–80 mesh. The

Table 4.2 Mesh sizes of stationary phase supports in GLC.

US sieve mesh size	Opening range, nominal (mm)
40– 60	0.42–0.25
60– 80	0.25–0.18
80–100	0.18–0.15
100–120	0.15–0.125

Table 4.3 Support treatments of stationary phases.

AW	Acid-washed
NAW	Non-acid-washed (untreated form)
AW-DMCS	Acid-washed and dimethyldichlorosilane-treated
HMDS	Hexamethyldisilazane-treated.
HP	High-performance (high-quality, acid-washed and dimethyldichlorosilane-treated)

finer particles are harder to pack uniformly and the columns will have higher back-pressures but higher efficiencies.

The loading of the liquid phase on the support material is usually expressed as a percentage (w/w) of the support phase. Although, in many cases, one support can be replaced by another of similar grade, because of the different densities of the support materials (Table 4.1) the proportion of liquid phase may need to be adjusted to give the same total amount of liquid phase on the column, and hence the same overall retention. Chromosorb G is 2.5 times heavier than Chromosorb W and thus a correspondingly lower w/w loading would be needed (i.e. 12% on Chromosorb W is equivalent to 5% on Chromosorb G).

For many chromatographic applications the support materials are chemically treated by acid washing (AW), which removes metallic impurities and and basic groups (Table 4.3). The remaining interactions caused by free silanol groups on the surface of the support can be largely eliminated by silylation with dimethyldichlorosilane (AW-DMCS) (Figure 4.2).

Silylated materials should always be used if the proportion of stationary phase is less than 3% of a polar liquid or 5% of a non-polar liquid. With higher percentages the liquid phase itself will completely coat the surface of the support, masking any silanol groups. Selected and purified AW-DMCS treated materials are available as high-performance (HP) grades or the specifically inert support materials Gas Chrom Q and Chromosorb 750, which have been designed for use in the analysis of sensitive steroids, pesticides and pharmaceutical samples. Some support materials are sold deliberately as untreated or non-acid-washed (NAW) for use with acid-sensitive liquid phases. A lower degree of activation can be obtained by reaction of the untreated support with hexamethyldisilazane (HMDS), but this is not used widely.

Both chemically treated and untreated support materials are available pregraded from a wide range of suppliers and few chromatographers would nowadays carry out their own treatment reactions or grading.

A few other materials have been used as chromatographic support materials including glass beads, charcoal (Carbopak B and C, see section 4.2.e) and

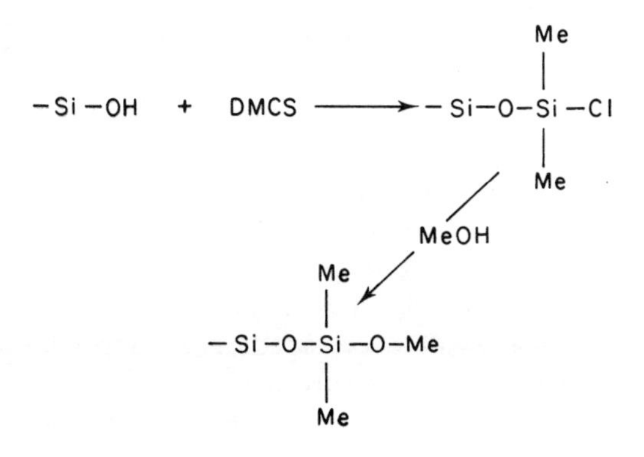

Figure 4.2 Silylation reaction to deactivate the surface of support material or glass column tubing using dimethyldichlorosilane (DMCS).

Chromosorb T (Teflon beads). Although the last of these is extremely inert, with virtually no tailing effects, the surface is so non-wetting that it is very difficult to coat with a uniform layer of liquid phase.

b. Open-tubular columns

Unlike packed columns, open-tubular columns contain no support material and the liquid phase is coated as a film on the wall of the column (wall-coated open-tubular WCOT). This has the advantage that these columns have very low back-pressures and thus for the same overall pressure drop can be much longer. However, in order to increase the efficiency compared to packed columns, the open-tubular column must be very narrow so that the mass transfer effects in the gas phase do not significantly contribute to band spreading. The columns are made from narrow-bore capillary tubing and are often referred to as capillary or Golay columns after their originator. Currently the term "open-tubular" is often preferred as it is the openness of the tube and the lack of a support material, rather than the dimensions, that are responsible for many of the useful properties.

When open-tubular columns were first introduced (Chapter 1), they were usually made from stainless steel, but this interacts with many polar samples and restricted the applications to non-polar samples, such as petroleum products. Because open-tubular columns contain only a small proportion of liquid phase compared to their cross-sectional area, they have a much lower sample capacity than packed columns. Sample introduction was a problem as the columns could be easily overloaded. Their application was therefore very restricted and until the late 1970s their use remained a specialist area. The first development in the creation of the modern open-tubular column was the ready availability of commercially prepared columns made of rigid narrow-bore borosilicate glass. These columns were fragile and the coating procedures, which required a number of surface pretreatments, were often poorly reproducible. However, these columns were very attractive because they had much higher efficiencies than packed columns. Because the surface of the tubing was relatively unreactive they could also be used for a wide range of samples. In practice, considerable operator skill was needed to fit the brittle columns into a chromatograph to give maximum efficiency. The ends of the tubing had to be straightened to fit into the injectors and detectors or additional couplings were needed. Few instruments were available that had been specifically designed to handle the low carrier gas flow rates of narrow open-tubular columns and with appropriately designed injectors and detectors. In many cases a packed column instrument was modified by the addition of a new injector port.

Before the columns were coated with the liquid phase the internal surface of the glass was etched to ensure a uniform layer and good spreading of the liquid. A number of different processes were used and the development of this step was a major topic of research. The liquid film was coated onto the column by either a static (evaporation of a solvent) or a dynamic process (forced through as a plug), and eventually developments in these methods enabled manufacturers to achieve consistent and uniform liquid film layers.

To increase the sample capacity of open-tubular columns, wider-bore columns (0.5 mm) were introduced in which the liquid film was coated onto a thin layer of an inert diatomaceous support material spread on the column wall (support-coated open-tubular, SCOT). This gave a greater surface area for the liquid phase but these columns have now virtually all been replaced by wide-bore columns with thick liquid films.

An important development was to replace the brittle borosilicate glass with flexible fused silica or quartz tubing, which could be bent without breaking. The technology was a spin-off from the development of fused silica optical fibres for telecommunications and gave chromatography a relatively cheap source of column material. Fused silica is almost 100% pure silica with only traces (a few parts per million) of metallic impurities. These columns are robust and easy to use. Being flexible, they can readily be fitted into injectors or detectors, and this removed a major source of user resistance. The same columns would also fit all chromatographs. The increased interest in open-tubular columns also led to improvements in instrument design and the production of specific capillary column chromatographs as the mainstream of most instrument ranges. The coating of the columns with the liquid phase still causes some difficulties and problems are still encountered with some liquid phases. The surface has to be modified to give a high surface deactivation and to increase the wettability, particularly for polar liquid phases. A further important step was the introduction of columns with a liquid phase that has been "chemically bonded' (in practice usually by crosslinking between a small proportion of vinyl groups) by photolytic or free-radical reactions with possibly some binding to the column wall, so that the liquid phase is permanently held in the column. This results in columns with a very low bleed and a very deactivated surface, so that even polar samples, such as acids or amines, show no tailing (Figure 4.3). These bonded liquid phases are extremely robust and the columns can be washed with organic solvents to remove residual contaminants without affecting the stationary phase. Even if the column is accidentally overloaded with sample solvent, the uniformity of the liquid phase will not be damaged, which was always a fear with the older unbonded liquid phases. As yet, only a limited range of different bonded liquid phases are available but as these increase it is likely that the next few years will see the almost complete replacement of unbonded liquid films by cross-linked phases. Many of these phases are closely related to non-bonded liquid phases but small differences in selectivity are sometimes found.

The fused silica columns are coated with an external polyimide layer, which protects the external surface from scratching and prevents the column from cracking. This layer often gives the column a brown or black colour. Care must be taken not to use the columns at high temperature (>350°C) or this coating may be pyrolysed and the column will then become fragile. More rececently an aluminium film has been used to give a more thermally stable coating enabling work up to 425°C.

Open-tubular columns are available in a wide range of column internal diameters (0.1–0.75 mm), liquid film thicknesses (0.1–5.0 μm) and column lengths (5–50 m). The highest efficiencies are obtained by using the smallest diameters (0.1–0.15 mm) but the sample capacities are very low. Generally,

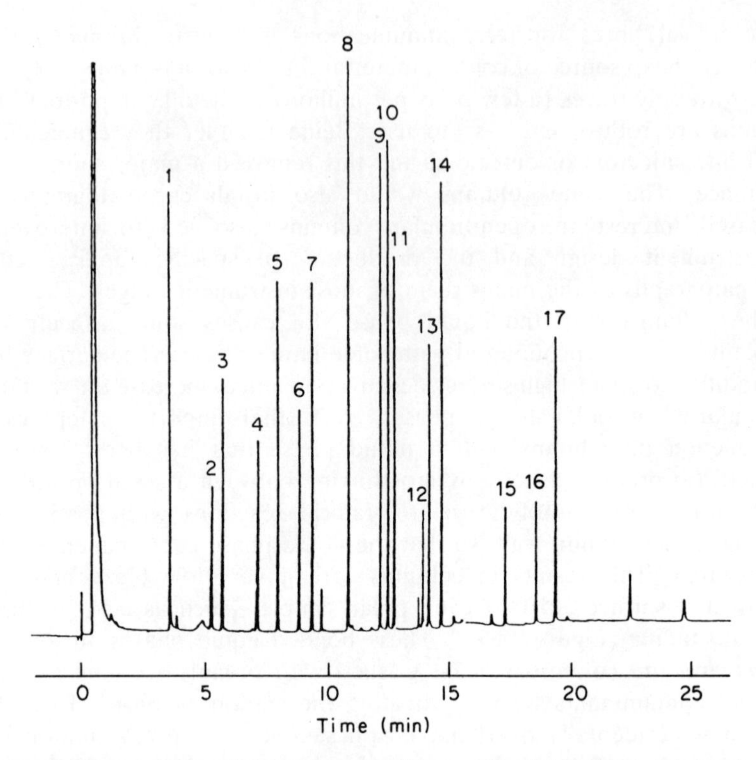

Figure 4.3 Separation of a complex drug mixture on bonded SE-54 methylphenyl-silicone open-tubular column 25 m × 0.25 mm i.d. using programmed temperature vaporisation injection. Initial injector temperature 65°C rapidly programmed to 400°C after sample has been injected. Column temperature 100°C for 5 min then programmed at 10°C min^{-1} to 300°C. Carrier gas helium and flame ionisation detector. Drugs: 1, nicotine; 2, barbital; 3, methyprylon; 4, butalbutal; 5, meperidine; 6, caffeine; 7, lidocaine; 8, cyclizine; 9, methaqualone; 10, amitriptyline; 11, imipramine; 12, oxazepam; 13, codeine; 14, diazepam; 15, flurazepam; 16, haloperidol; 17, triazolam. (Reproduced by permission of Perkin Elmer Corp.)

as the internal diameter or the film thickness increases, the column sample capacity increases but the efficiency decreases. The largest sizes are the wide-bore open-tubular columns (0.53–0.75 mm i.d. and 1–5 μm thick films), which have similar capacities and efficiencies to packed columns (Table 4.4 and Figure 4.4). Columns with larger internal diameters would be too rigid to handle easily. A common size for an analytical column is 25 m long, 0.22 mm internal diameter and a film thickness of 0.25 μm, which should have a typical efficiency of 100 000–125 000 plates.

Table 4.4 Comparison of packed and open-tubular columns.

Column dimensions	Film thickness	Column efficiency	Sample capacity (ng)
Open-tubular SE 30 columns			
0.25 mm × 60 m	(0.35 μm)	150 000	50–100
0.32 mm × 60 m	(1 μm)	126 000	400–500
0.50 mm × 60 m	(1 μm)	78 000	1000–2000
0.75 mm × 60 m	(1 μm)	49 500	10 000–15 000
Packed SE 30 column			
2 mm × 2 m	(10%)	4000	20 000

(Reproduced by permission of Supelco Inc., Bellefonte, PA 16823)

Shorter columns of 10 or 5 m can be used to provide very rapid separations, enabling analyses to be carried out in a much shorter time than on a packed column but with a similar efficiency. Columns with thicker films (1–3 μm) have completely replaced the older SCOT columns. At present some non-bonded open-tubular columns are still available but these are likely to be completely replaced by bonded columns in the future.

At one stage there was a great interest in the production of glass open-tubular columns in the individual laboratory. However, it is probable that with the introduction of fused silica columns and bonded phases most chromatographers will purchase precoated open-tubular columns because both these advances enable the manufacturer to provide a highly consistent and reliable product at a reasonable cost. Because the high efficiency of open-tubular columns usually enables almost all the components of any sample to be separated, only two or three stationary liquid phases are needed to handle almost all possible samples. In contrast, with packed columns a large number of different liquid phases are needed to provide a range of selectivities, in order to compensate for the lower efficiencies.

4.2 STATIONARY PHASES

Generally the same or very similar liquid stationary-phase materials are used for both packed and open-tubular columns. The selection and comparison of liquid phases is probably one of the most confusing areas for the newcomer to GLC. Over 300 different phases are widely available and 700–1000 phases have been reported in the literature. However, many of these phases are virtually the same material produced by different manufacturers or supply houses but given a different code or tradename. Most phases can therefore

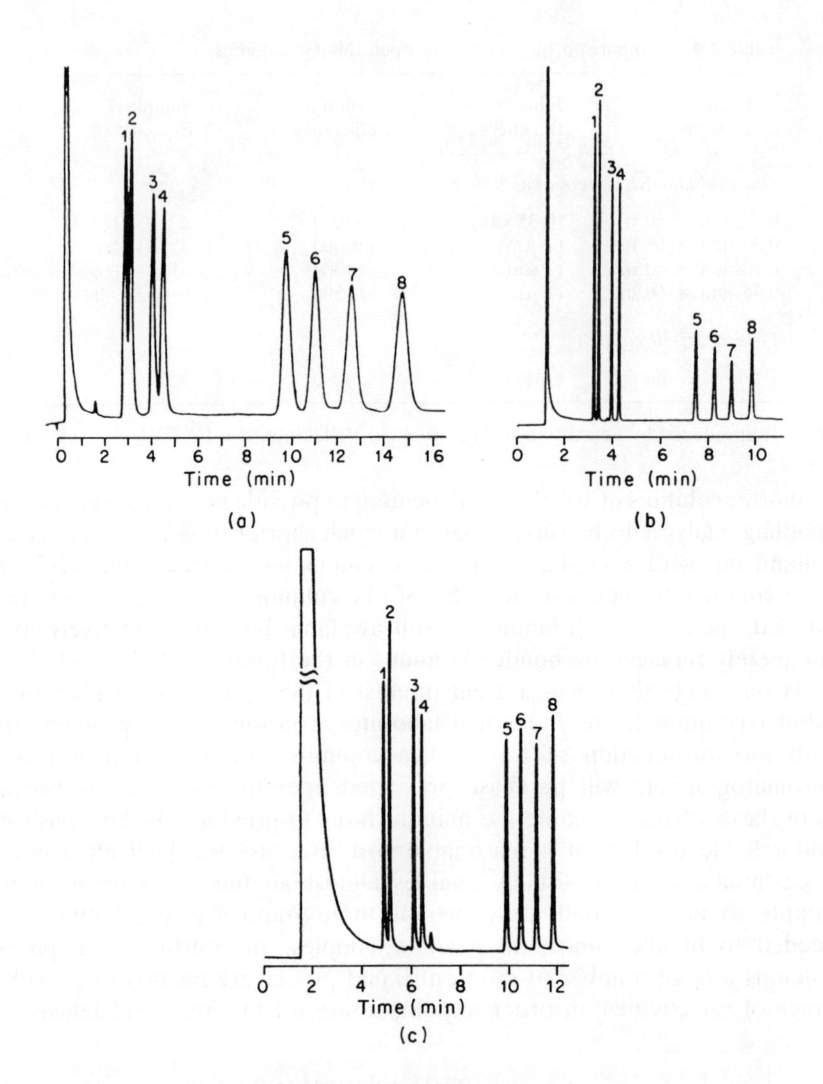

Figure 4.4 Comparison of (a) a packed column and (b) narrow-bore and (c) wide-bore capillary columns for the separation of alditol acetates. Samples: 1, rhamnitol; 2, fucitol; 3, ribitol; 4, arabinitol; 5, mannitol; 6, galactitol; 7, glucitol; 8, inositol. (a) 3% SP-2340 on 100–120 mesh Supelcoport 6 ft × 2 mm i.d. Column temperature 225°C. Carrier gas nitrogen 20 ml min^{-1}. (b) SP-2330 fused silica open-tubular column 15 m × 0.25 mm i.d. Film thickness 0.20 μm. Column temperature 240°C. Carrier gas helium. (c) SP-2330 wide-bore open-tubular column 30 m × 0.75 mm i.d. Film thickness 0.20 μm. Column temperature 200°C for 2 min then 4°C min^{-1} to 250°C. Carrier gas helium. (*Supelco Reporter*, 1985, 4, 10. Reprinted by permission of Supelco Inc., Bellefonte, PA 16823.)

be grouped into a limited number of typical types with related properties and structures. In practice, only a limited number of phases are widely used for packed columns and an even more restricted selection for open-tubular columns. Few of the early liquid phases were initially designed for GLC but were produced as insulators, lubricants, or detergents, although now specifically purified grades or specially synthesised materials are used.

The general criteria for a liquid phase to be suitable for GLC are that it should be chemically stable and unreactive towards the sample, involatile and stable to thermal decomposition. As the temperature rises, one of these last two criteria will usually fail and each liquid phase has a maximum recommended temperature (MRT), which should not be exceeded in operation. Care must be taken if the packing in the column extends into the injection port as this is often at a higher temperature than the oven. Some phases are solids at room temperature and therefore also have a lower temperature limit, as they must be a liquid for efficient separations. The liquid phase should also be a good solvent for the sample to ensure symmetrical peaks and this is usually an important criterion in the selection of a phase for a particular analysis (see Section 4.3).

The different liquid phases can be broadly divided into non-polar, polar and special phases. These differ in their selectivity, which is their ability to retain compounds of different structural types and thus to separate the components of a mixture. The non-polar phases contain no groups capable of hydrogen bonding or dipole interaction so that analytes are primarily eluted according to their volatilities. The elution order therefore varies as the boiling points of the analytes. The polar phases contain halogen, hydroxyl, nitrile, carbonyl, or ester polar functional groups and analytes containing polar groups will interact more strongly than non-polar analytes. Retentions will therefore depend on a combination of volatility and polar–polar interaction. By changing from a phase of one type to a phase of the other, large differences in relative retention can often be obtained for analytes containing different functional groups. These changes enable the selectivity and resolution of a separation to be optimised. As will been seen later, these interactions can be quantified and phases with similar or different properties identified.

Precoated support materials for packed columns can be purchased either as loose packing material or already packed into columns. Alternatively, many chromatographers coat the support material in their own laboratory and pack the columns at minimal material cost (Section 4.4). For an industrial laboratory this process can be time-consuming, and unless care is taken the results from column to column may not be reproducible. In these cases externally produced columns may be more cost-effective for routine analyses.

Table 4.5 Hydrocarbon liquid phases in GLC.

Hydrocarbon	Temperature range (°C)
Squalane ($C_{30}H_{62}$)	0–130
Apiezon L ($-(CH_2)_n-$)	50–280
Apiezon M ($-(CH_2)_n-$)	50–280
Apolane (C_{87} hydrocarbon)	30–280

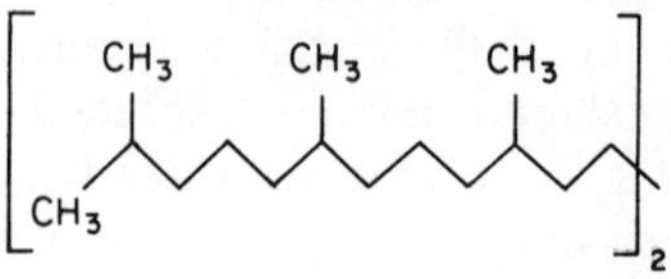

Squalane

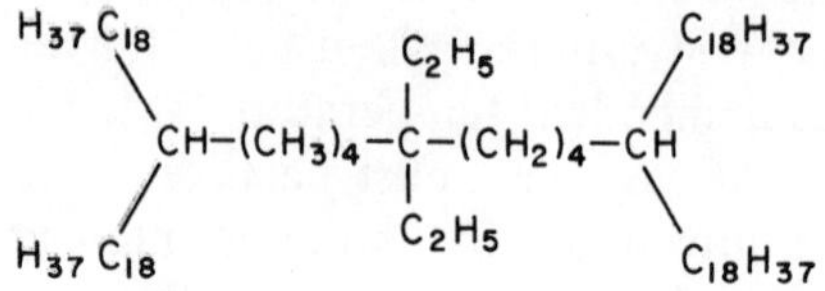

Apolane

Almost all open-tubular columns are purchased ready-coated and, unless large numbers of columns are required, in-house production of columns is unlikely to be viable. With bonded liquid phases the stability and lifetime of open-tubular columns has increased and the real cost per assay has significantly decreased despite the higher initial cost.

a. *Non-polar liquid phases*

i. Hydrocarbon phases

In order to obtain a hydrocarbon liquid phase with the necessary low volatility, either discrete large hydrocarbons, such as squalane (obtained by the hydrogenation of the natural triterpene squalene), Kováts' synthetic Apolane C_{87} hydrocarbon, or mixtures of long-chain *n*-alkanes such a Apiezon L have been used (Table 4.5). Although squalane is not widely used because it has a low temperature limit (MRT 140°C), it is important in GLC because it is the least polar phase in common use. It is therefore used as a reference point for the polarity of liquid phases (see section 4.3) and has the advantage that as a single compound its purity is easily defined.

At a high liquid-phase loading these hydrocarbons have found limited application for the analysis of low-boiling hydrocarbons (Figure 4.5) and other non-polar samples.

None of this group of phases has found application for open-tubular columns. They are all susceptible to oxidation by traces of oxygen in the

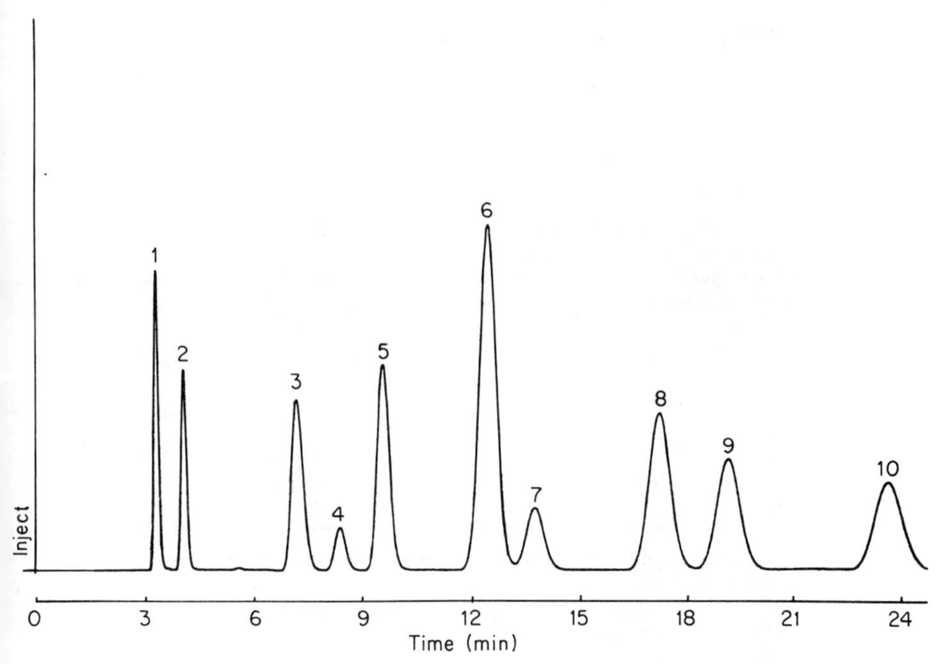

Figure 4.5 Separation of a mixture of straight-chain, branched and cyclic hydrocarbons. Column 15% squalane on Chromosorb P 80–100 mesh, 2 m × $\frac{1}{8}$ inch o.d. Column temperature 160°C. Carrier gas nitrogen at 6 ml min^{-1}. Flame ionisation detector. Peaks: 1, isopentane; 2, n-pentane; 3, cyclopentane; 4, 3-methylpentane; 5, n-hexane; 6, 2,4-dimethylpentane; 7, benzene; 8, cyclohexane; 9, 3-methylhexane; 10, n-heptane. (Reproduced by permission of Perkin Elmer Limited from *Perkin Elmer applications sheet*, 496-0771. © Perkin Elmer, 1972.)

carrier gas at high temperatures. Some perfluoroalkanes have also been examined as liquid phases but, although non-polar, they have limited applicability because they are so non-polar that they are poor solvents for most analytes.

ii. Alkylsilicone liquid phases

Polymers based on a silicon–oxygen–silicon backbone (Si–O–Si) are thermally very stable and form the basis of many stationary phases in GLC. They were originally prepared as insulation liquids for transformers. Most can be used up to 300°C without significant decomposition or volatilisation, particularly in the grades purified for GLC.

The most important group are the non-polar dimethylsilicones, which include the phases SE-30, OV-1 and OV-101, and these have been widely

Table 4.6 Dimethylsilicone liquid phases. -(-Si(Me)$_2$-0-)- polymers.

Dimethyl silicone	Temperature range (°C)
Liquid phases	
SE-30 Silicone gum rubber	50–350
MS 200/12 500 Silicone fluid	20–250
OV-1 (gum)	100–350
OV-101 (fluid)	0–350
CP-Sil 5	50–350
SP-2100 (fluid)	0–350
Open-tubular bonded phases	
BP-1	-320
CP-Sil 5 CB	0–325
Durabond DB-1	−50–320
SPB-1	−60–350

Trademarks and abbreviations of liquid phases (also used in subsequent tables):

BP	Scientific Glass Equipment bonded phases
CP	Chrompack
DC	Dow Corning
Durabond	J and W
OV	Ohio Valley
SE	General Electric
SP	Supelco
SPB	RSL Belgium/Alltech
Superox	RSL Belgium/Alltech

used for a range of moderate to non-polar samples (Table 4.6). These phases include closely related materials from different manufacturers or suppliers, all with effectively identical selectivities. For packed columns the less viscous liquid silicones, such as OV-101, are preferred as they give higher efficiencies, whereas viscous gums, such as OV-1, have been used for open-tubular columns because they give a more even coating on silica or glass surfaces. However, in this case they are largely being replaced by crosslinked dimethylsilicones, such as BP-1 and SPB-1.

These dimethylsilicone materials are very popular as general-purpose phases for the separation of low-volatility samples and are used in many areas. The OV-101 open-tubular column has been used as one of two standard phases in a recent large compilation of retention indices for the identification of analytes by GLC (see Chapter 6).

Table 4.7 Methylphenyl silicone liquid phases.
Methylphenylsilicones are polymers of –(Si–C(Me)(Ph)–O–)– groups. Dimethyldiphenyl silicones contain combinations of –O–Si$(Me)_2$–and –O–Si$(Ph)_2$–groups.

Methylphenyl silicone	Proportion of groups		Temperature range (°C)
	Phenyl	Methyl	
Liquid phases			
SE-52	5	95	50–300
SE-54	5	94 (+1% vinyl)	50–300
CP-Sil 8	5	95	−25–350
DC 550	25	75	20–250
OV-3	10	90	0–350
OV-7	20	80	0–350
OV-11	35	65	0–350
OV-17	50	50	0–350
OV-22	65	35	0–300
OV-25	75	25	0–350
SP-2250	50	50	0–375
Open-tubular bonded phases			
BP-5	5	95	-320
CP-Sil 8 CB	5	95	−25–350
Durabond DB-5	5	95	−50–320
Dimethyldiphenyl silicone	**Proportion of groups**		**Temperature range (°C)**
	Phenyl	Methyl	
Liquid phase			
OV-61	33	67	0–350
OV-73	5.5	94.5	50–350
Open-tubular bonded phases			
SPB-5	5	94 (+1% vinyl)	−60–320
SPB-20	20	80	0–300
SPB-35	35	65	0–300

A closely related series of liquid phases is based on the methylphenylsilicones in which different proportions of the methyl groups are replaced by phenyl groups (Table 4.7). The substitution can be either random or the liquid phase can be a copolymer of dimethyl and diphenylsilicone groups. In some phases, small proportions of vinyl groups are included for crosslinking. The methylphenylsilicones often give better peak shapes than the dimethylsilicones, particularly for analytes containing polar groups or aromatic rings (Figure 4.6). Increasing the proportion of phenyl groups

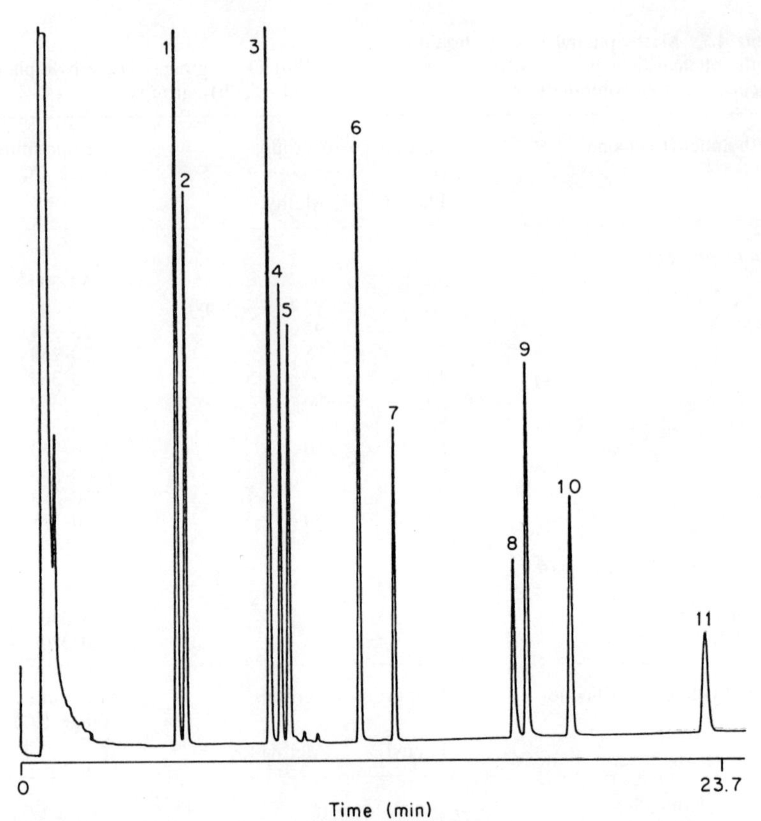

Figure 4.6 Separation of a mixture of polar phenols on a chemically bonded methylphenylsilicone (bonded OV-17) wide-bore open-tubular column 15 m × 0.53 mm i.d. Column temperature isothermal at 100°C for 1 min then temperature programmed from 100°C to 210°C at 8°C min^{-1}. Helium carrier gas. Peaks: 1, 2-chlorophenol; 2, phenol; 3, 2,4-dimethylphenol; 4, 2-nitrophenol; 5, 2,4-dichlorophenol; 6, 4-chloro-3-methylphenol; 7, 2,4,6-trichlorophenol; 8, 2,4-dinitrophenol; 9, 4-nitrophenol; 10, 2-methyl-4,6-dinitrophenol; 11, pentachlorophenol. (Reproduced by permission of Quadrex Corp.)

increases the retention of polycyclic aromatic compounds and polarisable groups. The most popular of this group of phases are the 50:50 methylphenylsilicones, OV-17 and SE-52, and the corresponding open-tubular bonded phase, BP-5.

b. Polar liquid phases

The main disadvantage of the non-polar phases is that, particularly as packed columns, they are often poor solvents for polar samples and give tailing or

Table 4.8 Halogen-substituted dimethylsilicone liquid phases.

Chloro-substituted dimethylsilicone			Temperature range (°C)
SP-400 Chlorophenylmethylsilicone			0–350
	Proportion of groups		
Trifluoroalkyl-substituted dimethylsilicone	Trifluoropropyl	Methyl	Temperature range (°C)
Liquid phases			
QF-1 (FS 1265)	50	50	0–250
OV-202 (fluid)	50	50	0–275
OV-210 (fluid)	50	50	0–275
OV-215 (gum)	50	50	50–275
SP-2401	50	50	0–275
Open-tubular bonded phase			
DB-210	50	50	45–240

distorted peaks. Over the years a number of polar liquid phases have therefore been developed that provide better separations. These materials have a range of different interactions, so the selectivity of a particular separation can be optimised. Many are known by tradenames and at first sight the number can be confusing. However, although most suppliers list large numbers of these phases, in reality very few are in widespread use. In some cases, the phases represent column materials employed in standard methods in the early days of GLC. These are only retained as it would be too costly to replace them with a newer material because of the time needed to revalidate and restandardise the method. Nowadays, if possible, all monomeric liquid phases should be avoided unless a direct comparison with an established method is required. Many of these phases have low temperature limits and can be chemically reactive towards analytes containing particular functional groups.

i. Substituted silicone liquid phases

The silicone skeleton has also been used to prepare substituted phases containing the polar trifluoromethyl (Table 4.8) or cyano groups (Table 4.9). Both materials have good chemical and thermal stability. By incorporating different proportions of the polar groups, liquid phases can be produced

Table 4.9 Cyano-substituted dimethylsilicone liquid phases.

Cyano-substituted dimethylsilicone	Proportion of groups			Temperature range (°C)
	Cyanopropyl	Methyl	Other	
Liquid phases				
XE-60		75	25 (cyanoethyl)	0–250
OV-105	5	95		20–250
OV-225	25	50	25 (phenyl)	20–250
OV-275	100 (dicyanoallyl)			20–275
OV-1701	7	85	7 (phenyl) 1 (vinyl)	20–235
CP-Sil 58	50		50 (phenyl)	50–275
CP-Sil 71	75		25 (phenyl)	50–275
CP-Sil 84	90		10 (phenyl)	50–275
CP-Sil 88	100			55–240
SP-2340	100			25–275
SP-2330	95		5 (phenyl)	25–250
SP-2300	50		50 (phenyl)	25–275
SP-2310	75		25 (phenyl)	25–275
Open-tubular bonded phases				
BP-10	7	86	7 (phenyl)	–270
BP-15	25	50	25 (phenyl)	–260
BP-75	10 (dicyanoallyl)			–250
CP-Sil 19 CB	7	85	7 (phenyl) 1 (vinyl)	50–300
CP-Sil 43 CB Cyanopropyl methylphenyl silicone	25	50	25 (phenyl)	50–225
Durabond DB-225	25	50	25 (phenyl)	40–220

with a wide range of polarities, which have been used for both packed and open-tubular columns.

The trifluoromethyl group has a high dipole moment, strong electron-acceptor properties and a particular affinity for analytes containing carbonyl and nitro groups. The most widely used materials are OV-210, the equivalent OV-202, which is a low-viscosity fluid for packed columns, and OV-215, a corresponding gum for open-tubular columns.

The cyano group is strongly electron-attracting and interacts with π-bonded groups, such as olefins, phenyl rings, carbonyl groups and esters. These phases can be particularly polar and the dicyanoallylsilicone OV-275 is regarded as the most polar liquid phase generally available. This phase also has the advantage that as a packed column it can be used to resolve *cis* and *trans* fatty acid methyl esters (Figure 4.7), which are not normally resolved on polyester liquid phases. However, this material is unsuitable for the separation of alkanes as they are poorly soluble and give poor peak

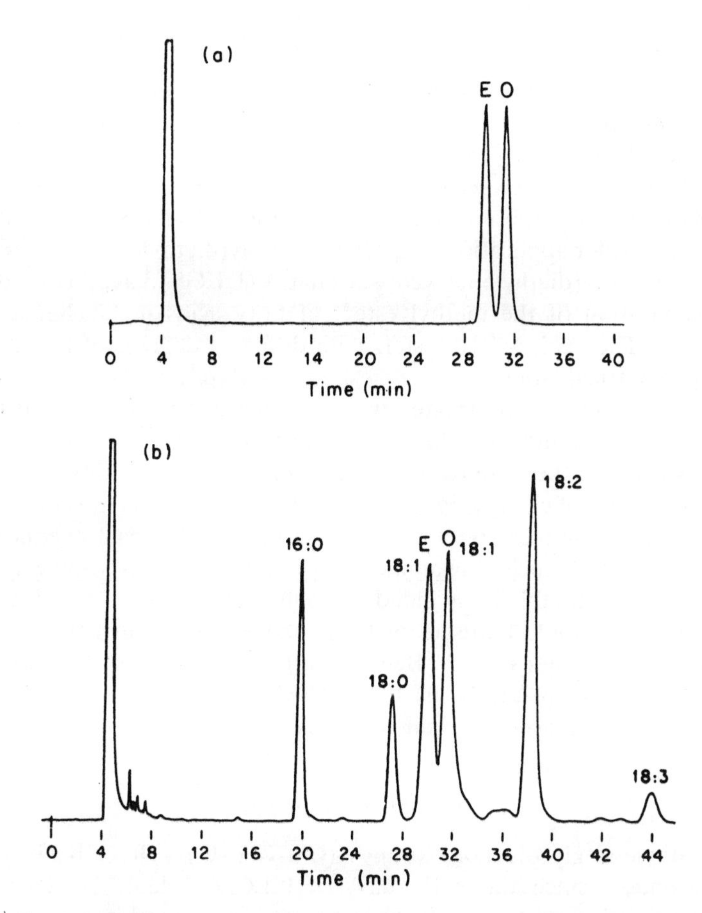

Figure 4.7 Separation of *cis*/*trans* isomers of fatty acid methyl esters: (a) standards esters; (b) a typical margarine. Peaks: E, methlyl elidate (*trans* 18:1); O, methyl oleate (*cis* 18:1). Column packed with 15% OV-275 on Chromosorb P AW-DMCS 100–120 mesh 20 ft × $\frac{1}{8}$ inch at 220°C. Carrier gas nitrogen 10 ml min^{-1}. (*Lipid Reporter*, 1976, **1**, 3. Reprinted with permission of Supelco, Inc., Bellefonte, PA 16823.)

shapes. Because of concern that cyano groups can be hydrolysed to amides, the carrier gas should be dried before use.

Cyano columns are becoming popular as alternatives to the thermally less stable polyester phases for the separation of fatty acid methyl esters, OV-275 replacing diethyleneglycol succinate (DEGS) and CP-Sil 58 replacing ethyleneglycol adipate (EGA). The corresponding crosslinked phases are becoming widely adopted as the typical polar stationary phase for open-tubular columns.

ii. Ester liquid phases

In the past a few high-molecular-weight monomeric esters (Table 4.10) were used as liquid phases but generally they have relatively low temperature limits (dinonyl phthalate, MRT 150°C). They can therefore only be used for volatile samples, such as simple esters or solvent mixtures. Polymeric esters (Table 4.10) prepared by the condensation of diacids and diols have found a much wider application. In particular, poly(diethylene glycol adipate) (DEGA) and poly(diethyleneglycol succinate) (DEGS) have been adopted for the separation of the methyl esters of fatty acids in biochemical and food studies. They can resolve esters with different degrees of unsaturation but not geometrical isomers.

These phases have to be treated with more care than the silicone phases and precautions should be taken to remove oxygen and water from the carrier gas because they can degrade thc esters. Reproducibility can also be a problem because of changes in the selectivity depending on the conditioning of the columns. Samples containing groups that might react chemically or exchange with esters, including amines, isocyanates, hydroxyls and carboxylic or Lewis acids, should be avoided. Usually a non-acid-washed (NAW) support material should be used for these phases, as residual traces of acid may hydrolyse the ester groups. Because of their relatively low temperature limits these phases are rarely used for open-tubular columns and as noted above are being replaced by cyanosilicones.

iii. Polyether liquid phases

The polyethylene glycols (polyoxiranes–$(CH_2CH_2\text{-}O)_n$–) have been popular as liquid phases since the early days of GLC (Table 4.11). They were originally manufactured as surfactants and a range of different materials with molecular weights from 200 to 20 000 (Carbowax 200 to Carbowax 20M) are available. Recently the series has been extended to higher-molecular-weight materials with Superox 0.1, Superox 0.6 and Superox 4.0, with molecular weights of 100 000, 600 000, and 4 000 000, respectively. These new materials have a higher purity, a particularly low content of trace metal ions, and increased stability but similar temperature limits. They are gums and were primarily intended for open-tubular columns and GLC–MS applications, where reduced column bleed is important, but are likely to be displaced by the recent introduction of crosslinked Carbowax phases (Figure 1.7).

Carbowax 20M is one of the most popular liquid phases and is widely used for the separation of polar and non-polar samples (Figure 4.8). It has

Table 4.10 Ester stationary phases for GLC.

Ester	Temperature range (°C)
Monomeric liquid phases	
Dinonyl phthalate	20–150
Didecyl phthalate	10–175
Polymeric liquid phases	
Butane-1,4-diol succinate	50–200
Diethylene glycol succinate (DEGS)	20–200
Diethylene glycol adipate (DEGA)	0–200
Reoplex 400 (polyester) (polypropyleneglycol adipate)	20–220
Ethylene glycol adipate (EGA)	100–200
Ethylene glycol succinate	100–200

Table 4.11 Polyether stationary phases for GLC.

Ether	Average molecular weight	Temperature range (°C)
Liquid phases		
Polyethyleneglycols		
Carbowax 400	400	0–125
Carbowax 600	600	20–125
Carbowax 1000	1000	40–150
Carbowax 20M	20 000	60–300
Superox 20M	20 000	60–300
Superox 0.1	100 000	–275
Superox 0.6	600 000	65–300
Superox 4.0	4000 000	65–300
CP-Wax 51	40 000	60–280
CP-Wax 600M	600 000	60–280
CP-Wax 4000M	4000 000	60–280
Polypropyleneglycol	2000	20–150
Ucon oil LB 550X polypropylene (90%)–polyethylene (10%) glycol	1400	20–200
OV-330 phenylsilicone–Carbowax copolymer		30–250
Open-tubular bonded phases		
BP-20 Bonded Carbowax 20M		–250
CP-Wax 52 CB		30–275
Supelcowax 10		50–280

also been adopted as the standard polar phase for retention index identification studies (see Chapter 6). Closely related materials based on poly(propylene glycol) are also available. They are often employed for the analysis of simple alcohols and for many years were used in the standard method for the determination of ethanol levels in blood from "drunken" motorists.

All the polyol phases are particularly susceptible to contamination and care should be taken to remove oxygen and moisture from the carrier gas and to avoid analytes, such as silylation reagents, that could react chemically with the terminal hydroxyl groups of the polymer.

c. *Specialised liquid phases*

A number of specialised phases have been developed to separate particular groups of analytes or for use with specialised analytical techniques. As well as the examples that follow, they include liquid-crystal phases and mixtures of inorganic salts, but these will not be discussed in detail.

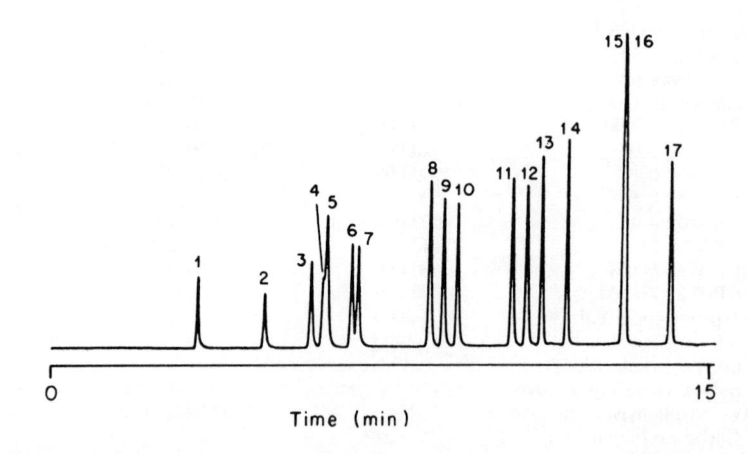

Figure 4.8 Separation of C_1 to C_5 alcohols on a DB-WAX+ bonded Carbowax column 30 m × 0.53 mm i.d. Helium carrier gas. Oven temperature 40°C for 3 min then programmed to 105°C at 7.5°C min^{-1}. Flame ionisation detector. Peaks: 1, acetaldehyde; 2, methyl acetate; 3, ethyl acetate; 4, methanol; 5, 2-methyl-2-propanol; 6, 2-propanol; 7, ethanol; 8, 2-methyl-2-butanol; 9, 2-butanol; 10, 1-propanol; 11, 2-methyl-1-propanol; 12, 3-pentanol; 13, 2-pentanol; 14, 1-butanol; 15, 2-methyl-1-butanol; 16, 3-methyl-1-butanol; 17, 1-pentanol. (Reproduced by permission of J. & W. Scientific, Inc.)

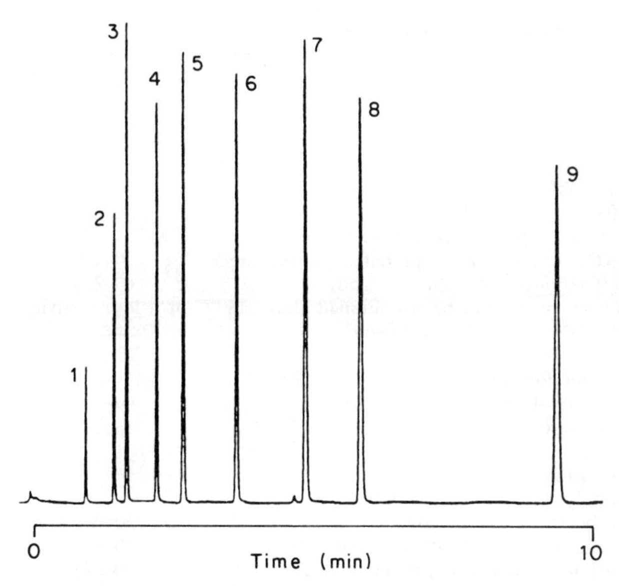

Figure 4.9 Separation of C_1 to C_7 free fatty acids on OV-351 open-tubular column 30 m × 0.25 mm i.d. Column temperature 145°C. Flame ionisation detector. Peaks: 1, acetic acid; 2, propionic acid; 3, isobutyric acid; 4, butyric acid; 5, isovaleric acid; 6, valeric acid; 7, isocaproic acid; 8, caproic acid; 9, heptanoic acid. (Reproduced by permission of J. & W. Scientific.)

i. Phases for carboxylic acids

On many packed columns, free carboxylic acids give badly tailing peaks because of interaction with the column or support materials, although this effect is less of a problem on bonded open-tubular columns. The separation of the C_1 to C_7 free carboxylic acids be can carried out successfully on Carbowax 20M impregnated with terephthalic acid (Carbowax 20M–TPA) or on columns of FFAP (Free Fatty Acid Phase), which are the product of the reaction of Carbowax 20M with nitroterephthalic acid (Figure 4.9 and Table 4.12). These columns have better temperature stability than Carbowax 20M but almost identical selectivity and can also be used for the separation of neutral compounds.

ii. Phases for basic compounds

Amines often interact very badly with silica support materials but can be separated on a column coated with Carbowax 20M to which a small amount of potassium hydroxide (2%) has been added during the coating process

Table 4.12 Specialised phases for GLC.

Specialised phase	Temperature range (°C)
Liquid phases for carboxylic acids	
Carbowax 20M–TPA (terephthalic acid)	60–250
FFAP (free fatty acid phase) (Carbowax 20M–nitroterephthalic acid)	50–275
OV-351 (Carbowax–nitroterephthalic acid polymer)	50–250
SP-1000 (Carbowax–nitrophthalic acid)	60–220
(Note: FFAP should not be conditioned above 180°C or it loses activity. Aldehydes are permanently retained)	
Liquid phases for bases	
Carbowax 20M–KOH (1–2%) (not commercially available)	60–250
High-temperature phases	
Dexsil copolymers with carborane	
300 Carborane methylsilicone	50–450
400 Carborane methylphenylsilicone	20–400
410 Carborane methyl-(β)-cyanoethylsilicone	20–375

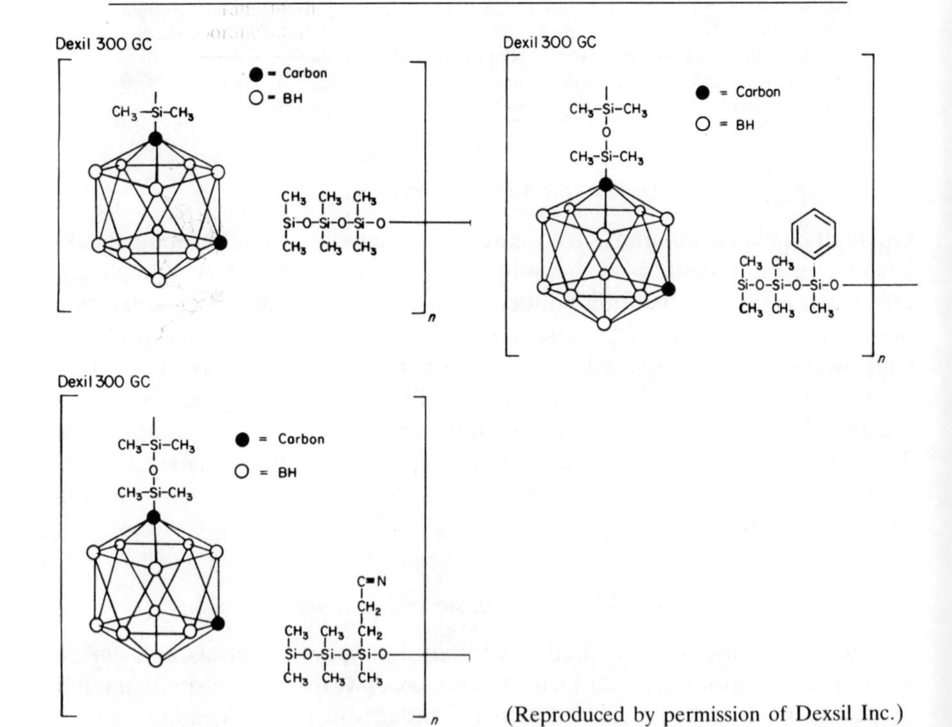

(Reproduced by permission of Dexsil Inc.)

(Table 4.12). Again the selectivity for neutral compounds is unchanged and the resulting columns have similar properties to Carbowax 20M. Volatile amines can also be analysed using porous polymer columns (see later).

iii. High-temperature phases

For coupled GC–mass spectrometry, a liquid phase with a very low bleed rate is desirable to avoid contamination of the mass spectrometer. This is one of the applications of the Dexsil liquid phases, which are based on substituted silicone–carborane copolymers (Table 4.12). Because of the stabilising effect of the carborane, these phases can be used to over 350°C with minimal bleeding.

d. Chiral phases

Of particular recent interest has been the development of liquid phases that can achieve the separation of enantiomers. Most chromatographic liquid phases are achiral and the separation environment is therefore symmetrical, so that enantiomeric compounds will have the same retention times and be unresolved. In order to achieve a difference in the properties of the column towards enantiomers, the liquid phase itself must be chiral. Alternatively, the analyte must be derivatised with an optically active chiral reagent to give a pair of diastereoisomers, which can be separated on an achiral column (see Chapter 6).

A number of chiral phases have been developed that incorporate amino-acid derived chiral centres (Table 4.13). These include monomeric materials based on lauryl esters and liquid phases derived from cyano-bonded siloxane polymers, including König's material and Chirasil-Val. The latter column has the ability to separate ester acyl derivatives of amino acids (Figure 4.10). The D-isomer of each amino acid is eluted before the L-isomer. Further more versatile and widely applicable phases are likely to be produced in the future.

e. Gas–solid phases and polymer phases

i. Gas–solid stationary phases

Gas–solid chromatography has never been very popular except in a few limited application areas but there are a number of separations, particularly of inorganic and organic gases, in which solid stationary phases are preferred.

Table 4.13 Chiral stationary phases.

Open-tubular columns
Chirasil-L-Val (L-valine-t-butylamide on carboxymethylsiloxane)
Chirasil-D-Val (D-valine-t-butylamide on carboxymethylsiloxane)
XE-60-*S*-valine-*S*-(α)-phenylethylamide
SP-300 (*N*-n-lauroyl-*N*-L-valine-t-butylamide)

L-valine-t-butylamide on carboxymethylsiloxane

XE-60-*S*-valine-*S*-α-phenylethylamide

The main phases are based on silica gel, charcoal, or molecular sieves (Table 4.14).

Molecular sieves 5 Å are often used for the analysis of gases because of their unique ability to separate oxygen and nitrogen (see Figure 5.5). Precautions must be taken to remove water from the carrier gas or the column will become deactivated. Carbon-based materials have been used directly for the analysis of gases, particularly sulphur gases and the oxides of nitrogen.

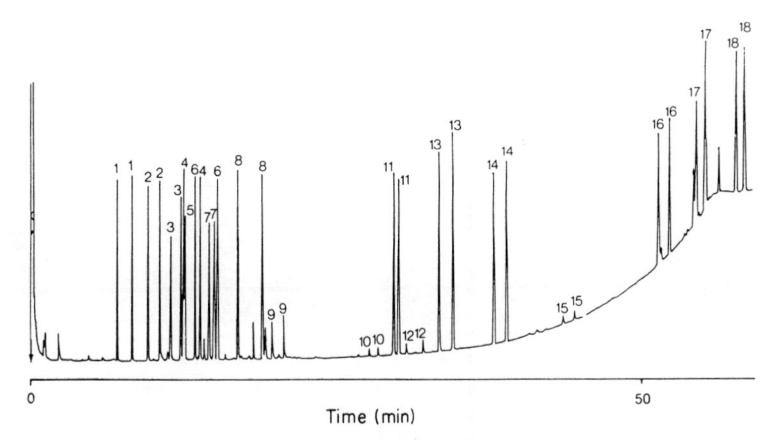

Figure 4.10 Chiral separation of *N*-pentafluoropropionyl isopropyl esters of D- and L-amino acids on Chirasil-L-Val open-tubular column 25 m × 0.22 mm. Nitrogen carrier gas. Column temperature isothermal at 80°C for 3 min then programmed at 3°C min^{-1} to 190°C. Flame ionisation detector. (D-isomer elutes first in each case). Peaks: 1, DL-alanine; 2, DL-valine; 3, DL-threonine; 4, DL-*allo*-isoleucine; 5, glycine; 6, DL-isoleucine; 7, DL-proline; 8, DL-leucine; 9, DL-serine; 10, DL-cysteine; 11, DL-aspartic acid; 12, DL-methionine; 13, DL-phenylalanine; 14, DL-glutamic acid; 15, DL-tyrosine; 16, DL-ornithine; 17, DL-lysine; 18, DL-tryptophan. (Reproduced with permission from *Chrompack application note*, No. 29.)

Table 4.14 Gas–solid stationary phases.

Molecular sieves 5 Å	Pore size 4–5 Å Separates O_2 and N_2 Carbon dioxide and water adsorbed
Molecular sieves 13X	Pore size 8–10 Å Separation of freons
Alumina	Retains unsaturated olefins
Silica gel	Retains carbon dioxide
Chromasil 310	Sulphur gases
Chromasil 330	C_1–C_3, RSH and sulphur gases
Charcoal-based materials	
Carbosieve G (replaces B)	
Carbosieve SII (spherical)	Permanent and sulphur gases, nitrogen oxides
Spherocarb	15 Å pore size C_1–C_4, stack gases, SO_2, nitrogen oxides
Open-tubular column	
PLOT columns	Porous-layer open-tubular columns
Al_2O_3/KCl	C_1–C_{10} hydrocarbons
Molsieve 5 Å	Permanent gases, oxygen/nitrogen
PoraPLOT Q	Similar properties to Porapak Q

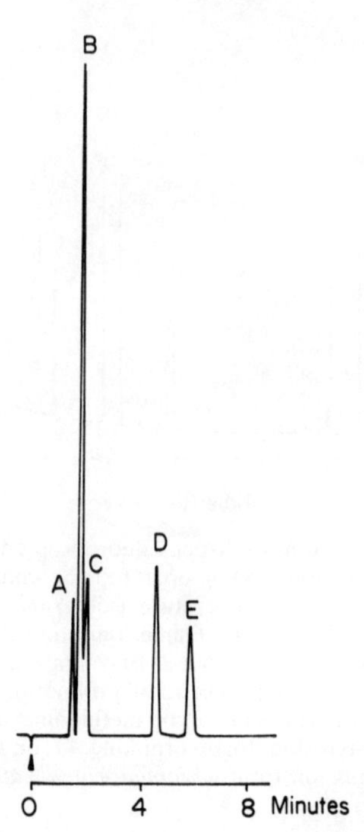

Figure 4.11 Separation of gases from soil on combination of columns (a) Chromosorb 101 packed column 6 ft × $\frac{1}{8}$ inch at 40°C then flow switched to molecular sieve 5Å column. 6 ft × $\frac{1}{8}$ inch at 40°C. Peaks: A. CO_2; B. N_2O; 3. O_2; D. N_2; E. CH_4. (Reproduced by permission of Varian Associates.)

Open-tubular gas–solid columns in which a layer of a porous solid stationary phase is deposited on the wall of the columns (porous-layer open-tubular, PLOT) have been developed using molecular sieves or alumina for the separation of gases and highly volatile analytes (Table 4.14).

ii. Porous polymer phases

It has been shown that beads of porous polymers can also be used as stationary phases. They behave much more like liquids than solids and have a very high retention power. They are very popular for the separation of volatile samples and gases (Figure 4.11) and have often replaced methods formerly carried out on solid phases. The polymers are mainly based on polystyrene/divinylbenzene polymers or acrylates (Table 4.15). The two main groups are the Chromosorb Century series and Porapak phases. Each set includes materials with a wide range of properties and special characteristics.

Table 4.15 Polymer column materials. Temperature limits range from 190 to 250°C.

Name	Material (Use)	Density (g cm^{-3})	Surface area (m^2 g^{-1})
Porapak P	Non-polar polystyrene–divinylbenzene (Glycols/ROH)	0.26	100–200
Porapak P-S	Non-polar silanised (RCHO)	0.26	100–200
Porapak Q	Slight–moderate polarity divinylbenzene–ethylvinylbenzene (Hydrocarbons, oxides of nitrogen)	0.34	500–600
Porapak Q-S	Slight–moderate polarity (Fatty acids/amines)	0.34	500–600
Porapak R	Moderately polar polyvinylpyrrolidone (Esters)	0.32	450–600
Porapak S	Moderately polar polyvinylpyridine (Normal and branched alcohols)	0.35	300–450
Porapak N	Polar polyvinylpyrrolidone	0.41	250–350
Porapak T	Very polar ethyleneglycol dimethylacrylate (Formaldehyde in water)	0.39	225–350
Chromosorb 101	Non-polar styrene–divinylbenzene (Fatty acids, alcohols, carbonyl compounds)	0.3	<50
Chromosorb 102	Slightly polar styrene–divinylbenzene (Alcohol, permanent gases)	0.29	300–400
Chromosorb 103	Non-polar/basic crosslinked polystyrene (Amines, basic compounds, amides)	0.32	15– 25
Chromosorb 104	Very polar acrylonitrile–divinylbenzene (Nitriles, nitroalkanes, inorganic gases)	0.32	100–200
Chromosorb 105	Moderately polar polyaromatic (Formaldehyde, acetylene, adsorbent)	0.34	600–700
Chromosorb 106	Non-polar crosslinked polystyrene (Gases, C_2–C_5 alcohols)	0.28	700–800
Chromosorb 107	Polar crosslinked acrylic ester (Formaldehyde, sulphur gases, polar compounds)	0.30	400–500
Chromosorb 108	Polar crosslinked acrylic (Gases, alcohols, aldehydes)	0.30	100–200
Tenax-GC	2,6-Diphenyl-*p*-phenylene oxide (General-purpose polar compounds)	375	
Tenax-TA	(Trapping agent for organic volatiles)	0.16	35

For example, Chromosorb 103 is designed for the analysis of basic amines and Porapak T is used for the determination of formaldehyde in water.

Another polymeric material that has been employed in gas chromatography is Tenax GC (poly-2, 6-diphenyl-*p*-phenylene oxide). It can be used as a stationary phase but is probably more commonly employed as a trapping agent because of its high capacity for organic compounds and low capacity for water. The trapped components can then be released into the injection port of a GLC by heating.

4.3 COLUMN SELECTION AND THE CLASSIFICATION OF LIQUID PHASES

a. Column selection

In selecting a liquid phase and column system for a GLC separation, the operator has to balance a number of factors. The main influences are the nature of the analyte of interest compared with other components that may be present in the sample and the complexity of the sample. The availability of equipment will determine whether a packed or open-tubular column can be used. Often the choice will be a matter of the experience of the operator coupled with a knowledge of methods reported in the literature for the same or similar compounds. Frequently, a range of systems may be possible for a particular analysis, but usually only with demanding samples will the choice be critical.

Although open-tubular columns offer a number of advantages, particularly of high efficiency and increased sensitivity, packed columns are still suitable for many samples. Packed columns are generally easier to use because they require less operator skill and can tolerate larger samples and less efficient injection techniques. They would therefore probably be the method of choice for relatively simple samples with a limited number of components, such as the quality control of a known compound, purity checks on solvents, or monitoring a chemical reaction. As the peaks are broader they are often easier to quantify, whereas peak heights and shapes from a open-tubular column can be critically dependent on the injection technique. Although most separations can be carried out on liquid stationary phases packed on inert supports, columns packed with porous polymers or solid stationary phases are preferable for very volatile samples and gases.

Open-tubular columns are now available in a wide range of column dimensions and film thicknesses, which overlap packed columns in sample capacity and robustness. Consequently, wide-bore columns with thick films can directly replace many packed column separations. In addition, bonded stationary-phase materials offer a column stability and ease of cleaning that are very attractive particularly for impure samples. The high efficiency of narrow-diameter open-tubular columns enables many separations to be achieved. The high resolving power also means that only two or three different liquid phases are needed to obtain sufficient selectivity for most samples, unlike packed columns when a wide range of liquid phases are needed. The newer crosslinked column materials are also extremely inert and show much fewer distortions with polar analytes than packed or non-bonded open-tubular columns.

In any laboratory a limiting factor in the choice of system may be the equipment available. Many older gas chromatographs are badly designed for narrow-bore open-tubular columns and much of the potential efficiency may be lost by inappropriate injection ports or detectors. Almost all current instruments are primarily intended for use with open-tubular columns and some may even be unsuitable for packed column operation.

In selecting the liquid phase to be used for a particular application, with either an open-tubular or packed column, the initial choice is governed by the general rule of *LIKE WITH LIKE*. Thus the first choice for the separation of samples containing esters should be an ester liquid phase such as DEGS (or cyanopropyl silicones); for alcohols, a Carbowax 20M or Superox phase; for alkanes, an Apiezon L column, methyl-, or phenyl silicone phase; and for ketones and aldehydes, a cyano silicone column (both π-bonded C=X systems). These relationships usually ensure a good solubility of the analyte in the liquid phase and therefore good peak shapes. If a phase with very different properties to the analyte is used, such as the separation of polar alcohols on Apiezon L or alkanes on a poly(propylene glycol) phase, the peaks will inevitably show distortion or tailing.

The retention of an analyte on a liquid phase is governed by its volatility and the nature of the interactions that take place. Three principal modes have been identified. The strongest interaction is provided by orientation forces or permanent dipole–dipole interactions. Less polar analytes or liquid phases may interact by Debye forces or dipole–induced dipole interactions; and non-polar analytes are retained only by London or dispersion forces. The stronger the forces that are present, the more a particular analyte will be retained.

Normally if there is a choice, the liquid phase with the highest maximum recommended temperature should be used. Many of the non-silicone-based phases may be restricted to the separation of more volatile compounds (see Tables). However, the availability of a particular phase or packing already within the laboratory, even if not the ideal choice, may enable an adequate separation to be carried out without delay or avoid the need to prepare a special column.

Many samples are mixtures of compounds containing a range of functional groups and the initial selection of a column may not satisfy all the compounds. It such cases the best approach is to avoid the more extreme polar or non-polar phases. For example, although OV-101 is an excellent phase for the separation of hydrocarbons, it is very non-polar and the slightly less non-polar methyl phenyl silicone phases are often more suitable for mixtures of non-polar and polar samples. Open-tubular columns can frequently be successfully used with a wider range of molecular types than the corresponding packed column, as specific interactions of the column are greatly reduced. Even very polar compounds such as free carboxylic acids and amines can often be analysed on non-polar phases, such as BP-1. In many cases the initial selection of a liquid phase will not achieve complete resolution of the components of the sample. It will then be necessary to examine alternative liquid phases to alter the selectivity of the separation.

With packed columns it is also important to use a suitable support material. For non-polar analytes Chromosorb P supports are often used, but for polar samples Chromosorb W and Chromosorb G are preferred. To obtain similar retentions the loading of the liquid phases on the different supports may need to be different because of their different densities (see Section 4.1.a.i). If loadings lower than 5% are being used, the support material should be silylated, particularly if it is coated with a non-polar liquid phase.

For most laboratories a fairly limited set of packed columns will enable most separations to be carried out. Hawkes has suggested columns coated with OV-1, OV-17, DEGS, OV-275, OV-210 and Carbowax 20M [1] and similar sets have been proposed by other authors. With open-tubular columns, two liquid phases, a chemically bonded dimethylsilicone and a chemically bonded Carbowax 20M column, should enable almost all mixtures to be resolved. In addition, a cyano silicone column may be desirable for less volatile polar analytes. If work with more volatile samples is expected, the more retentive porous polymers and solid stationary phases must not be forgotten.

b. Classification of liquid phases

Because of the large number of commercially available liquid phases, it is important to be able to identify those materials which are closely related or are significantly different. If a particular phase is unsuitable for a separation, it will then be possible to choose an alternative with different selectivity properties. In addition, it is frequently desirable to repeat a method reported from another laboratory based on a liquid phase that is not readily available. This can often be a problem as many virtually identical phases are frequently sold under different labels and with different tradenames. Differences between columns, in the proportion of the liquid phase, the support material, or column dimensions, are less important than the nature of the liquid phase as they primarily alter the amount of liquid phase in the column. This alters the overall retention rather than the selectivity. The operator can compensate for these effects by altering the length of the column or the loading or by changing the temperature of the oven to speed up or slow down the analysis, although this may alter the selectivity.

In order to be able to classify liquid phases, it is first essential to have a scale of measurement. Retention times are no good as they are so dependent on the temperature, carrier gas flow rate and column dimensions. In 1956 Kováts proposed the use of a retention index scale in which the retention of analytes on GLC is interpolated between the retentions of the *n*-alkanes, whose retention indices (*RI*) are defined as their carbon number × 100 (i.e. hexane = 600). Because it is a relative retention scale, the values are virtually independent of flow rates and column dimensions (for a fuller discussion see Section 6.2.d).

Based on this scale, Rohrschneider measured the retention indices of a series of five test compounds, namely benzene, ethanol, butan-2-one, nitromethane and pyridine, as a guide to the selectivity of different liquid phases (Table 4.16) [2, 3]. The standards were chosen to reflect different interactions of the liquid phase; benzene is characteristic of aromatics and olefins; ethanol is characteristic of alcohols, nitriles, acids and halides; butanone is characteristic of ketones, ethers, esters and epoxides; nitrome-

Table 4.16 Test compounds for GLC stationary phases.

Rohrschneider test compounds (used at 100°C)		McReynolds test compounds (used at 120°C)		Retention index on squalane
Benzene	x	Benzene	x'	653
Ethanol	y	1-Butanol	y'	590
2-Butanone	z	2-Pentanone	z'	627
Nitromethane	u	1-Nitropropane	u'	652
Pyridine	s	Pyridine	s'	699
		2-Methyl-2-pentanol	H	690
		1-Iodobutane	J	818
		2-Octyne	K	841
		1,4-Dioxane	L	654
		cis-Hydrindane	M	1000

thane is characteristic of nitro and nitrile groups; and pyridine is characteristic of bases and heterocyclic compounds. The retentions in each case are a combination of the properties of the standard and the interaction with the liquid phase. The retention index value for each standard on the liquid phase under study was compared with that for the same analyte measured on a squalane column. The differences in the retention index values, expressed as (Δ *RI*)/100, were used as a guide to the interaction properties of the liquid phases. The retention index differences of the standards are often referred to respectively as x, y, z, u and s from the symbols used in the original theoretical discussion. Squalane was chosen as the reference liquid phase because it was the least polar liquid base readily available and its polar interactions could be regarded as negligible. The liquid phases constants were measured at 100°C using columns with a 20% loading to reduce any effect due to the support material. Liquid phases with similar Rohrschneider constants are considered to have comparable retention and selectivity properties. Two liquid phases with different constants should demonstrate different selectivity properties [4].

This method for classifying liquid phases has been largely superseded by a closely related method developed by McReynolds [5]. He proposed the use of an expanded set of test compounds, namely benzene, 1-butanol, 2-pentanone, 1-nitropropane, pyridine, 2-methyl-2-pentanol, 1-iodobutane, 2-octyne, 1,4-dioxane and *cis*-hydrindane, with the measurement of the retention indices at 120°C using a column with a 20% loading. The column constants are again derived by comparison of the retention indices with those on a squalane column and are expressed as Δ retention index values. In this case the symbols x', y', z', u', s', H, J, K, L and M are used, respectively.

The wider range of test compounds was chosen to reflect additional interaction factors but frequently only the first five values, x' to s', are quoted. A list of some of the more popular phases is given in Table 4.17. The similarity of the selectivities of the different methylsilicones, SE-30, OV-101, OV-1 and SP-2100, is clear. The different silicone oils will therefore usually give the same elution order for a sample mixture. However, a particular methylsilicone phase may still give a poor chromatogram because the efficiency of the separation will depend on the viscosity and loading, and the different materials may have different temperature limits depending on their purity. In contrast, the differences between the cyanopropylsilicones OV-225 and OV-275 are very marked.

Often the McReynolds constants for the first five test compounds (designated x' to s') are added together to give an overall polarity value (Σ). These cumulative totals have been used by some manufacturers to

Table 4.17 Comparison of McReynolds constants for a selection of widely used liquid phases.

Liquid phase	MRT (°C)	McReynolds constants				
		x'	y'	z'	u'	s'
Squalane	0–130	0	0	0	0	0
Apiezon L	50–280	32	22	15	32	42
Apolane C_{87}	30–280	21	10	3	12	25
Carbowax 20M	60–225	322	536	368	572	510
Diethylene glycol adipate	0–200	378	603	460	665	658
Diethylene glycol succinate	0–200	496	746	590	837	835
Didecyl phthalate	10–175	136	255	213	320	235
Dinonyl phthalate	20–150	83	183	147	231	159
Dexsil 300	50–450	41	80	103	148	96
Dexsil 400	20–400	72	108	118	166	123
Ethyleneglycol adipate	100–200	372	576	453	655	617
Ethyleneglycol succinate	100–200	537	787	643	903	889
FFAP	60–275	340	580	399	602	627
OV-1 Dimethylsilicone gum	100–350	16	55	44	65	42
OV-101 Dimethylsilicone fluid	0–350	17	57	45	67	43
OV-11 Methylphenylsilicone	0–350	102	142	145	219	178
OV-17 Methylphenylsilicone	0–350	119	158	162	243	202
OV-210 Trifluoromethylsilicone	0–275	146	238	358	468	310
OV-225 Cyanopropylmethylsilicone	20–250	228	369	338	492	386
OV-275 Dicyanoallylsilicone	25–275	629	872	763	1106	849
OV-351 Carbowax–nitroterephthalic acid	50–250	335	552	382	583	540
Polypropyleneglycol	20–150	128	294	173	264	226
QF-1	0–250	144	233	355	463	305
SE-30 Methyl silicone	50–350	15	53	44	64	41
SE-52 Methylphenyl silicone	50–300	32	72	65	98	67
SP-400 Chloromethyl silicone	0–350	32	72	70	100	68
SP-2100 Methyl silicone	0–350	17	57	45	67	43
Superox 0.6	65–300	311	560	359	565	497
XE-60 Cyano silicone	0–250	204	381	340	493	367
Ucon LB-550-X (propylene glycol)	20–200	118	271	158	243	206

McReynolds constants are as follows: x', benzene; y', 1-butanol; z', 2-pentanone; u', 1-nitropropane; s', pyridine.

provide polarity classifications and listings. However, although the two phases didecyl phthalate (Σ = 1159) and polypropyleneglycol (Σ = 1083) have similar overall polarities, examination of the individual values show marked differences reflecting different selectivities.

McReynolds constants have also been reported for a number of bonded liquid phases on open-tubular columns, and some small but significant differences have been found between the bonded phases and the comparable non-bonded liquid phase, e.g. OV-17 and the bonded SP-2250, both 50:50

Table 4.18 McReynolds constants for bonded open-tubular columns.

Column	McReynolds constants				
	x'	y'	z'	u'	s'
SPB-1 Methylsiloxane (100)	4	58	43	56	38
SPB-5 Methylphenylsiloxane (95:5)	19	74	64	93	62
SPB-20 Methylphenylsiloxane (80:20)	67	116	117	174	131
SPB-35 Methylphenylsiloxane (65:35)	101	146	150	219	180
CP-Sil 5 CB Methylsiloxane	15	50	41	60	36
CP-Sil 8 CB Methylphenylsiloxane (95:5)	33	68	59	98	65
CP-SiL 19 CB Methylphenylsiloxane (50:50)	82	170	157	236	160
CP-Sil 43 CB Cyanopropylmethylsiloxane	227	373	336	489	398
CP-Wax 51 CB Carbowax 20M	299	503	345	541	478

McReynolds constants are as follows: x'. benzene; y', 1-butanol; z', 2-pentanone; u', 1-nitropropane; s', pyridine.

methylphenylsilicones (Table 4.18). These effects may be partly because on an open-tubular column the liquid film layer is very thin and part of the underlying silica surface may be exposed. Related tables of retention values have been prepared for porous polymer columns, enabling the different materials to be compared with one another and with the liquid phases.

As well as the original paper and a number of subsequent reviews and additional surveys, compilations of the McReynolds constants are available from many sources (see Bibliography), and often manufacturers and chromatography supply houses include a listing in their catalogues. These can be invaluable in trying to identify equivalent phases to those reported in literature methods. A selection of these catalogues and their selectivity listings should form part of the basic reference library of every chromatographer.

4.4 COLUMN PACKING AND TESTING

Many laboratories still prepare their own stationary phases for packed columns from liquid phases and support materials. Alternatively, precoated stationary phases can be purchased, which are prepared to the customer's requirements. These can then be readily packed into columns with little equipment or time required. However, as noted earlier, very few laboratories prepare their own open-tubular columns and bonded phases are almost all commercially prepared. Preparing a packed column falls into two stages: firstly, preparing the stationary phase and, secondly, packing it into the column.

a. *Preparation of the stationary phase*

The simplest method to prepare a stationary phase is to dissolve the required weight of the liquid-phase material in an organic solvent and mix it with a weighed amount of support material. Usually the support material is obtained pretreated by silanisation or acid washing. Care should be taken that the support is not overloaded; Chromosorb G should have a loading of less than 5% but other phases can accept higher loadings (see Table 4.1). The tables of liquid phases in suppliers' catalogues usually include the most suitable solvent, typically methanol, acetone, or chloroform. Some solvents should be avoided with particular phases. For example, solvents containing carbonyl groups, such as acetone, are unsuitable for QF-1 and related materials. The slurry of solvent and packing material is then placed in a flask and the solvent removed on a rotary evaporator until a dry powder remains. However, the tumbling action of the evaporator, as well as giving an even distribution of the liquid phase, can also break up the support material. Friable materials such as Chromosorb W AW-DMCS can suffer from this effect as the fractures will reveal fresh surfaces, which will contain untreated active sites, and these may interfere with the chromatography.

There has been concern that this method often gives columns that were not reproducible between laboratories. This problem was investigated in detail by the Analytical Methods Committee of the Royal Society of Chemistry. They developed a method in which the liquid phase is dissolved in a specified limited volume of the solvent, which is then mixed with the support material. The two are lightly mixed to give a slightly damp mixture but with no free solvent. This is then placed in a sealed container overnight and finally left open in a fume hood for the solvent to evaporate. The standing is important as it appears to promote an even distribution of the liquid phase. The method has so far been described in detail for methyl silicone [6] and Carbowax 20M [7] stationary phases. Using these methods, columns prepared in different laboratories were found to have virtually identical retention and selectivity characteristics for a series of sesquiterpene test compounds.

b. *Column packing*

Before packing the stationary phase into the columns, the column tubing should be well cleaned. Glass columns are usually silanised by treatment with a solution of 1% dimethyldichlorosilane (DMCS) in toluene for 10 min, washing with methanol and then acetone, and followed by drying.

The detector end of the column is attached to a vacuum line and the stationary phase is poured into the column in small portions, with vibration to ensure a uniform packing. Once the column is filled, the vacuum is released and an inert carrier gas is passed through the column with further vibration until the bed is stable. The pressure is allowed to drop and the entrance of the column is plugged with silanised glass wool. The packing process is much easier to monitor with glass columns as any irregularities are easily seen.

Finally the column is placed in a column oven but the exit of the column is not attached to the detector. With a carrier gas flowing through the column, the oven temperature is then slowly raised to near the maximum recommended temperature of the liquid phase to condition the column. This process removes any volatile residues and traces of solvent that might interfere with subsequent analyses or contaminate the detector. More specialised conditioning routines are used for some phases, such as the porous polymers, and the manufacturer's instructions should then be followed.

Finally the column should be labelled with the liquid phase, loading, support material and MRT. A record should be maintained for each column giving the preparation and conditioning treatment.

c. *Column testing*

Once the column has been conditioned it can be connected to the detector and is ready for use. At this point the performance of the column should be tested. At intervals during use, the test should be repeated to ensure that the efficiency and selectivity have not altered through aging or contamination. In addition, method-specific tests may be used as part of any assay protocol.

Desirably a test sample should contain a mixture of compounds that would behave in a similar way to the expected sample and could be used for both qualitative and quantitative evaluation. A well retained component can be used to test the efficiency of the column to avoid variations due to extra-column effects. If *n*-alkanes are included, the retention indices (see later) of the test compounds can be determined. Because of the wide variation in samples, columns and conditions in gas chromatography, there are few standard mixtures but each chromatographer usually prepares his own test samples.

In 1971 Grob and Grob proposed a mixture containing neutral, acidic, and basic compounds for the evaluation of open-tubular columns and Grob *et al.* subsequently refined it in 1978 (Table 4.19) [8]. The *n*-alkanes and esters in the mixture were intended for the evaluation of retention and efficiency as Trennzahl numbers. 1-Octanol and 2,3-butanediol interact with adsorption sites, and 2,6-dimethylphenol and 2,6-dimethylaniline are included to test the acidity and basicity of the column. A particularly severe test is provided by the very polar components, dicyclohexylamine and 2-ethylhexanoic acid, and the peak for the latter is often distorted by interaction with the column (Figure 4.12). On many non-polar columns this acid will give a badly fronted peak if the surface is not totally inert. The base dicyclohexylamine

Table 4.19 Grob test mixture for open-tubular columns [8] (each approximately 40 mg l^{-1}).

Decane	2,6-Dimethylphenol
Undecane	Dicyclohexylamine
1-Octanol	2-Ethylhexanoic acid
Nonanal	Methyl dodecanoate
2,3-Butanediol	Methyl undecanoate
2,6-Dimethylaniline	Methyl decanoate

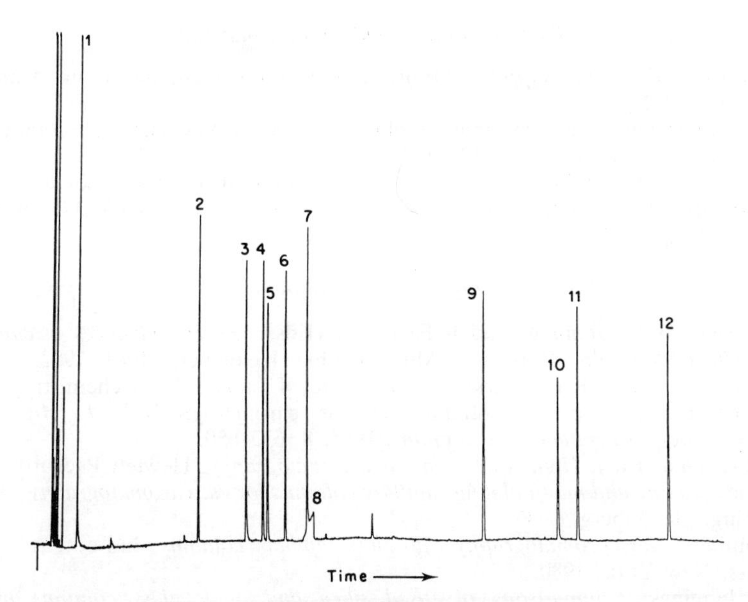

Figure 4.12 Separation of Grob activity test mixture on a chemically bonded dimethylsilicone open-tubular column BP-1 25 m × 0.22 mm i.d. Film thickness 0.25 μm. Column temperature programmed from 40°C to 145°C at 1°C min^{-1}. Flame ionisation detector. Peaks: 1, 2,3-butanediol; 2, n-decane; 3, 1-octanol; 4, nonanal; 5, 2,6-dimethylphenol; 6, n-undecane; 7, 2,6-dimethylaniline; 8, 2-ethylhexanoic acid; 9, C_{10} methyl ester; 10, dicyclohexylamine; 11, C_{11} methyl ester; 12, C_{12} methyl ester. (Reproduced with permission from Scientific Glass Engineering.)

will detect free silanol groups. Although this mixture is widely used for open-tubular column evaluation, it is not ideal as a single solution because some of the components will interact with one another, and on storage additional components may be formed.

BIBLIOGRAPHY

General

L.S. Ettre and E.W. March, "Efficiency, resolution and speed of open tubular columns as compared to packed columns", *J. Chromatogr.*, 1974, **91**, 5–24.

G. Guiochon, "Comparison of the performances of the various column types used in gas chromatography", in J.C. Giddings and R.A. Keller (Eds.), *Advances in chromatography*, Vol. 8, Marcel Dekker, New York, 1969, **8**, pp. 179–270.

Packed columns and support materials

Chromosorb. Diatomite supports for gas–liquid chromatography, Johns Manville, Denver, 1981.

D.M. Ottenstein, "The chromatographic support in gas chromatography", *J. Chromatogr. Sci.*, 1973, **11**, 136–144.

N.C. Saha and J.C. Giddings, "Comparative column efficiencies of some common solid supports in gas liquid chromatography", *Anal. Chem.*, 1965, **37**, 830–835.

Open-tubular columns

W. Bertsch, W.G. Jennings and R.E. Kaiser (Eds.), *Recent advances in capillary gas chromatography*, Vols. 1–3, Alfred Hüthig, Heidelberg, 1981, 1982.

L.G. Blomberg and K.E. Markides, "The role of organosilicon chemistry in the preparation of capillary columns for gas chromatography", *J. High Res. Chromatogr., Chromatogr. Commun.*, 1985, **8**, 632–650.

R.R. Freeman (Ed.), *High resolution gas chromatography*, Hewlett Packard, 1981.

K. Grob, *Making and manipulating capillary columns for gas chromatography*, Alfred Hüthig, Heidelberg, 1986.

W. Jennings, *Gas chromatography with glass capillary columns*, 2nd ed., Academic Press, New York, 1980.

W.G. Jennings, *Comparisons of fused silica and other glass columns in gas chromatography*, Alfred Hüthig, Heidelberg, 1981.

M.L. Lee, F.J. Yang and K.D. Bartle, *Open tubular column gas chromatography. Theory and practice*, John Wiley, New York, 1984.

R.C.M. de Nijs and J. de Zeeuw, "Aluminium oxide-coated fused-silica porous-layer open-tubular column for gas–solid chromatography of C_1–C_{10} hydrocarbons", *J. Chromatogr.*, 1983, **279**, 41–48.

J.G. Nikelly (Ed.), *Advances in capillary chromatography*, Alfred Hüthig, Heidelberg, 1986.

Stationary liquid phases

G.E. Baiulescu and V.A. Ilie, *Stationary phases in gas chromatography*, Pergamon, New York, 1975.
J.K. Haken, "Developments in polysiloxane stationary phases in gas chromatography", *J. Chromatogr., Chromatogr. Rev.*, 1984, **300**, 1–77.
J.A. Yancey, "Liquid phases used in packed gas chromatographic columns. Part I. Polysiloxane liquid phases", *J. Chromatogr. Sci.*, 1985, **23**, 161–167.
J.A. Yancey, "Liquid phases used in packed gas chromatographic columns. Part II. Use of liquid phases which are not polysiloxanes", *J. Chromatogr. Sci.*, 1985, **23**, 370–377.

Stationary solid phases and porous polymer phases

S.B. Dave, "A comparison of the chromatographic properties of porous polymers", *J. Chromatogr. Sci.*, 1969, **7**, 389–399.
O.L. Hollis, "Separation of gaseous mixtures using porous polyaromatic polymer beads", *Anal. Chem.*, 1966, **38**, 309–316.
W.R. Supina and L.P. Rose, "Comparison of Porapak and Chromosorb porous polymers", *J. Chromatogr. Sci.*, 1969, **7**, 192.

Chiral liquid phases

R. Charles, U. Beitler, B. Feibush and E. Gil-Av, "Separation of enantiomers on packed columns containing optically active diamide phases", *J. Chromatogr.*, 1975, **112**, 121–133.
H. Frank, G.J. Nicholson and E. Bayer, "Rapid gas chromatographic separation of amino acid enantiomers with a novel chiral stationary phase", *J. Chromatogr. Sci.*, 1977, **15**, 174–176.
H. Frank, G.J. Nicholson and E. Bayer, "Chiral polysiloxanes for resolution of optical antipodes", *Angew. Chem., Int. Ed.*, 1978, **17**, 363–365.
W.A. König, *The practice of enantiomer separation by capillary gas chromatography*, Alfred Hüthig, Heidelberg, 1987.
W.A. König, I. Benecke and S. Sievers, "New results in the gas chromatographic separation of enantiomers of hydroxy acids and carbohydrates", *J. Chromatogr.*, 1981, **217**, 71–79.
T. Saeed, P. Sandra and M. Verzele, "$(GC)^2$ separation of the enantiomers of proline and secondary amines", *J. High Res. Chromatogr., Chromatogr. Commun.*, 1980, **3**, 35–36.
R.W. Souter, *Chromatographic separations of stereoisomers*, CRC Press, Boca Raton, FL, 1985.

Column and liquid phase selection

Column selection guide, GC Bulletin 723E, Supelco, Bellefonte, PA, 1978.
R.A. Keller, "Phase selectivity in gas and liquid chromatography", *J. Chromatogr. Sci.*, 1973, **11**, 49–59.
M.S. Klee, M.A. Kaiser and K.B. Laughlin, "Systematic approach to stationary phase selection in gas chromatography", *J. Chromatogr.*, 1983, **279**, 681–688.

H. M. McNair, *Column selection in gas chromatography*. ACS Audio Course, American Chemical Society, Washington, DC, 1982 (cassette tape/manual).

J.A. Yancey, "Liquid phases used in packed gas chromatographic columns. Part III. McReynolds' constants, preferred liquid phases, and general precautions", *J. Chromatogr.Sci.*, 1986, **24**, 117–124.

REFERENCES

1. S. Hawkes, D. Grossman, A. Hartkopf, T. Isenhour, J. Leary and J. Parcher, "Preferred stationary liquids for gas chromatography", *J. Chromatogr. Sci.*, 1975, **13**, 115–117.
2. L. Rohrschneider, "Die Vorausberechnung von Gas Chromatographischen Retentionszeiten aus Statistisch ermittelten 'Polaritäten'", *J. Chromatogr.*, 1965, **17**, 1–12.
3. L. Rohrschneider, "Eine Methode zur Charakterisierung von Gaschromatographischen Trennflüssigkeiten", *J. Chromatogr.*, 1966, **22**, 6–22.
4. W.R. Supina and L.P. Rose, "The use of Rohrschneider constants for classification of GLC columns", *J. Chromatogr. Sci.*, 1970, **8**, 214–217.
5. W.O. McReynolds, "Characterization of some liquid phases", *J. Chromatogr. Sci.*, 1970, **8**, 685–691.
6. Analytical Methods Committee, "Application of gas–liquid chromatography to the analysis of essential oils. Part VIII. Fingerprinting of essential oils by temperature programmed gas–liquid chromatography using methyl silicone stationary phases", *Analyst*, 1981, **106**, 448–455.
7. Analytical Methods Committee, "Application of gas–liquid chromatography to the analysis of essential oils. Part VII. Fingerprinting of essential oils by temperature programmed gas–liquid chromatography using a Carbowax 20M stationary phase", *Analyst*, 1980, **105**, 262–273.
8. K. Grob Jr, G. Grob and K. Grob, "Comprehensive standardized quality control test for glass capillary columns", *J. Chromatogr.*, 1978, **156**, 1–20.

CHAPTER 5

DETECTORS FOR GAS–LIQUID CHROMATOGRAPHY

5.1 GENERAL CRITERIA

a. *Introduction*

The elution of the analytes from the GLC column is monitored by the detector, whose performance largely defines the sensitivity and selectivity of the analysis. A large number of different types of GLC detectors have been described but most are of only academic or experimental interest. The remainder can be divided into a group of five established detectors, most of which will be found in any chromatography laboratory, and a number of less important but commercially available detectors with more specialised applications. In addition, gas–liquid chromatographs can act as coupled sample separation and inlet systems for more sophisticated spectroscopic instruments, such as mass spectrometers or Fourier transform infrared (FTIR) spectrometers, which are powerful analytical tools in their own right.

All the detectors produce an output which is proportional to the amount of analyte being eluted, although the response of different compounds in the same detector may differ. The output is recorded as a continuous trace of signal strength against time and the peak areas can be measured either electronically using an integrator (Chapter 15) or manually from a chart recorder. Manual measurements are usually only used for approximate

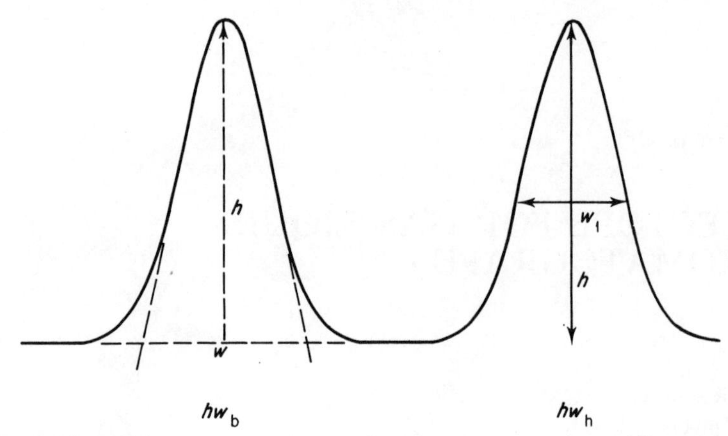

Figure 5.1 Estimated measurement of peak areas. Area = w_b h/2 or more usually = w_h h.

quantification using an estimated area calculated as peak height × width at half-height (w_h) (Figure 5.1). This method assumes that the peak is a triangle, whereas peak shapes are normally closer to a Gaussian curve. However, if the peaks from standards and from samples are examined under the same conditions there will be a constant proportional relationship between the estimated and true areas enabling the former to be used to establish calibration curves.

b. Detector performance

The operation and applicability of the different chromatographic detectors can be compared in the following terms.

Response

This is the magnitude of the electrical signal from the detector per unit mass of analyte, and is expressed in different ways depending on the mode of action of the detector.

Sensitivity

Often given as the limit of detection, this is the smallest amount of analyte that gives a peak height twice (sometimes defined as three times) the noise level of the background signal (Figure 5.2). It is thus the smallest peak that can be positively discriminated from the background noise. A second guide

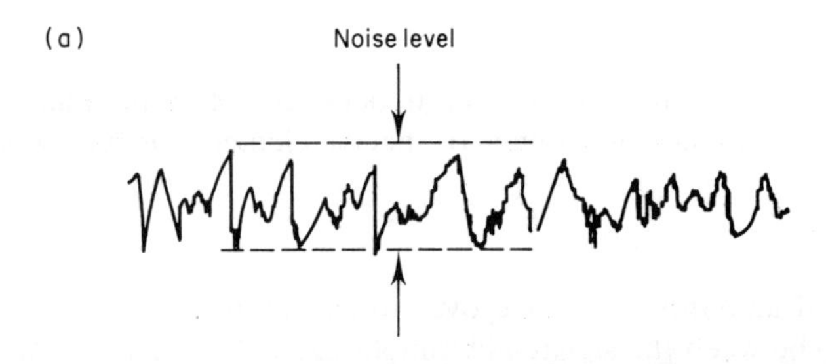

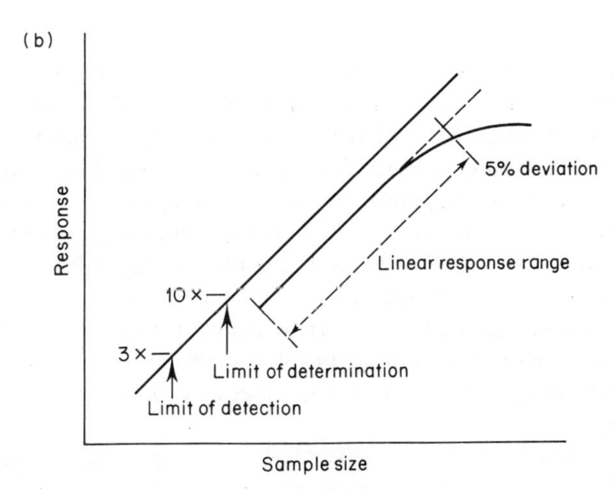

Figure 5.2 Determination of limit of detection. (a) Measurement of background noise level (b) Calibration curve for chromatography detector. Limit of detection (LOD) = 3 × background noise. Limit of determination = 10 × background noise. The linear range runs from the limit of determination to the point of 5% deviation from a linear relationship between signal and sample size.

to the sensitivity is the limit of determination, which is based on the smallest peak that can be confidently quantified with accuracy. It is usually defined as the amount of analyte that gives a peak with a height 10 times the background noise level.

Selectivity

This is the differences in the response of the detector to different compounds or different types or groups of compounds.

Specificity

This is used when the discrimination between different analytes can be related to the presence or absence of specific elements or functional groups.

Linear range

The range of amounts of analyte over which the detector will give a linear relationship between the signal and sample size is known as the linear range. It is calculated as the range from the limit of determination up to the point at which the relationship has a 5% deviation from linearity (Figure 5.2b).

Response time

This is a measure of the speed with which the detector can respond to changes in the output of the column. It is determined as the time for the detector signal to reach 63% of true value following a sharp change. The response is usually limited by the electronics of the amplifier in the detector or in the recorder or integrator. In a well designed modern system the response time should not be a problem, but older equipment, originally designed for the broad peaks from packed columns, may deliberately dampen the signal to reduce background noise. This can mean that the output will not able to respond rapidly to the narrow peaks from open-tubular columns (often less than 10 s wide). Similar problems can be found with older chart recorders and sharp peaks may be erroneously broadened or shortened electronically.

c. *Principal groups of detectors*

The detectors used in GLC can be divided into two principal groups based on the type of response.

i. Bulk property detectors

These detectors measure a bulk property of the carrier gas stream as it is eluted from the column and respond to changes in that property caused by the presence of the analyte. The response at any particular moment therefore depends on the concentration of analyte in the carrier gas and on the difference between its value for the bulk property and that of the carrier gas, and can be expressed as mV mg ml^{-1}. With this mode of detection, the total area under a peak is dependent on the flow rate of the carrier gas,

Table 5.1 Comparison of principal GLC detectors.

Detector	Sensitivity[a]	Linear range
TCD	10^{-7} g ml^{-1}	10^3–10^4
FID	10^{-12} g s^{-1}	10^7
ECD	10^{-16} moles ml^{-1} (lindane)	10^3–10^4
TID	10^{-14} g(N) s^{-1} (azobenzene)	10^3–10^5
	10^{-15} g(P) s^{-1} (tributyl phosphate)	
FPD	10^{-10} g(S) s^{-1} (thiophene)	10^{3}[b]
	10^{-12} g(P) s^{-1} (tributyl phosphate)	10^5
Hall	10^{-13} g(Cl) s^{-1}	10^2–10^5
	10^{-12} g(N) s^{-1}	10^4
	10^{-12} g(S) s^{-1}	10^4

[a] For the compounds indicated in the case of selective detectors.
[b] Following linearisation of the response.

because a low flow rate will cause the sample to remain for longer in the detector cell and hence give broader peaks with the same height than a fast flow rate. However, in most modern instruments, the flow can be regarded as constant and the response is directly related to the sample size.

ii. Mass flow detectors

These detectors give little or no signal for the carrier gas but only respond to the analytes. The response depends on the mass of the sample being eluted from the column at any point, and is expressed as mV mg^{-1}. The response is independent of the concentration and the peak area therefore depends only on the total mass being eluted from the column, irrespective of the flow rate.

Of the large number of detectors in routine use (Table 5.1), most of the work is carried out with the five main detectors:

Thermal conductivity detector	TCD
Flame ionisation detector	FID
Electron capture detector	ECD
Thermionic ionisation detector	TID
Flame photometric detector	FPD

These can be regarded as the established detectors, and will be found in most well equipped analytical laboratories. There are also related detectors with similar underlying modes of action, such as the photoionisation detector (PID). The first two detectors, FID and TCD, are usually considered universal detectors as they will respond to most samples. The other three

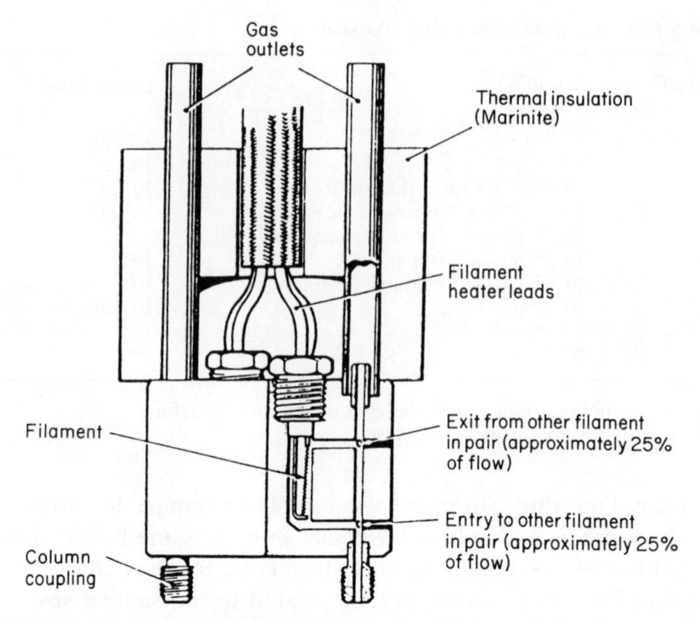

Figure 5.3 Thermal conductivity detector showing one of the four matched filaments. (Reproduced by permission of Philips Analytical from *An introduction to gas chromatography*, Pye Unicam, Cambridge, 1980, p. 9.)

detectors, ECD, TID and FPD, are the main selective detectors and give differential responses for analytes containing different groups. This category also includes other interesting commercially available detectors, such as the thermal energy analyser and electrochemical detectors, and these will be briefly considered later.

5.2 UNIVERSAL DETECTORS

a. *Thermal conductivity detector*

This is one of the earliest detectors used in GLC and is also known as the hot-wire detector, katharometer or catharometer detector, or in a slightly modified form as the thermistor detector. It is a bulk property detector, in which an electrically heated tungsten filament is placed in the carrier gas stream (Figure 5.3). The rate at which heat from the filament is transferred to the surrounding metal block depends on the thermal conductivity of the eluent gas. With just the carrier gas flowing through the detector cell, a thermal equilibrium will be established. When the analyte is eluted, the

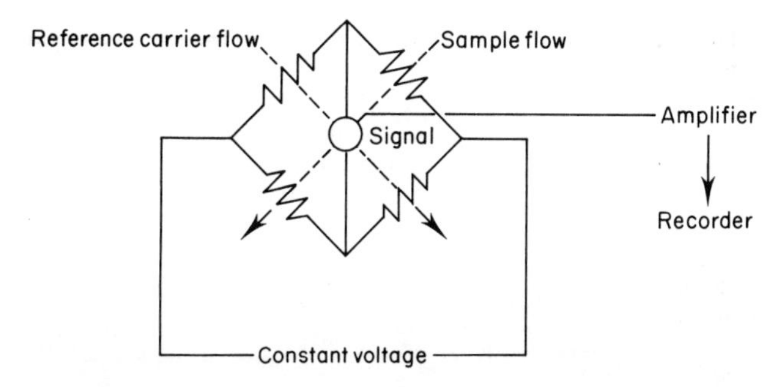

Figure 5.4 Electronic circuit for thermal conductivity detector: Wheatstone bridge circuit showing the positions of the four filaments, two in the eluent stream from the column and two in a reference carrier gas stream.

thermal conductivity of the carrier gas stream will change and hence the rate of heat loss from the filament will alter. This will cause the temperature and thus the resistance of the filament to change. These changes in electrical resistance can be detected using a Wheatstone bridge circuit. Opposite sides of the bridge are placed in the column eluent stream and in a reference carrier gas stream, and the current flowing across the bridge is recorded (Figure 5.4). In newer instruments the current through the filament is adjusted electronically to keep its temperature constant and the change in applied potential is monitored. Care must be taken to flush any air from the detector before the current is applied to the filaments or they may be damaged by the oxygen or burn out.

The sensitivity of the detector depends on the magnitude of the difference between the thermal conductivity of the carrier gas and that of the carrier gas plus analyte. Nitrogen is not usually used as a carrier gas, because it has a thermal conductivity similar to that of most organic compounds and any change will be small (Table 5.2). Most compounds would therefore give only a poor response. Consequently, helium or hydrogen, which have much higher thermal conductivities, are frequently used as the carrier gas because the change when an organic analyte is eluted will be greater and hence the sensitivity will be increased. However, if the sample includes hydrogen or helium as an analyte, then nitrogen or argon can be used as the carrier gas to increase the sensitivity.

However, in all these cases the differences between the thermal conductivities of the analyte and carrier gases are small. Thus unless the concentration of the analyte is high, it will have only a minor effect on the total thermal conductivity. This detector is therefore relatively insensitive compared to typical ionisation detectors (see later). Its main advantage is

Table 5.2 Thermal conductivities of typical carrier gases and samples at 38°C.

Compound	Thermal conductivity (10^{-6} cal s^{-1} cm^{-1} $°C^{-1}$)
Carrier gases	
Helium	369
Hydrogen	459
Nitrogen	64
Samples	
Organic compounds	
Methane	86
Propane	45
Ethanol	37
Benzene	35
Inorganic compounds	
Water	45
Carbon dioxide	42

that it is a universal detector and will respond to any compound, irrespective of its structure, whose thermal conductivity differs from that of the carrier gas.

The main applications of the thermal conductivity detector are for the determination of those compounds which give a poor or negligible response with the other established detectors. In particular, it is the standard detector for inorganic gases, such as water, carbon dioxide, oxygen, nitrogen, carbon disulphide, carbon tetrachloride and hydrogen (Figure 5.5). It can also be used for the separation of highly halogenated or oxygenated organic compounds or in preparative separations when large samples are being analysed. It is very suitable for portable instruments as only a carrier gas and electrical power are required for its operation. A major disadvantage, in addition to the low sensitivity, is that this detector can only be used for isothermal separations. Programming the oven temperature would cause variations in the carrier gas temperature and hence baseline drift.

The closely related gas density balance detector works on a very similar principle but in this case the hot filaments are isolated pneumatically from the carrier gas flow. This design protects the filaments from potentially corrosive analytes, but this detector is now little used except for some specialised applications in gas analysis.

b. Flame ionisation detector

The flame ionisation detector is the most important detector in GLC and is the standard detector fitted to almost all chromatographs (Figure 5.6). It

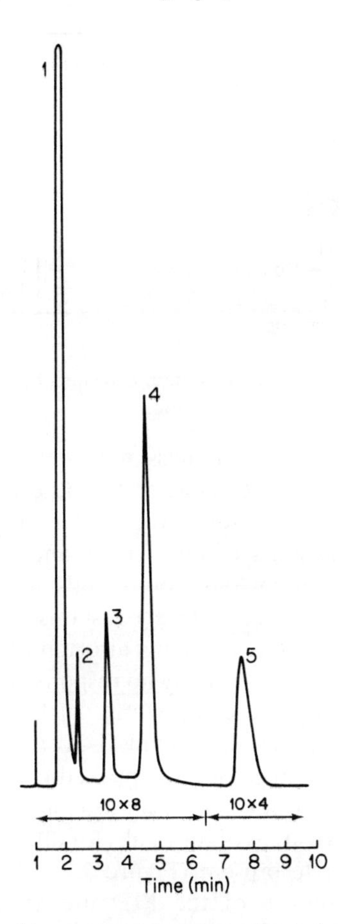

Figure 5.5 Detection of inorganic gases using a thermal conductivity detector. Unusually the separation uses argon as the carrier gas so that both hydrogen and nitrogen can be detected in the sample. The column was packed with 5 Å molecular sieve in order to achieve the separation of nitrogen and oxygen. Peaks: 1, H_2; 2, O_2; 3, N_2; 4, CH_4; 5, CO. (Reproduced by permission of Philips Analytical from P.J. Ridgeon, *Gas chromatography separations*, 3rd ed., Pye Unicam, Cambridge, 1976, p. 31.)

operates by mixing hydrogen with the eluent carrier gas and burning it at a jet in a supply of air. The ionisation in the flame is determined by applying a potential of 50–170 V from the jet to a collector. The current that flows across the flame is amplified and recorded. Typically the hydrogen–air flame has a very low background current of 10^{-13} to 10^{-14} A, but when an organic compound is eluted into the flame, the current increases to 10^{-12} to 10^{-6} A.

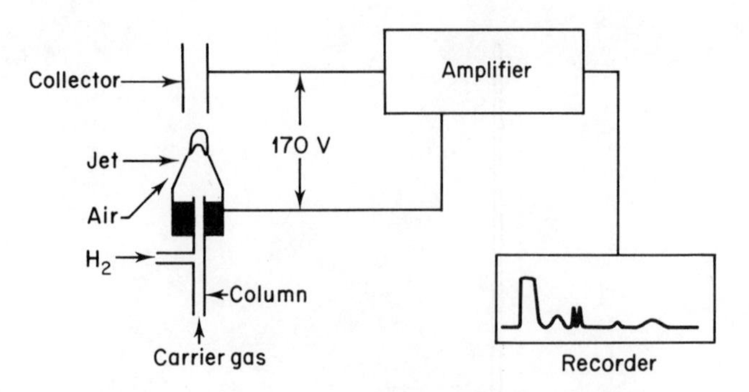

Figure 5.6 Flame ionisation detector—basic design.

The detector thus responds as a mass flow detector measuring only the analyte and gives no significant response to the carrier gas.

Almost all organic compounds give similar responses, approximately proportional to the total mass of the carbon and hydrogen in the analyte (Table 5.3). The only anomalous compounds are the first members of homologous series, which give reduced responses. Generally compounds with a large proportion of oxygen atoms also tend to give lower responses. Unlike the TCD, the FID gives virtually no response for inorganic compounds or totally halogenated organic compounds, such as carbon tetrachloride. The linear range for organic compounds is very long ($x10^7$) and covers almost the entire working range of most packed columns. It is therefore the universal detector of choice for almost all organic compounds and to a first approximation the relative areas of the peaks for different analytes correspond to the quantities of analyte present (Figure 5.7).

The stability and response of the FID are affected by the operating conditions (Figure 5.8). For the optimum response, the flow of hydrogen (typically 30–40 ml min^{-1} for a packed column) should be approximately 1.2 times the flow rate of the carrier gas. Large deviations from this ratio can give an unstable flame, which produces a noisy response or is extinguished as the solvent peak is eluted. The air supply is less critical as long as sufficient air is present to provide full combustion (typically >300 ml min^{-1}). Under normal conditions the FID is very stable and can give high sensitivity for most organic compounds with a limit of detection about 10^{-12} g s^{-1}. The disadvantages of this detector are the need for three gases, carrier, air and hydrogen, and the absence of any response to inorganic compounds. Although it is destructive, if a second method of detection is required, the column effluent can usually be split, with a proportion of the flow going simultaneously to the detectors in parallel.

Table 5.3 Response of typical compounds in the flame ionisation detector relative to hexane = 100.

Compound	Relative response
Methane	97
Ethane	97
Propane	98
Hexane	100
Ethylene	102
Acetylene	107
1-Hexene	99
Benzene	112
Toluene	107
Ethylbenzene	103
Methanol	23
Ethanol	46
Propanol	60
Butanol	66
Pentanol	71
Butanal	62
Acetone	49
Butan-2-one	61
Cyclohexanone	72
Formic acid	1
Acetic acid	24
Propionic acid	40
Hexanoic acid	63
Octanoic acid	65
Methyl acetate	20
Ethyl acetate	38
Butyl acetate	55
Aniline	75
Dibutylamine	75

(Reproduced by permission of Varian Associates from McNair and Bonelli, *Basic gas chromatography*, p. 142.)

In the the early days of GLC, radioactive sources were also used to produce ionisation. These included β-irradiation from a strontium-90 source, which was used in the argon ionisation detector and the ionisation cross-section detector, but these detectors can all now be regarded as obsolete.

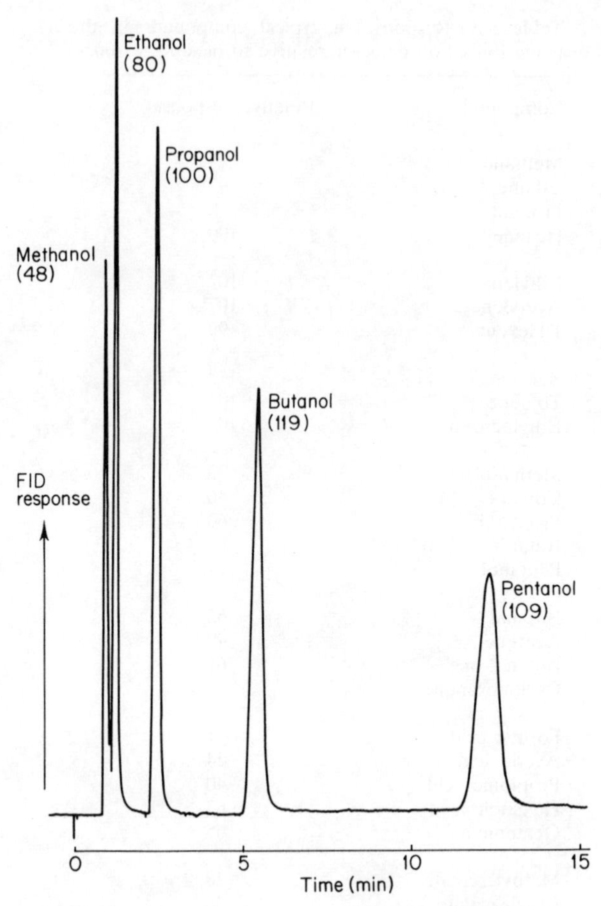

Figure 5.7 Separation of a mixture of equal volumes of the homologous alkanols, methanol to pentanol. The different responses of the alkanols (measured as relative peak areas compared to propanol) in the flame ionisation detector are given in parentheses. These show the reduced detector response for methanol and ethanol compared to the higher homologues. The separation was carried out using a 10% polypropyleneglycol column 1 m × 3 mm i.d. at 45°C.

c. *Photoionisation detector*

The photoionisation detector is closely related to the FID, but instead of using a flame, the analytes are ionised (Equation 5.1) by passing the eluent into a chamber illuminated by a powerful lamp (Figure 5.9):

$$R + h\nu \longrightarrow R^+ + e^- \tag{5.1}$$

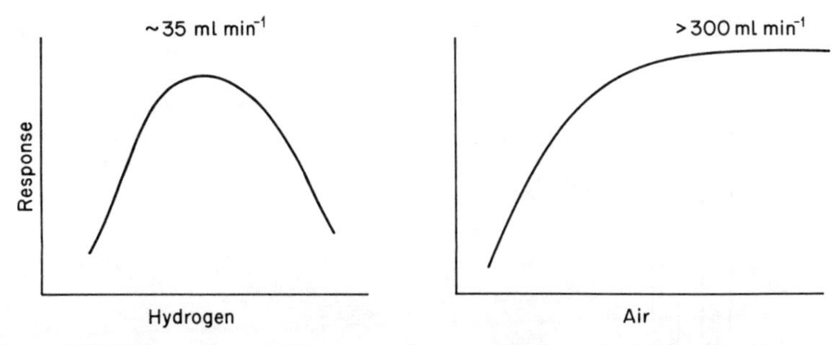

Figure 5.8 Effect of air and hydrogen flows to the flame ionisation detector on the response.

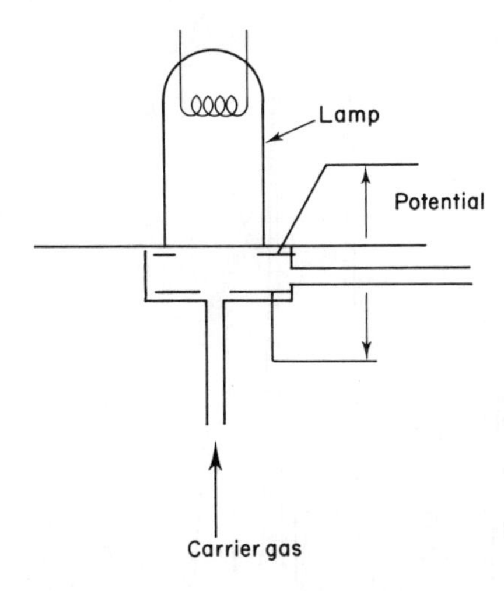

Figure 5.9 Photoionisation detector design.

The extent of ionisation is detected as a current by placing a potential across the cell. The response and sensitivity are similar to the FID, but as might be expected there is an enhanced response for unsaturated and aromatic compounds, which absorb ultraviolet light. This selectivity can enable the peaks from unsaturated or aromatic compounds, such as benzene or toluene, to be identified in a complex hydrocarbon mixture (Figure 5.10).

On using lamps with different energies, emitting either 9.5, 10.2 or 11.7 eV photons, the responses of the lower members of homologous series alter. The higher energies can enable some compounds that have a poor FID response, such as formaldehyde, to be easily detected. However, the PID

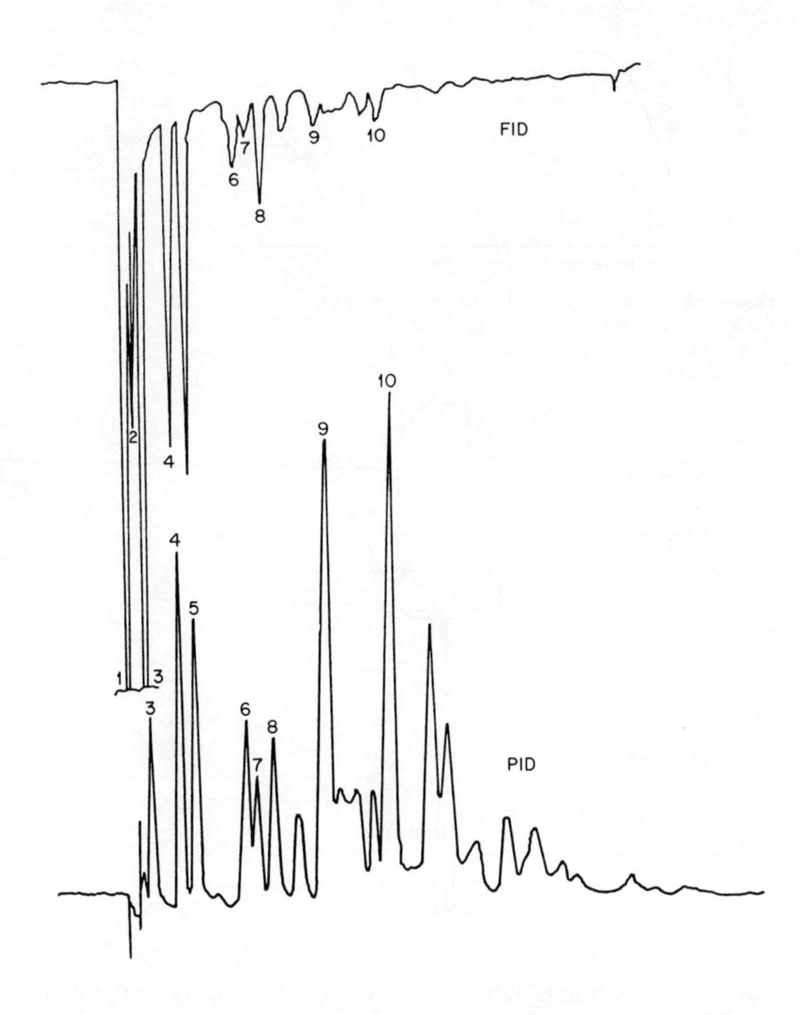

Figure 5.10 Separation of a light hydrocarbon synthetic natural gas feedstock using a flame ionisation detector (FID) and a photoionisation detector (PID; 10.2 eV lamp) in parallel. Column Porapak Q column programmed from 0°C to 190°C at 8°C min^{-1}. Nitrogen carrier gas. Note the high response for the aromatic components benzene (9) and toluene (10) from the photoionisation detector. (Reproduced by permission of Elsevier Science Publishers from *J. Chromatogr.*, 1978, **158**, 171–180.)

is more expensive than the FID and as yet has not gained wide acceptance. It is potentially a very suitable detector for portable GLC systems.

5.3 SELECTIVE AND SPECIFIC DETECTORS

Selective detectors are those which respond differently to compounds containing different elements or functional groups. In recent years, there has been a noticeable increase in the use of the more specific detectors, which only respond to compounds containing a specific or defined range of groups or elements. These are particularly useful for the analysis of complex mixtures, where a universal detector may be swamped by a multitude of signals. A selective detector, which will respond to only a few of the components, will often enable the compound of interest to be specifically quantified and identified. The response of an analyte in a selective detector can also confirm the presence of a particular type of group and thus aid identification. In addition, most selective detectors also offer increased sensitivity compared to the universal FID.

a. *Electron capture detector*

The electron capture detector (ECD) is the most widely used selective detector because of its high sensitivity for many compounds of environmental interest. Its operation has been the subject of much detailed study, which has been comprehensively reviewed (see Bibliography). It was one of the earliest detectors to be proposed for GLC as the "electron affinity detector". Nowadays a number of different designs are used by the different manufacturers but in each case the basic concept is of a chamber containing a radioactive β-emitter source (Figure 5.11). The chamber is heated above the column temperature to prevent condensation of the sample. The source is usually ^{63}Ni (typically 8–15 mCi) which has a half-life of 85 years and an energy of 0.067 MeV. More rarely ^{3}H (100–250 mCi), as a titanium–tritium foil, is used, which has a half-life of 12.5 years and an energy of 0.018 MeV. The latter source is only thermally stable up to 225°C and it should not therefore be used above 200°C, so that its application is severely limited. In contrast, nickel sources can be used safely up to 350°C, although their sensitivity is slightly lower.

As the carrier gas enters the chamber, it is ionised to give free electrons, (Equation 5.2),

$$\text{Carrier} + \beta \longrightarrow \text{Carrier}^{+} + 2e^{-} \qquad (5.2)$$

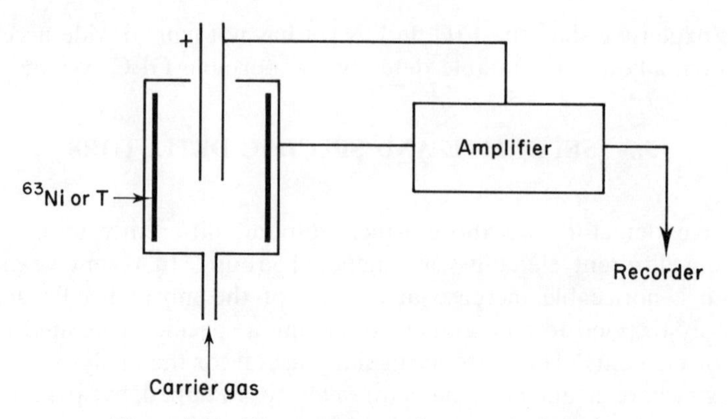

Figure 5.11 Electron capture detector—basic design.

which are monitored at the anode by applying a potential of about 50 V across the chamber. This background or standing current (approximately 10^{-9} A) is amplified and passed to the recorder (Figure 5.12a). If an analyte molecule that is capable of capturing electrons is eluted, when it collides with a free electron either an non-dissociative or dissociative reaction can occur depending on the analyte (Equation 5.3).

$$
\begin{aligned}
AB + e^- &\longrightarrow AB^- && \text{non-dissociative} \\
&\longrightarrow A^{\cdot} + B^- && \text{dissociative} \quad (5.3)
\end{aligned}
$$

These reactions are thermally controlled and increases in the temperature will increase the dissociative mechanism and change the response. Ideally the detector temperature should be optimised for each analyte but it must also be held above the column temperature to prevent condensation of the analyte. Highly mobile free electrons have thus been replaced by slowly moving ionised molecules or radicals. This increases the possibility of a recombination collision taking place with positively charged carrier gas ions, lowering the current across the cell (Figure 5.12a). The detector response therefore depends on the concentration of the analyte in the eluent.

Compounds that do not capture electrons will have no effect on the current (unless they are themselves ionised by the β-radiation, which only rarely happens). This detector typically responds to halogenated, unsaturated and aromatic organic compounds and to organometallic compounds but not to simple alcohols, ethers, or other saturated groups (Table 5.4).

The reduction in the standing current is displayed as a positive peak by reversing the output from the amplifier and setting the standing current signal to the baseline (Figure 5.12b). The magnitude of the signal from an ECD is limited because, once the original standing current is fully quenched,

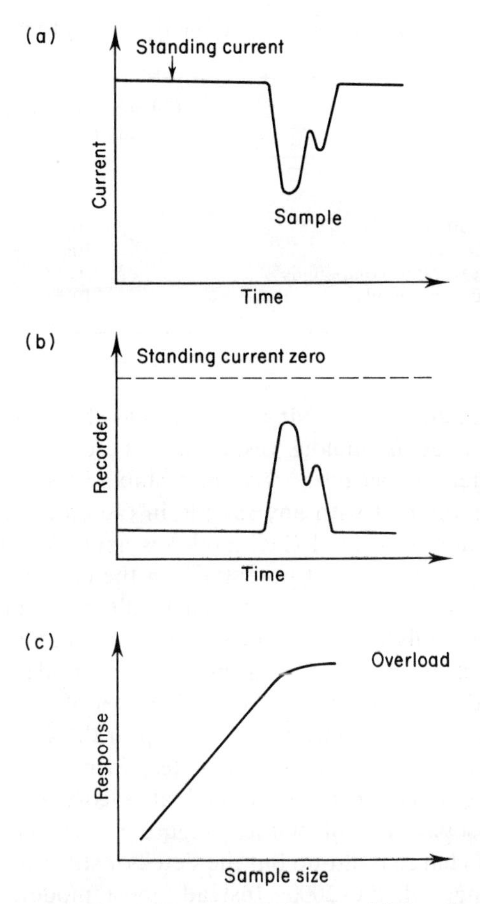

Figure 5.12 Operation of a electron capture detector. (a) Standing current from ionisation of the carrier gas, and effect on standing current of the elution of an electron capturing analyte. (b) Signal as displayed after inversion and offset of the standing current. (c) Response curve on overloading the detector.

any further increase in sample size can produce no further change (Figure 5.12c). Overloading the detector therefore produces flat-topped peaks. The sensitivity of the ECD is so high that this effect will occur with quite small sample sizes, near the limit of detection of the FID detector.

Nitrogen can be used as the carrier gas but the sensitivity of the detector is enhanced by using argon containing 5–10% of either methane or carbon dioxide as a quench gas. Some detectors are specifically designed to give the best results with one carrier gas or the other. The quench gas reduces the mean speed of the free electrons and therefore increases the possibility

Table 5.4 Selectivity and response of different types of analytes in the electron capture detector.

Analytes	Enhanced response compared to FID response
Organohalogens	10^4
Organometallic compounds	10^3
Aromatic compounds	10^2
Conjugated unsaturated compounds	10
Non-conjugated compounds	10^{-4}

of interaction with the analyte. It also increases the dynamic range of the detector and removes anomalous responses. These would occur with pure argon as the carrier gas because it give metastable ions on ionisation, which would themselves interact with any sample in the eluent to give a signal.

When a continuous potential (DC mode) is applied to the detector cell, there is a build-up of a "contact potential" on the electrode surfaces and of a "space charge" in the gas phase, which result in anomalous peaks with electron capturing analytes. It has been shown that better peak shapes and reproducibility can be obtained by using a pulsed mode of operation. The potential across the cell is applied as short pulses of 0.5–1 μs. This allows the concentration of free electrons to build up in the cell and increases the probability of their interaction with analytes, increasing the sensitivity of the detector. The pulses can be applied at a constant pulse frequency, typically with a separation of 500 μs (Figure 5.13a). This method avoids most problems of charge build-up but the detector still has a limited dynamic concentration range of 200–1000. Instead, most modern instruments are designed to operate in a constant current mode in which the pulse frequency is varied, typically between 1 kHz and 1.4 MHz, so that there is a constant current across the cell (Figure 5.13b). This method gives a longer linear range, up to 10^4, and the detector is reported to show less susceptibility to contamination.

Many ECD chambers have a significant internal volume (0.2–0.3 ml). If an open-tubular column with a low flow rate or if hydrogen or helium are used as the carrier gas, a make-up gas of nitrogen or argon/methane is blended with the eluent after the column so that a standing current can be formed and to reduce the effective dead volume of the detector by increasing the flow rate through the chamber.

In practical operation the purity of the carrier gas is critically important. Traces of oxygen, air and water are electron capturing and quench the

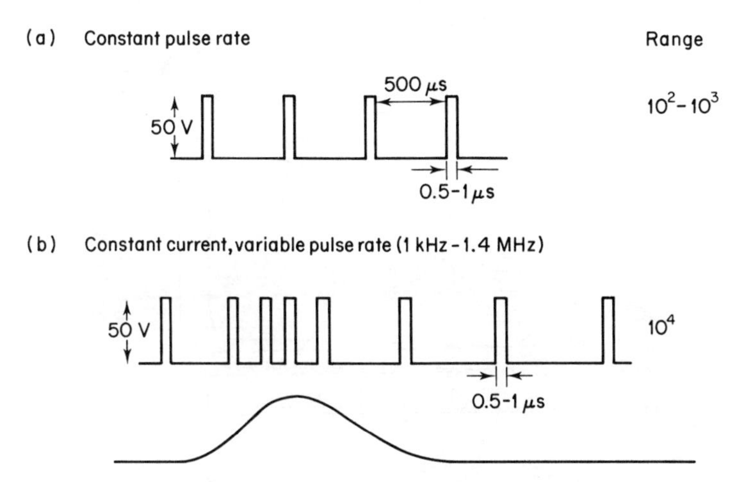

Figure 5.13 Different modes of pulsed operation of an electron capture detector: (a) constant pulse rate; and (b) constant current operation.

standing current. They must therefore be scrupulously removed from the carrier stream to a level below 10 ppm by using molecular sieve and charcoal traps. Any carrier gas lines must be metal rather than plastic as air can diffuse through the latter. Column bleed is also a problem and low percentages of liquid phases should be used and halogenated phases (i.e. OV-210) or thermally unstable phases must be avoided. The ECD is also very sensitive to contamination by analyte impurities or even from solvent vapours, such as dichloromethane and chloroform, in the laboratory atmosphere. Obviously halogenated compounds must not be used as solvents, even at an earlier stage in the sample preparation, as even residual traces left in the injected sample can deactivate the detector for prolonged periods. Contaminated ^{63}Ni detector cells can often be cleaned by heating to 350°C or can be rinsed with solvent, taking radiochemical precautions to monitor any activity. Even if grossly contaminated, detector cells should not be dismantled or mechanically cleaned in any way, except by the manufacturer. Because of the radioactivity a sealed source licence is normally required to purchase an ECD and the gas chromatograph should be monitored at intervals to test for radioactive contamination.

Despite these problems, the ECD is widely used, but in order to achieve consistent results and maximum sensitivity, considerable care, experience and expertise are needed. The popularity of the ECD is mainly due to its high sensitivity to organohalogen compounds, which include many compounds of environmental interest, including pesticides (Figure 5.14) such as DDT,

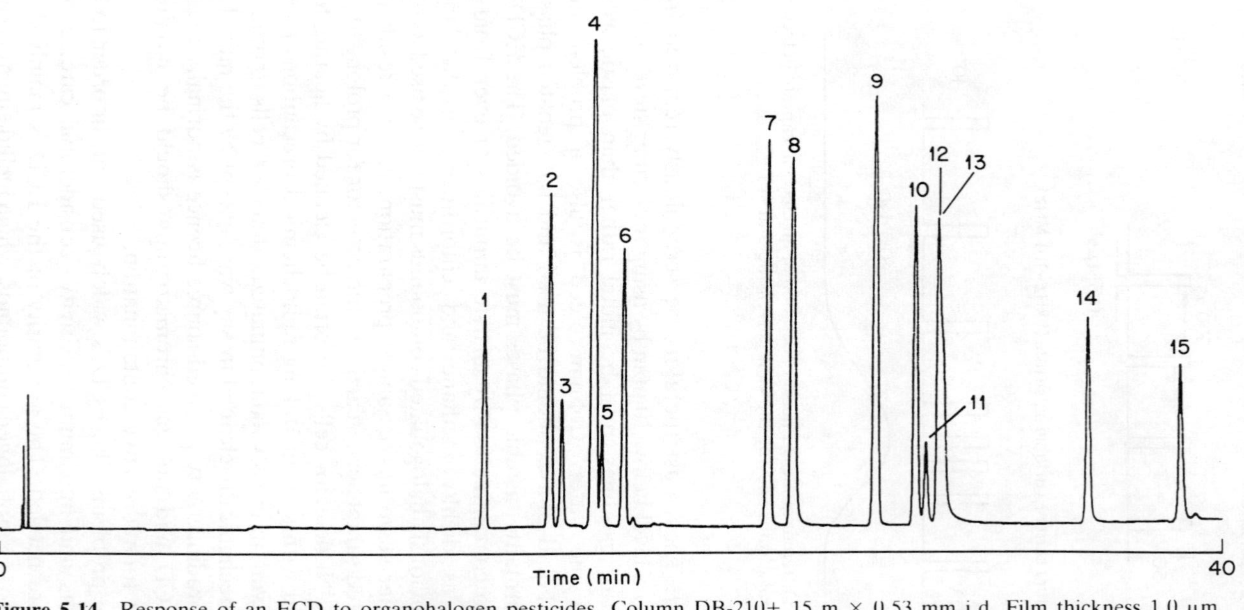

Figure 5.14 Response of an ECD to organohalogen pesticides. Column DB-210+ 15 m × 0.53 mm i.d. Film thickness 1.0 μm. Carrier gas helium. Column temperature 100°C to 220°C at 3°C min^{-1}. Injection of approximately 600 pg of each pesticide. Peaks: 1, α-BHC; 2, lindane (γ-BHC); 3, heptachlor; 4, aldrin; 5, β-BHC; 6, δ-BHC; 7, heptachlor epoxide; 8, *p,p'*-DDE; 9, dieldrin; 10, endrin; 11. *p,p'*-DDD; 12, β-endosulfan; 13, *p,p'*-DDT; 14, endrin aldehyde; 15, endosulfan sulphate. (Reproduced by permission of J. & W. Scientific.)

herbicides, polychlorinated biphenyls (PCBs) and halogenated solvents. Lindane, which is often used as a test sample, can usually be detected at the 0.1–1 pg level. The ECD has also found application for the selective determination of diacetyl in beer and organolead compounds present in petrol as antiknocking agents.

Quantitation is a problem with the ECD because the response is very dependent on the structure of the analyte. Each compound needs individual calibration and the linearity should always be tested over the operational range. Because of the short linear range, care must be taken to avoid overloading the detector or completely quenching the standing current. In many cases ECD and FID detectors are used in parallel to give both universal and selective responses. The carrier gas will often be split between ECD and FID in the ratio of 1:10 or 1:100, otherwise the FID may be below its detection limit, when the sample sizes are small enough to be within the linear range of the ECD.

b. Thermionic ionisation detector

The thermionic ionisation detector (TID) is often also known as the N/P detector (NPD) or alkali flame ionisation detector (AFID). It is a modified flame ionisation detector, which can be tuned by chemical or electronic adjustments to give a negligible response as a universal detector but a high sensitivity specifically for compounds containing nitrogen and/or phosphorus atoms. It is also possible to adjust it to give specificity towards other elements, but these modes are rarely used.

Although this detector has been known for many years, it has only recently become of widespread interest as commercial detector designs have improved and the reproducibility and stability have increased. The original detectors were based on a conventional flame ionisation detector in which an alkali salt or alkaline-earth salt had been placed in the flame. This gave an enhanced response for compounds containing phosphorus as well as the normal FID responses. Later designs placed the salt as a collar around the base of the jet in the burner but this still gave both an FID and TID response. Subsequently a stacked two-flame design was developed with the lower collector giving an FID response and the upper detector a specific response for compounds containing nitrogen or phosphorus atoms. This avoided the quenching of the second flame, which was sometimes observed in a single-flame detector when large quantities of unresponsive samples were being eluted.

Most present-day commercial detectors are based on a three-electrode design introduced by Kolb and Bischoff [1] (Figure 5.15) and its refinement

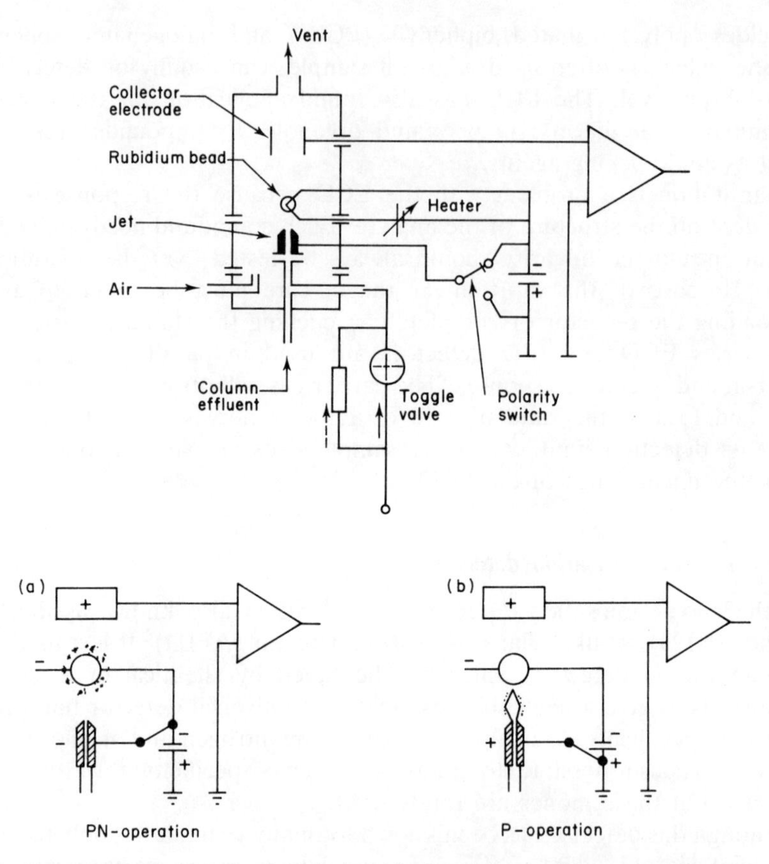

Figure 5.15 Design of a thermionic ionisation detector. (Reproduced from the *Journal of Chromatographic Science* by permission of Preston Publications, A Division of Preston Industries, Inc.) (a) Conditions optimised for nitrogen-phosphorus response with low hydrogen flow and electrically heated bead to give plasma around bead. (b) Conditions optimised for phosphorus compounds with higher hydrogen flow and bead heated by flame (Reproduced by permission of Perkin Elmer)

by Patterson [2]. A silica or ceramic bead, impregnated with a rubidium or caesium salt and containing an electrical heating element, is placed between the jet and the collector of an FID. By adjusting the potential applied to the jet, the bead and the collector, the selectivity and the response of the detector can be altered.

In the most useful mode, the flow of hydrogen is reduced to 1–2 ml min^{-1}, which is insufficient to form a flame. The bead is heated electrically until bright red (about 600–800°C). The hydrogen then burns in a plasma around

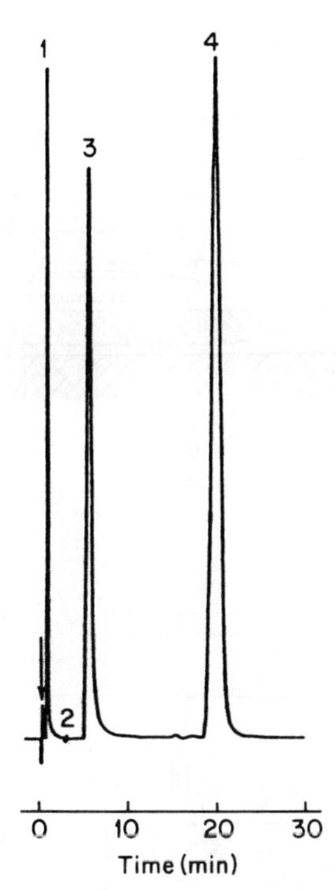

Figure 5.16 Selective detection using a thermionic detector. High response for nitrogen and phosphorous containing peaks; 1, pyridine (49.5 ng); 3, nitrobenzene (337 ng); and 4, tri-n-butyl phosphate (22.5 ng) compared to 2, n-hexadecane (50 000 ng). (Reproduced by permission of Chrompack)

the heated bead (Figure 5.15a). Compounds containing nitrogen or phosphorus atoms give a response and their sensitivity (10^{-13} to 10^{-14} g s^{-1}) is similar to that from an FID. However, the normal FID response is totally suppressed and virtually no signal will be recorded for compounds containing only carbon, hydrogen and oxygen (Figure 5.16). This mode of the detector therefore provides an important and highly selective method for the detection of many basic drugs of therapeutic importance in simple extracts of body fluids.

If the hydrogen flow is increased to a similar level as in an FID, a flame is formed, but the jet is maintained at a positive potential only compounds containing phosphorus atoms will give a signal (Figure 5.15b). However, except for the organophosphorus pesticides, such as malathion, few phosphorus compounds are of general interest and this mode has limited application.

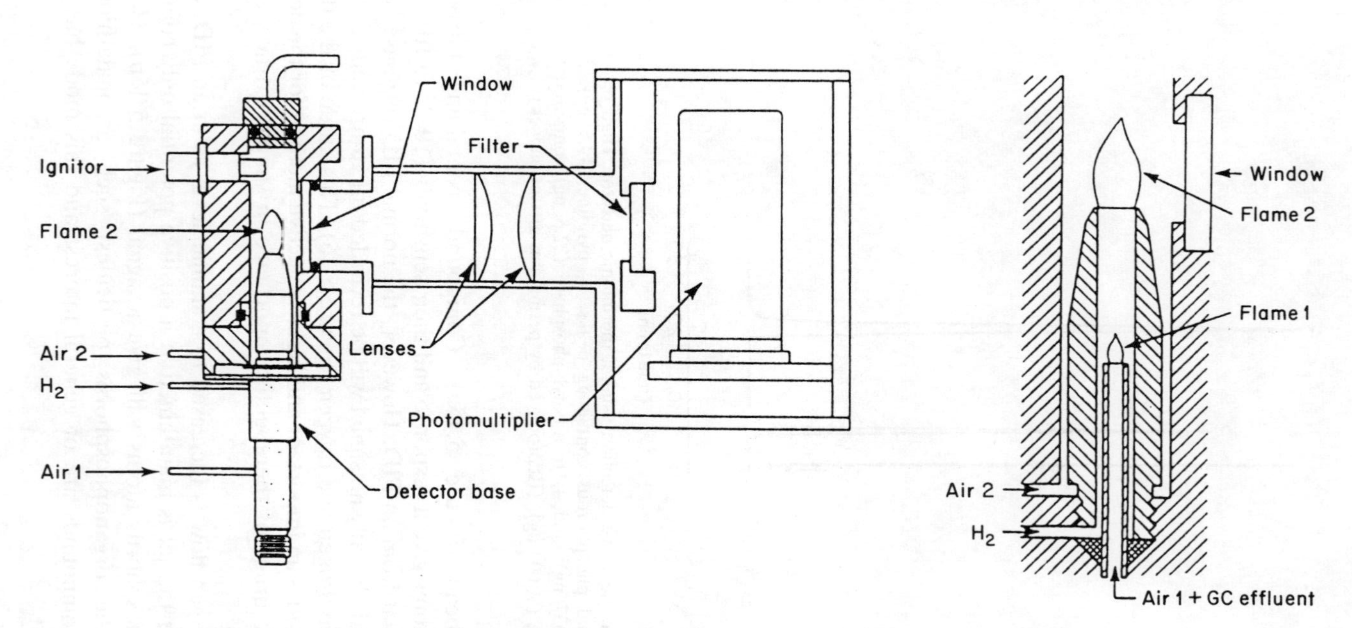

Figure 5.17 Flame photometric detector. (Reproduced by permission of Varian Associates.)

Most alkaline and alkali salts can be used to give a response but the more volatile compounds are rapidly lost from the detector and the reproducibility of the response is poor. Longer lifetimes and improved stabilities have been obtained by using silica or ceramic beads impregnated with rubidium sulphate or caesium salts. The mechanism of detection using these detectors is not clear. Different proposals have been made but none fully explains the observed responses. This is partly because each different design of detector appears to give slightly different responses for different compounds.

The TID has a wide linear response (10^4) and the response for compounds containing N or P atoms is typically 10^4 greater than for hydrocarbons.

c. *Flame photometric detector*

In the flame photometric detector (FPD) the carrier gas is mixed with hydrogen and the analyte is burnt in a relatively cool hydrogen-rich flame. Any emitted light is detected using a photomultiplier (Figure 5.17). By using the appropriate filter, only the emission with the wavelength characteristic of the particular element of interest will be detected to give a specific mass flow response for that element. In common with other emission spectroscopic techniques used in analytical chemistry, such as fluorescence, the detector has a very low background signal and is therefore very sensitive. The most commonly used modes are designed to give specific responses for compounds containing either phosphorus or sulphur atoms.

Analytes containing phosphorus atoms break down to form the excited HPO group, which emits light at 526 nm with a sensitivity of 4×10^{-11} g s^{-1}. However, as with the TID detector, the possible applications of this mode are limited. A higher selectivity and better reproducibility compared to the TID often mean that the FPD is preferred for the determination of pesticides such as methyl parathion (Figure 5.18b).

By using a 394 nm filter, analytes containing sulphur atoms can be detected by the emission of light from excited S_2^* molecules with a sensitivity of 2×10^{-11} g sec^{-1} (Figure 5.18a). The response of the detector varies approximately as the square of the amount of sulphur present in the sample. The exact relationship depends on the particular sample and detector, with typical values of about 1.84. This non-linear response causes problems with calibration and ideally the peak height of samples should be matched as closely as possible to standard samples. Alternatively a wide-range calibration curve should be prepared over the operating range. Some detectors provide electronic compensation based on an approximate correction factor to linearise the output signal, but these should only be used for rough

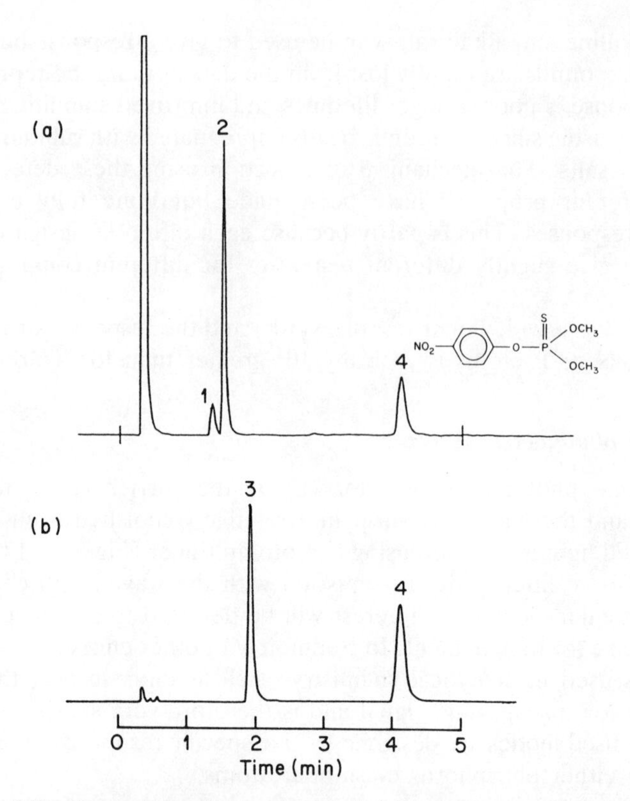

Figure 5.18 Comparison of the selectivity of flame photometric detector in different modes for mixture of sulphur and phosphorus containing compounds. Peaks: 1, pentadecane (4000 ng); 2, dodecanethiol (20 ng); 3, tributylphosphate (20 ng); 4, methyl parathion (20 ng). (a) The sulphur mode with detection of dodecanethiol and methyl parathion. (b) The phosphorus mode with the detection of tributyl phosphate and methyl parathion. Only a very small response is found for the excess of pentadecane (Reproduced by permission of Varian Associates.)

determinations. As the use of microcomputers increases, more sophisticated methods are likely to become available in the future.

The selectivity of the FPD for sulphur is 5×10^4 to 1×10^5 compared to CH compounds. This mode is of considerable interest in the petroleum industry, for the detection of sulphur compounds in crude and refined oils, and in the food industry, as many flavour and odour compounds are thiols or related compounds.

The FPD can also be tuned for other elements but these methods are rarely used routinely. A more general emission detector has also been developed, in which the sample is excited with a microwave helium plasma torch. The emission spectrum is resolved using a monochromator and can be used for the detection and quantification of carbon, oxygen, halogen, nitrogen and other atoms. It can therefore determine the relative elemental composition of any peak, but the detector is expensive and has not gained wide application.

d. Less commonly used detectors

i. Electrochemical methods

The Hall/Coulson electrolytic conductivity detectors (HECD) have been known since 1965 but have never proved very popular although they can specifically detect compounds containing either sulphur, nitrogen, or halogen atoms, depending on the operating conditions. The eluent from the GLC is passed into a reaction chamber at 600–1000°C under oxidative conditions, for the detection of sulphur atoms, or reductive conditions, for nitrogen and halogen atoms. The degradation products, respectively SO_2/SO_3, NH_3, or hydrogen halides, are absorbed in an aqueous solution and changes in the conductivity are monitored. The specificity is altered by changing the reaction conditions and by the use of selective scrubbers to remove interfering reaction products. In all cases the sensitivity is about 10^4–10^6 fold greater than for analytes without the specific groups.

The major impact of these detectors has been as selective detectors of sulphur in the petroleum industry, where the linear response range of 10^4 is preferred to the PFD. Some interest has been shown in recent years in their use as specific nitrogen detectors for the determination of drugs in biological samples or of herbicides in soils, as alternatives to the TID. However, the use of reaction solutions and the need to monitor the condition of absorbents have meant that they are relatively messy detectors to use and this has limited their acceptance.

ii. Thermal energy analyser

The thermal energy analyser (TEA) is a good example of a detector that satisfies a specialised but limited need. Because of concern about the carcinogenicity of *N*-nitrosamines, it is necessary to monitor their presence in foodstuffs. In the thermal energy analyser, the eluent carrier gas stream is passed to a pyrolysis chamber and the reaction products are mixed with ozone. Any N–NO groups will give $NO^{\cdot}$, which will be converted to excited

NO_2^* species. These will fall back to their ground state with the emission of light with a wavelength of 600 nm. This chemiluminescence can then be detected with a filter and photomultiplier (Equation 5.4):

$$R_2N{-}NO \xrightarrow{\text{heat}} NO^{\bullet} + O_3 \longrightarrow$$

$$NO_2^* \longrightarrow h\nu\ (\lambda = 600\text{nm}) + NO_2 \qquad (5{\cdot}4)$$

The reaction is highly specific and compounds lacking a *N*-nitroso group give no response. It is therefore much more selective than the TID, which would also respond to compounds with amino, nitro, or other nitrogen-containing groups.

iii. Other detectors

Other methods have also been used to monitor the output from a GLC. Those available commercially include, for example, the ultrasonic detector, which monitors the variation in the speed of sound across the carrier gas stream, and the far-ultraviolet detector, which measures the absorbance of light in the region of 120–150 nm. At these wavelengths all organic compounds will respond but helium can still be used as a carrier gas. Numerous further experimental detectors have been developed in research laboratories but most are less sensitive or selective than existing detectors.

5.4 COUPLED DETECTORS

Although universal and specific detectors can monitor the output from a gas chromatograph, they provide little guidance to the nature or structure of the analytes. Much more information can be obtained by coupling the output of gas chromatographs to mass or infrared spectrometers or to other detection devices that are analytical methods in their own right. These so-called hyphenated techniques, such as GC–MS or GC–FTIR, have become more important in recent years as the technology of interfacing has improved and more advanced microelectronics in the spectrometers have enabled spectra to be measured rapidly on the flowing carrier gas stream.

a. *GC–human nose*

The human nose is a very specialised detector, which can be used to examine the odour of the eluent from the column. This organoleptic method has an important role in flavour and perfumery laboratories for the assessment of the individual components in complex natural and synthetic samples and to identify the active constituents and subtle changes in quality and composition. However, it is difficult to train staff for this work and reproducibility is hard to maintain. Results are often therefore qualitative descriptions but the high sensitivity of the human nose enables studies to be carried out on very small quantities.

b. *GC–radiochemical detectors*

GC–radiochemical detection using a flow-through Geiger counter is a valuable technique in metabolic and biosynthetic studies to monitor radiolabelled tracers of drugs or their metabolites. However, the sensitivity of the detector can be limited because the residence time of the analyte in the detector is short. Alternative detectors convert all organic compounds in the carrier gas stream to carbon dioxide to aid counting.

c. *GC–atomic absorption spectroscopy*

Organometallic compounds or volatile metal complexes can be detected by passing the carrier gas into the flame of an atomic absorption spectrometer, which can be set for selective detection of specific metallic elements. This approach has been used to identify the structure of metal complexes in environmental speciation studies and for the determination of some organometallic drugs. If it is necessary to determine which metal is present, the carrier gas can be passed to an inductively coupled plasma emission spectrometer, which can carry out a multi-element determination.

d. *GC–infrared spectroscopy*

Because the infrared spectrum of a compound is characteristic of the whole molecule, by comparing it with standard spectra it can be used for positive identification of an analyte. Until recently, infrared spectrometers were not used widely because of the time required to measure a spectrum, typically 3–15 min using a conventional scanning spectrometer. This was so long that the sample would have completely passed through the cell before the spectrum had been recorded. Some work could be done using a fixed-wavelength spectrometer but this method is limited to monitoring known compounds or for searching for particular functional groups characteristic of the compounds of interest. In both cases the carrier gas had to be transferred to the spectrometer using a heated tube to avoid condensation of the components.

Advances in instrument design and the use of Fourier transform (FT) algorithms to convert a rapidly measured interferogram into a spectrum now enable a complete infrared spectrum to be scanned in 0.3 s and so are compatible with even the narrow peaks from open-tubular separations. The sensitivity of the detector has also greatly increased partly because an FTIR spectrometer measures the whole spectrum simultaneously rather than just examining one wavelength at a time as in a scanning spectrometer. Flexible open-tubular columns can also be used to carry the eluent all the way to the detector cell without the need for separate transfer lines and couplings.

Full infrared spectra from 5000 to 600 cm^{-1} can now be measured on every significant component even in a complex mixture (Figure 5.19). As the spectra are obtained in the gas phase, the bands are usually sharper than in thin-film or solid spectra. Libraries of GC–FTIR standard spectra are becoming available as both printed spectra and digital computer records.

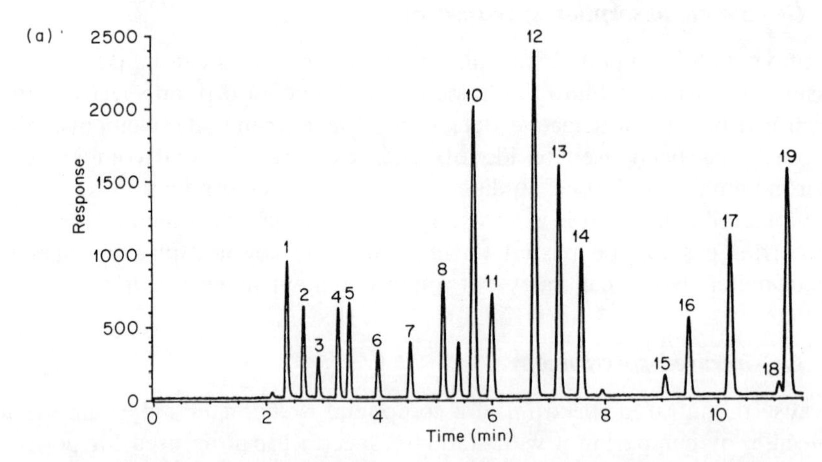

Figure 5.19 (a) Total infrared response chromatogram of mixture of industrial solvents detected using an FTIR detector. Separated on a open-tubular column. Peaks: 1, MeOH; 2, methyl formate; 3, EtOH; 4, acetone; 5, i-PrOH; 6, CH_2Cl_2; 7, n-PrOH; 8, MEK; 9, s-BuOH; 10, EtOAc; 11, i-BuOH; 12, i-PrOAc; 13, nitropropane; 14, 1,4-dioxane; 15, toluene; 16, mesityl oxide; 17, diacetone-OH; 18, *m*-xylene; 19, cyclohexane. (Reproduced with permission of Hewlett Packard. *Application brief*, IRD-86-3.)

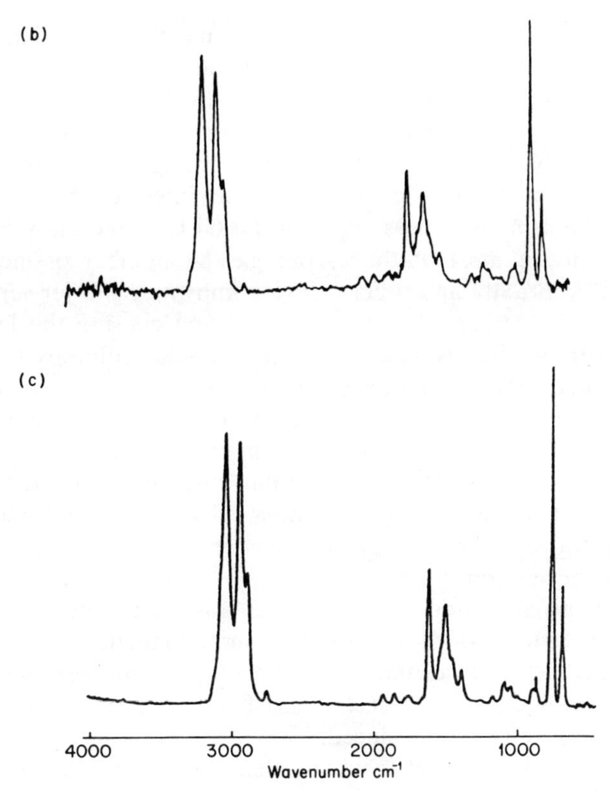

Figure 5.19 (cont.) (b) Infrared spectrum of the peak No. 18 assigned to *m*-xylene. (c) The reference spectrum of *m*-xylene spectrum for comparison.

As FTIR spectra are obtained in a digital format, comparisons can be automatically carried out by a computer to identify unknown components. However, compared to other detectors, FTIR spectrometers and their associated computer systems are still relatively expensive but are invaluable for the identification of the unknown components of complex natural products, pollution samples, or synthetic mixtures.

e. *GC–mass spectrometry*

i. Interfacing to the mass spectrometer

Because a mass spectrum can provide information on the molecular formula and structure of a sample, coupled with a high sensitivity, the mass spectrometer is one of the most valuable of analytical detection techniques. However, if the carrier gas from a packed column is passed directly to the source of a mass spectrometer, the pressure in the sampling area will rise. This will result in complete loss of the spectrum. Normally the source pressure must be very low, 10^{-6} torr (mm Hg), and for many years considerable effort was employed to devise an interface that would transfer as much of the sample as possible from the GLC to the mass spectrometer but would remove most of the carrier gas. A number of methods were studied and eventually an all-glass device known as the jet separator was widely adopted as being the most efficient and causing the least sample loss. However, in the last few years open-tubular columns have become widely adopted and these have much lower carrier gas flow rates. Combined with improvements in the vacuum pumps used in the mass spectrometers, it is now possible to maintain a sufficient vacuum even if all the carrier gas from the open-tubular column is passed to the mass spectrometer. The flexible open-tubular column can be carried directly into the ion source, eliminating the need for sample transfer lines, or for an interface or separator.

Helium is usually used as the carrier gas as it interferes least with the mass spectrum. Both quadrupole and magnetic/electrostatic sector spectrometers with high scanning rates have been used. Very recently newer designs of mass spectrometer, including FT mass spectrometers and the ion-trap detector, have been introduced. These can scan a spectrum very rapidly with increased sensitivity and should be ideal as mass spectroscopic detectors. However, all these mass spectroscopic systems are expensive and would be primarily used for research studies rather than routine determinations. However, in some areas involving the separation of complex samples, such as environmental pollution studies or the dope and drug testing of race horses and athletes, they are routinely used to ensure positive identification of any trace components in the sample.

ii. Modes of operation

The mass spectrometer can be used in a number of scanning modes with different sensitivities and selectivities.

Total ion monitor

The total ion current (TIC), sometimes referred to as the total ion monitor (TIM), passing through the mass spectrometer, due to the ionisation of the analyte, can be used to monitor the total sample chromatogram as a very expensive universal detector.

Scanning mode

The most common and informative mode of operation is for the mass spectrum to be scanned at frequent intervals (typically every 0.5–2 s) across the full mass range (30–500 mass units) and for the spectra to be recorded on a magnetic disc. The individual spectra from each peak can be subsequently examined and used to identify the analytes (Figure 5.20). Usually a minicomputer or large microcomputer is needed to store the spectra. This can also hold a library file of standard spectra, which can be searched automatically to identify unknown peaks.

Single/multiple ion monitoring mode

In the scanning mode only one mass (out of the full range) is being monitored at any moment, so that a relatively small proportion of the total signal is detected. This problem is avoided in the FTMS system or ion-trap spectrometers as they accumulate the ions before detection. If only a few ions are of particular interest, the mass spectrometer can continuously monitor a single ion (single ion monitoring, SIM) or small group of 2–8 mass/charge ratios (multiple ion monitoring, MIM). By concentrating on a few ions a greater proportion of each one will be detected and the sensitivity can be increased by up to a thousand fold. The ions chosen would be characteristic of the analyte of interest. With MIM the ratio of the peaks can also be measured and if the signals increase in the expected proportions the presence of a particular analyte can be indicated. This method can be very specific and is particularly useful for the study of complex mixtures such as environmental samples when a known pollutant is being monitored.

5.5 CRITERIA FOR THE ADOPTION OF A NEW DETECTOR

Frequently, new detectors are offered to the chromatographer, based on new concepts or major modifications of old detectors. In assessing their value the potential user must ask if they can either (i) satisfy a specific unsatisfied demand for a particular analysis or (ii) be significantly more sensitive or more selective than an established detector. Because of development costs it is highly unlikely that any new detector will be simply cheaper to build or run than existing detectors.

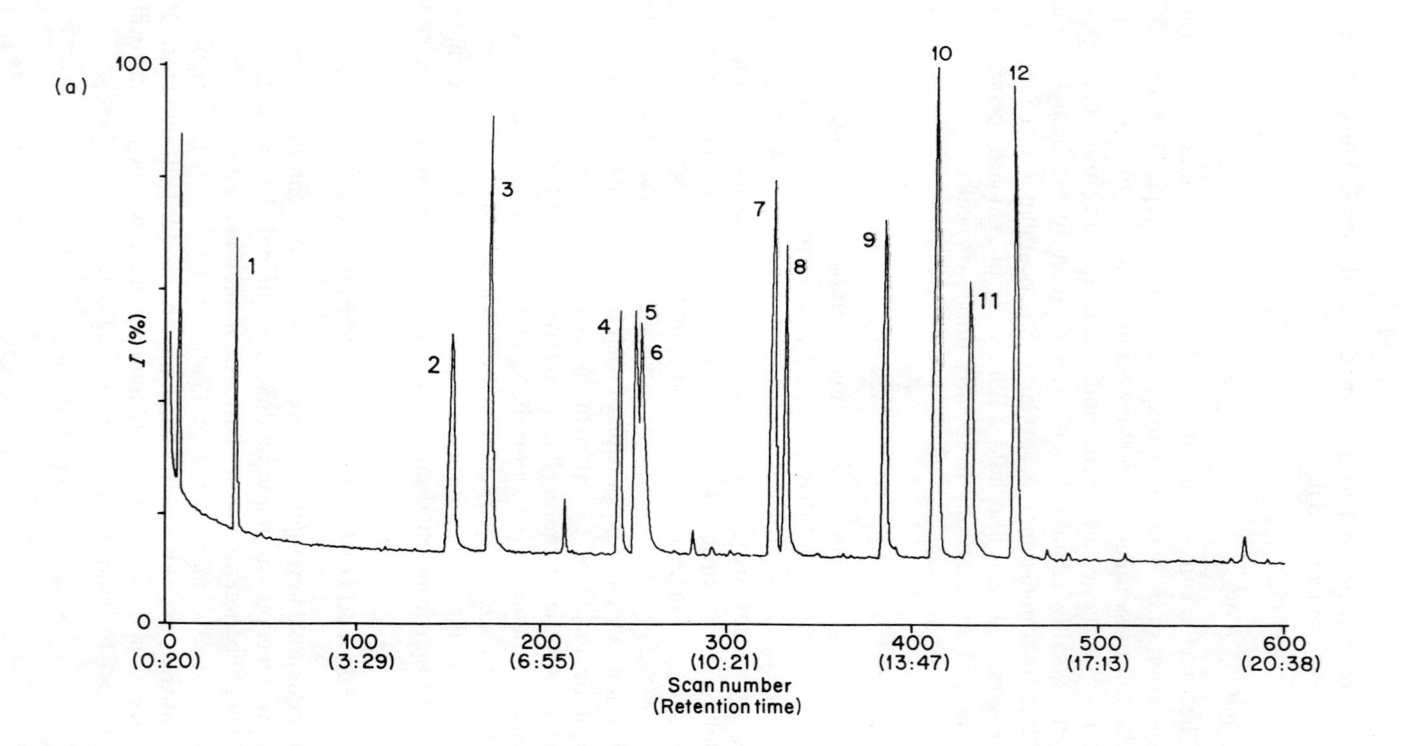
(a)
100
I (%)
0
0
(0:20)
100
(3:29)
200
(6:55)
300
(10:21)
400
(13:47)
500
(17:13)
600
(20:38)
Scan number
(Retention time)
1
2
3
4
5
6
7
8
9
10
11
12

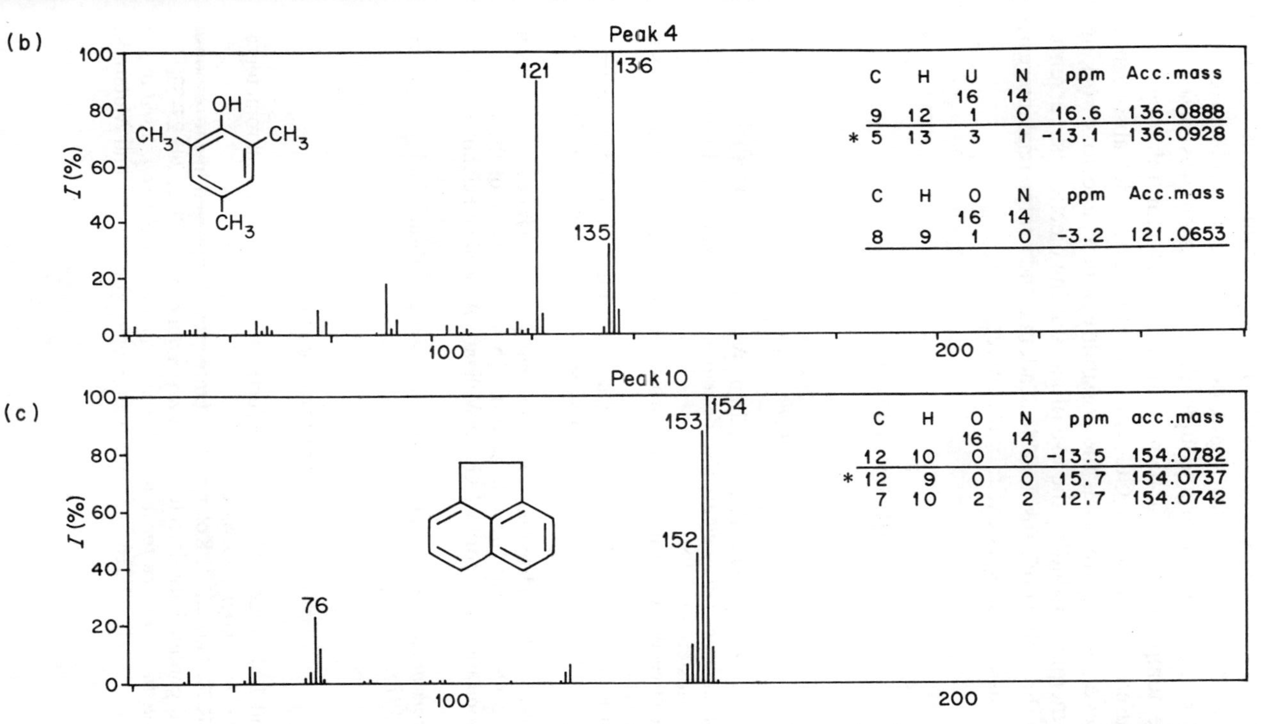

Figure 5.20 (a) Total ion current chromatogram of polarity test mixture detected using a mass spectrometric detector. Separation carried out on OV-1 open-tubular column 25 m programmed from 60°C to 150°C at 4°C min^{-1} with direct coupling to the mass spectrometer. Peaks: 1, 1,2-dimethylbenzene; 2, isooctanol; 3, 2,6-dimethylphenol; 4, 2,4,6-trimethylphenol; 5, 3,4-dimethylphenol; 6, 2-(1-methylethyl)phenol; 7, 2,6-dimethylundecane; 8, nicotine; 9, *N*-cyclohexylcyclohexanamine; 10, 1,2-dihydroacenaphthylene; 11, 1-dodecanol; 12, 2-methyltetradecane. (b) and (c) The mass spectra obtained from peaks 4 (2,4,6-trimethylphenol) and 10(1,2-dihydroacenaphthylene) respectively. (Reproduced by permission of VG Analytical Ltd from *VG micromass application notes*, No. 1, 1978.)

It is worth remembering that the gas chromatography literature describes many detectors that are now little used or have become extinct. These include the helium ionisation and argon detectors and the related cross-section detector, the flame thermocouple detector, which was based on changes in the temperature of a flame, the mass detector (although the name is now used for a form of GC–MS), and the ultraviolet absorbance detector. However, the future holds the possibility of new detectors working on different principles, including the use of gas sensing electrodes, currently rather slow but the subject of considerable academic and commercial interest, or fluorescence using lasers as excitation sources.

BIBLIOGRAPHY

General

D.J. David, *Gas chromatographic detectors*, John Wiley, New York, 1974.

M. Dressler, *Selective gas chromatographic detectors*, J. Chromatogr. Library, Vol. 36, Elsevier, Amsterdam, 1986.

L.S. Ettre, "Selective detection in column chromatography", *J. Chromatogr. Sci.*, 1978, **16**, 396–417.

S.O. Farwell, D.R. Gage and R.A. Kagel, "Current status of prominent selective gas chromatographic detectors: a critical assessment", *J. Chromatogr. Sci.*, 1981, **19**, 358–376.

A.J.C. Nicholson (Ed.), *Detectors and chromatography*. (Proceedings of an international conference held to celebrate the 25th anniversary of the invention of the flame ionisation detector by I.G. McWilliam), Australian Scientific Industry Association.

J. Ševčík, *Detectors in gas chromatography*, J. Chromatogr. Library, Vol. 4, Elsevier, Amsterdam, 1976.

Thermal conductivity detector

A.E. Lawson and J.M. Miller, "Thermal conductivity detectors in gas chromatography", *J. Gas Chromatogr.*, 1966, **4**, 273–284.

P.L. Patterson, R.A. Gatten, J. Kolar and C. Ontiveros, "Improved linear response of the thermal conductivity detector", *J. Chromatogr. Sci.*, 1982, **20**, 27–32.

Standard recommended practice for testing thermal conductivity detectors used in gas chromatography, ASTM E516–74, American Society for Testing and Materials, Philadelphia, 1974.

Flame ionisation detector

Standard recommended practice for testing flame ionization detectors used in gas chromatography, ASTM E594–77, American Society for Testing and Materials, Philadelphia, 1977.

Photoionisation detector

J.N. Driscoll, "Review of photoionisation detectors in gas chromatography: the first decade", *J. Chromatogr. Sci.*, 1985, **23**, 488–492.

J.N. Driscoll, J. Ford, L.F. Jaramillo and E.T. Gruber, "Gas chromatographic detection and identification of aromatic and aliphatic hydrocarbons in complex mixtures by coupling photoionization and flame-ionization detectors", *J. Chromatogr.*, 1978, **158**, 171–180.

M.L. Langhorst, "Photoionisation detector sensitivity of organic compounds", *J. Chromatogr. Sci.*, 1981, **19**, 98–103.

P. Verner, "Photoionisation detection and its application in gas chromatography", *J. Chromatogr., Chromatogr. Rev.*, 1984, **300**, 249–264.

Electron capture detector

R.J. Maggs, P.L. Joynes, A.J. Davies, and J.E. Lovelock, "The electron capture detector—a new mode of operation", *Anal. Chem.*, 1971, **43**, 1966.

P.L. Patterson, "Pulse-modulated electron capture detection with nitrogen carrier gas", *J. Chromatogr.*, 1977, **134**, 25–37.

C.F. Poole, "The electron-capture detector in capillary column gas chromatography", *J. High Res. Chromatogr., Chromatogr. Commun.*, 1982, **5**, 454–471.

Standard practice for the use of electron-capture detectors used in gas chromatography, ASTM E697–79, American Society for Testing and Materials, Philadelphia, 1979.

A. Zlatkis, C.K. Lee, W.E. Wentworth and E.C.M. Chen, "Constant current linearization for determination of electron capture mechanisms", *Anal. Chem.*, 1983, **55**, 1596–1599.

A. Zlatkis and C.F. Poole (Eds.), *Electron capture. Theory and practice in chromatography*, J. Chromatogr. Library, Vol. 20, Elsevier, Amsterdam, 1981.

Thermionic ionisation detector

V.V. Brazhnikov, M.V. Gur'ev and K.I. Sakodynsky, "Thermionic detectors in gas chromatography", *Chromatogr. Rev.*, 1970, **12**, 1–41.

A. Karmen and L. Giuffrida, "Enhancement of the response of the hydrogen flame ionization detector to compounds containing halogens and phosphorus", *Nature*, 1964, **201**, 1204–1205.

B. Kolb, M. Auer and P.Pospisil, "Reaction mechanism in an ionization detector with tunable selectivity for carbon, nitrogen and phosphorus", *J. Chromatogr. Sci.*, 1977, **15**, 53–63.

P.L. Patterson, "New uses of the thermionic ionization detector in gas chromatography", *Chromatographia*, 1982, **16**, 107–111.

P.L. Patterson, "Recent advances in thermionic ionization detection for gas chromatography", *J. Chromatogr. Sci.*, 1986, **24**, 41–52.

Flame photometric detector

S.S. Brody and J.E. Chaney, "Flame photometric detector. The application of a specific detector for phosphorus and for sulfur compounds—sensitive to subnanogram quantities", *J. Gas Chromatogr.*, 1966, **4**, 42–46.

T.J. Cardwell and P.J. Marriott, "Some characteristics of a flame photometric detector in sulphur and phosphorus modes", *J. Chromatogr. Sci.*, 1982, **20**, 83–90.

S.O. Farwell and C.J. Barinaga, "Sulfur-selective detection with the FPD: current enigmas, practical usage, and future directions", *J. Chromatogr. Sci.*, 1986, **24**, 483–494.

J.F. McGaughey and S.K. Gangwal, "Comparisons of three commercially available gas chromatographic-flame photometric detectors in the sulfur mode", *Anal. Chem.*, 1982, **52**, 2079–2083.

D.F.S. Natusch and T.M. Thorpe, "Element selective detectors in gas chromatography", *Anal. Chem.*, 1973, **45**, 1184A–1194A.

J. Ševčík and N.P. Thao, "The selectivity of the flame photometric detector", *Chromatographia*, 1975, **8**, 559–562.

Electrochemical detectors

D.M. Coulson, "Electrolytic conductivity detector for gas chromatography", *J. Gas Chromatogr.*, 1965, **3**, 134–137.

S. Gluck, "Performance of the model 700A Hall electrolytic conductivity detector as a sulfur-selective detector", *J. Chromatogr. Sci.*, 1982, **20**, 103–108.

R.C. Hall, "A highly sensitive and selective microelectrolytic conductivity detector for gas chromatography", *J. Chromatogr. Sci.*, 1974, **12**, 152–160.

Thermal energy analyser detector

D.H. Fine, D. Lieb and F. Rufeh, "Principle of operation of the thermal energy analyser for the trace analysis of volatile and non-volatile *N*-nitroso compounds.", *J. Chromatogr.*, 1975, **107**, 351–357.

D.H. Fine and D.R. Rounbehler, "Trace analysis of volatile *N*-nitroso compounds by combined gas chromatography and thermal energy analysis", *J. Chromatogr.*, 1978, **109**, 271–279.

Other detectors

T.R. Roberts, *Radiochromatography. The chromatography and electrophoresis of radiolabelled compounds*, J. Chromatogr. Library, Vol. 14, Elsevier, Amsterdam, 1978.

K.J. Skogerboe and E.S. Yeung, "Quantitative gas chromatography with analyte identification by ultrasonic detection", *Anal. Chem.*, 1985, **56**, 2684–2686.

GC–atomic absorption spectroscopy

L. Ebdon, S. Hill and R.W. Ward, "Directly coupled chromatography–atomic spectroscopy. Part 1. Directly coupled gas chromatography–atomic spectroscopy. A review", *Analyst*, 1986, **111**, 1113–1138.

GC-infrared spectroscopy

P.R. Griffiths, J.A. de Haseth and L.V. Azarraga, "Capillary GC/FT-IR", *Anal. Chem.*, 1983, **55**, 1361A–1387A.

A. Hanna, J.C. Marshall and T.L. Isenhour, "A GC/FT-IR compound identification system", *J. Chromatogr. Sci.*, 1979, **17**, 434–440.

W. Herres, *HRGC-FTIR: theory and applications*, Alfred Hüthig, Heidelberg, 1987.

K.S. Kalasinsky, "GC/FTIR applications in pesticide chemistry", *J. Chromatogr. Sci.*, 1983, **21**, 246–253.

D.R. Mattson and R.L. Julian, "Programming techniques for obtaining maximum sensitivity in real-time detection of GC effluents", *J. Chromatogr. Sci.*, 1979, **17**, 416–422.

R. Seelemann, "GC-FTIR coupling—a modern tool for analytical chemistry" *Trends Anal. Chem.*, 1982, **1**, 333–339.

GC-mass spectrometry

W.L. Budde and J.W. Eichelberger, *Organics analysis using gas chromatography/mass spectrometry*, Ann Arbor Science, Ann Arbor, 1979.

L.S. Ettre and W.H. McFadden (Eds.), *Ancillary techniques of gas chromatography*, John Wiley, New York, 1969.

B.J. Gudzinowicz and M.J. Gudzinowicz, *Analysis of drugs and metabolites by gas chromatography–mass spectrometry*, Vols. 1–7, Marcel Dekker, New York, 1977–1980.

W. McFadden, *Techniques of combined gas chromatography/mass spectrometry: applications in organic analysis*, John Wiley, New York, 1973.

G.M. Message, *Practical aspects of gas chromatography/mass spectroscopy*, John Wiley, New York, 1984.

R.L. Settine, J.A. Kisinger and S. Ghaderi, "Fourier transform mass spectrometry and its combination with high resolution gas chromatography", *Eur. Spectrosc. News*, 1985, **58**, 16–18.

G. Stafford, P.E. Kelley, J.E.P. Syka, W.E. Reynolds, and J.F.J. Todd, "Recent improvements in and analytical applications of advanced ion trap technology", *Int. J. Mass Spectrom. Ion Processes.*, 1984, **60**, 85–98.

REFERENCES

1. B. Kolb and J. Bischoff, "A new design of a thermionic nitrogen and phosphorus detector for GC", *J. Chromatogr. Sci.*, 1974, **12**, 625–629.
2. P.L. Patterson, R.A. Gatton and C. Ontiveros, "An improved thermionic ionization detector for gas chromatography", *J. Chromatogr. Sci.*, 1982, **20**, 97–102.

CHAPTER 6

SAMPLE IDENTIFICATION AND QUANTIFICATION IN GAS–LIQUID CHROMATOGRAPHY

6.1 SAMPLE PREPARATION FOR GAS–LIQUID CHROMATOGRAPHY

Generally sample preparation for GLC separations causes few problems. Volatile liquids and gases can be examined directly but more often both liquids and solids are dissolved in a volatile solvent to give a dilute solution. Unless the sample components are themselves low-boiling compounds, almost any solvent can be used, except that halogenated solvents must not be used with electron capture detection. The front of the solvent peak is often used as a guide to the column void volume.

Any insoluble materials should be removed by filtration or centrifugation and the sample solution should ideally contain no involatile components, such as proteins from blood or plasma, or polysaccharides, lignans and lipids from food products, plants materials, etc. These would be left on the top of the column as charred residues, which would cause column deterioration or a high background signal. It may therefore be necessary to separate the volatile compounds of interest from involatile components of the matrix before analysis. Solvent extraction or steam distillation will separate most non-polar analytes from involatile biological polymers and inorganic ions. Very polar compounds, which also tend to be involatile, such as amino acids or carbohydrates, can often be removed by filtering the sample through a short silica or alumina liquid chromatographic column. This also serves to remove insoluble materials. Not all matrices can be handled in these ways and some alternative techniques are given in Chapter 7, together with methods to improve the separation of analytes that are themselves involatile and thus could not normally be analysed by GLC.

6.2 SAMPLE IDENTIFICATION BY GAS–LIQUID CHROMATOGRAPHY

Gas chromatographic retention times can be used to confirm the identity of the components of a sample before quantification or as part of the identification of the unknown components in complex samples, such as essential oils, drug metabolites, or environmental pollution extracts. Both these approaches are possible because of the direct relationship between the retention time of an analyte and its distribution constant between the gas and stationary liquid phases. As this value is a fundamental property of the analyte, the retention time will be independent of the matrix of the injected sample. A pure analyte or the analyte as a component in a complex mixture will thus give peaks with the same retention time.

To identify a peak, either a direct comparison can be made between the analyte and an authentic standard on the same column, or the retention time can be related to the interaction of the analyte with the column and hence its structural features. Alternatively, the analyte can be related to a standard scale of retentions and hence can be compared with the retention values of standards obtained in other laboratories.

a. Identification by comparison with standard compounds

The most commonly employed technique for the identification of analytes is the direct comparison of the retention with the retention of an authentic standard sample (Figure 6.1). From the earlier discussion, it is clear that if two samples have different retentions under the same experimental chromatographic conditions, they must differ. Unfortunately the converse is not true. If a standard sample gives a peak with the same retention time as the analyte, the compounds may be the same or the similarity may only be a coincidence. Under most chromatographic conditions many compounds may have closely similar or identical retentions. Any apparent identification of an analyte by direct comparison should therefore be confirmed by repeating the comparison using a second chromatographic system. This should use a very different liquid phase, which will have a different interaction with the analytes. Usually, a pair of non-polar and polar stationary phases would be used. This ensures that if coincidence is obtained on both columns then the analyte and the standard compound must have both the

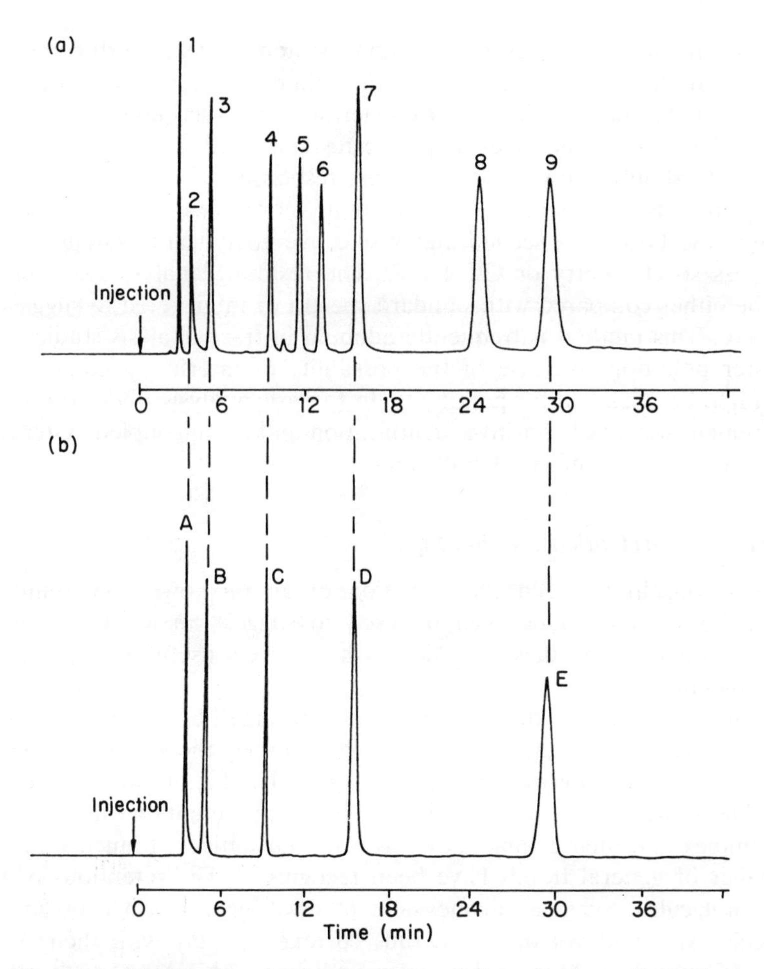

Figure 6.1 Identification of unknown compounds by comparison with standards. (a) Separation of "unknown" alkanols. (b) Separation of standard n-alkanols showing identifications on the basis of retention times. Peaks: A, methyl alcohol; B, ethyl alcohol; C, n-propyl alcohol; D, n-butyl alcohol; E, n-amyl alcohol. (Reproduced with permission of Varian Associates from McNair and Bonelli, *Basic gas chromatography*.)

same volatility *and* the same polar interactions. Typical liquid phases would be a non-polar dimethylsilicone oil, such as OV-101, and the polar polyether, Carbowax 20M. The best method is to chromatograph the sample and the standard individually and then examine a sample that has been spiked with the standard compound to ensure that the "unknown" and known compounds are separated under exactly the same conditions.

Clearly the ability of a chromatographic system to distinguish compounds will depend on the resolving power of the column and open-tubular columns are popular because of their high efficiencies and discriminating power. If the standard compound gives a specific response with a selective detector, the analyte should show a corresponding response.

A more positive identification of an unknown peak can be made by coupling the GLC to a second highly selective analytical technique such as GC–mass spectrometry or GC–FTIR. The resulting analyte spectrum can then be either compared with standard spectra or interpreted to suggest the structure. This method is frequently adopted in trace analysis studies, such as water pollution, because of the possibility of interfering co-extractives that might be mistaken for pollutants. In forensic analysis, there is a similar requirement for a very positive identification and again coupled systems are routinely used to confirm identifications.

b. Structure–retention relationships

The relationship between the retention of an unknown compound and related known compounds can be used to suggest an identification and an understanding of these relationships is often useful during method development.

Retention in GLC is directly related to the distribution constant, which is dependent only on the structure of the sample, the nature of the two phases and the temperature of the separation. The factors which determine the value of the distribution constant and hence the retention for any analyte are complex and their examination has been the subject of much study, but a number of general trends have been recognised. The retentions of non-polar molecules, such as alkanes and ethers, depend primarily on their molecular size and volatility and thus correlate closely with their boiling points (Figure 6.2). More polar compounds can also interact with a polar stationary phase by dipole–dipole interactions, but with non-polar columns, where no interaction can occur, the retentions are again broadly related to the boiling points. Thus a comparison of the retention times of an unknown analyte on polar and non-polar columns, relative to a standard compound, can give a guide to its polar interactions.

These interactions can also be related to the number and type of functional groups present in the molecule. The influence of the individual groups on the retention is cumulative and a particular group will have a corresponding effect in different compounds. However, because of the great range of experimental conditions that can be used in GLC, the prediction of retentions based on these theoretical aspects have not found wide application.

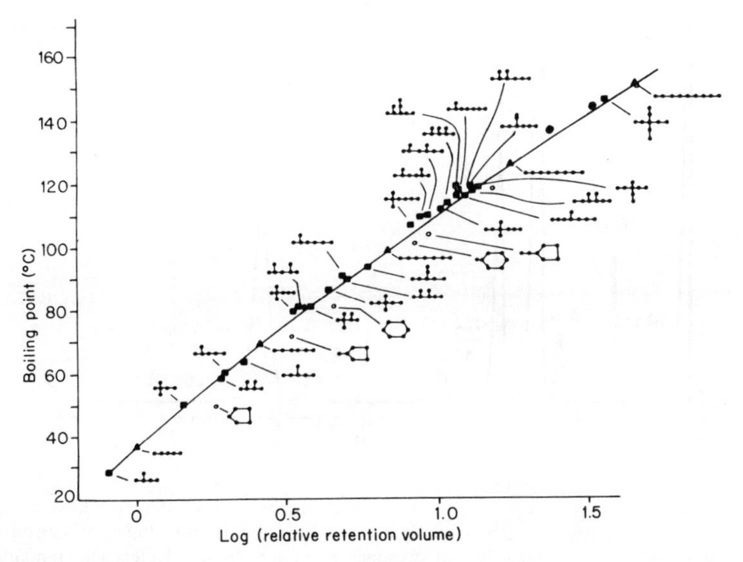

Figure 6.2 Relationship between logarithm of retention volume relative to n-pentane and boiling point for straight-chain, branched-chain and cyclic hydrocarbons. Stationary phase n-octadecane at 65°C. (Reproduced by permission of Chapman and Hall from Gurr and James, *Lipid biochemistry*, p.17.)

The one area in which use has been made use of these relationships is the comparison of homologues. The sequential addition of a methylene group causes a systematic increase in the retention, such that for the members of a homologous series there is a logarithmic relationship between the adjusted retention time and the number of carbon atoms C_n (Equation 6.1, A and B are constants).

$$\log t'_R = AC_n + B \tag{6.1}$$

As predicted, the effect of adding a methylene group is very similar for different series of homologous compounds, and if the retentions are compared using Equation 6.1 a similar slope for the relationship is found in each case.

This relationship forms the basis of the retention index system (see Section 6.2.d) but can also be used to predict the retention of homologues, if these are not available for direct examination. For example, although the even carbon number fatty acids (C_{14}, C_{16}, C_{18}, etc.) are readily available and their methyl esters can be examined, the odd carbon number acids are relatively rare. Using the linear relationship, their retentions can be predicted by interpolation (Figure 6.3). Similarly the introduction of a double bond into the fatty acids causes a systematic change in each case.

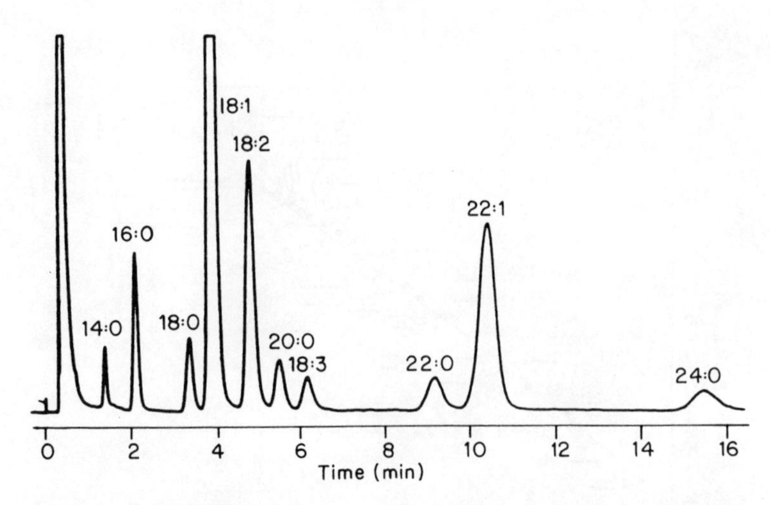

Figure 6.3 Separation of free fatty acid methyl esters showing logarithmic relationship between carbon number and adjusted retention times of saturated esters and incremental effect of presence of double bonds. Reference standard: RM-3 peanut oil type. (Reprinted with permission of Supelco, Inc., Bellefonte, PA 16823.)

c. *Relative retention standards*

Many chromatographic analyses, such as those used in quality control, may be repeated at intervals over a prolonged period. Positive identifications will therefore depend on the reproducibility of the retention times, which will depend on the reliability of the carrier gas flow, column temperature and the stability of the liquid phase. There are particular problems if a column has to be replaced as the proportion of liquid phase in the new column may differ slightly, as it will depend on the coating procedure and efficiency of the packing process. The column dimensions, both length and internal diameter, may also differ, altering the amount of stationary liquid phase used in the assay. In addition, during use, changes will slowly occur with time as the liquid phase may bleed from the column, particularly if it has been used near its maximum recommended temperature. However, it is often not practicable to measure the full set of possible standards with every batch of samples.

Except for temperature variations, most of these variable factors only alter the proportion of stationary liquid phase compared to the gas phase (the phase ratio) and thus the distribution constants and selectivity should be unaltered. If instead of using the absolute retention time, the retention of an analyte is recorded as a relative adjusted retention time compared to a standard compound, then more reproducible results should be obtained. The adjusted retention times of the sample and standard would both have been affected proportionally by any differences in the phase ratio. In practice, the slightly less reproducible relative unadjusted retention times are often used as these can be reported directly by many integrators. It is also highly desirable to use a standard sample routinely in most separation procedures as this method will compensate for day-to-day changes in flow rates and column ageing and will rapidly indicate leaking septa or column fittings.

If the standard compound can be chosen so that it does not interfere with the rest of the sample then it can be added as an internal standard to the sample before injection. This will also aid accurate quantification (see Section 6.3). The selection of a suitable standard compound can be very important. The ideal standard will have a closely related structure to the analyte of interest so that it will respond in a similar manner to any changes in the separation conditions. Its retention time should be similar to that of the analyte (usually longer) so that the ratio of the values is not very large or very small. Typically, homologues of the analyte or structural isomers are very suitable but care must be taken to ensure that these will not occur naturally in the sample, either as impurities or as metabolites. These might interfere with quantification and could mislead identification. For example in the determination of fatty acid methyl esters, unusual isomers, such as odd-numbered or iso- fatty acid methyl esters, or an ethyl ester could be used as an internal standard.

d. Retention indices

Although recording the retention of an analyte as a relative retention time compared to a standard compound can avoid many of the problems caused by small changes in the separation conditions, it is often still difficult to compare retentions measured in different laboratories, as each may use a different standard for each assay. Ideally, it would be desirable that any laboratory carrying out a GLC separation on a particular analyte would be able to obtain the same value for the retention on a specified separation system. It would then be possible for compilations of retention values to be accumulated, which could then be used to identify unknown samples. Comparisons could be carried out with a "standard" retention value rather than by direct comparison with an authentic standard, avoiding the need for each laboratory to hold large stocks of possible samples.

However, a single or small group of retention standards could not be used with all the possible conditions that might be employed in GLC applications. In 1958, Kováts proposed that an alternative method would be to use the n-alkanes as a series of universal standard compounds [1]. They are eluted over a wide range of separation conditions, are readily available in all laboratories, and are chemically very stable and unreactive. It was also suggested that, rather than calculate relative retentions compared to a single standard, the alkanes should be used to define a retention index scale. The retentions of analytes would then be determined by interpolation (Figure 6.4). The retention indices of the n-alkanes are defined as the number of carbon atoms $\times$ 100 (i.e. hexane = 600). The retentions index (RI) of an analyte is then calculated using Equation 6.2 in which t'_{Ra}, $t'_{\mathrm{R}n}$, and $t'_{\mathrm{R}n+1}$, respectively, are the adjusted retention times of the analyte, and of the n-alkanes containing n and n+1 carbons, which are eluted immediately before and after the analyte.

$$RI = n \times 100 + 100 \times \frac{\log t'_{\mathrm{Ra}} - \log t'_{\mathrm{R}n}}{\log t'_{\mathrm{R}n+1} - \log t'_{\mathrm{R}n}} \qquad (6.2)$$

Alternatively, a graphical method can be used in which the logarithms of the adjusted retention times are plotted against the number of carbon atoms in the n-alkanes (Figure 6.5). If the separation is carried out using temperature programming, the relationship between the carbon numbers and the retention times becomes linear and the retention indices can be calculated using a direct linear interpolation.

Because retention indices are based on an interpolated scale of retention measurements, as long as the same liquid phase and temperature are used, the values should be independent of the proportion of the liquid phase in the column. Retention indices measured in different laboratories under the same conditions should thus be directly comparable and the values have found extensive use for the comparison of data and as the basis of physicochemical measurements. However, because some interactions do take place between the analyte and the underlying support material, there are often small differences between retention indices measured in different laboratories.

In recent years, better methods have been described for the preparation of packed columns and there has been advances in the coating of open-tubular columns. As a result there has been a considerable improvement in the reproducibility of retention indices and a number of extensive compilations of retention indices have been published. These typically use columns coated with OV-101 or Carbowax 20M or their bonded equivalents and can be used to identify unknown analytes. Some of these compilations are available as computer databases so that they can be readily searched using a microcomputer. Compounds with particular combinations of retention indices

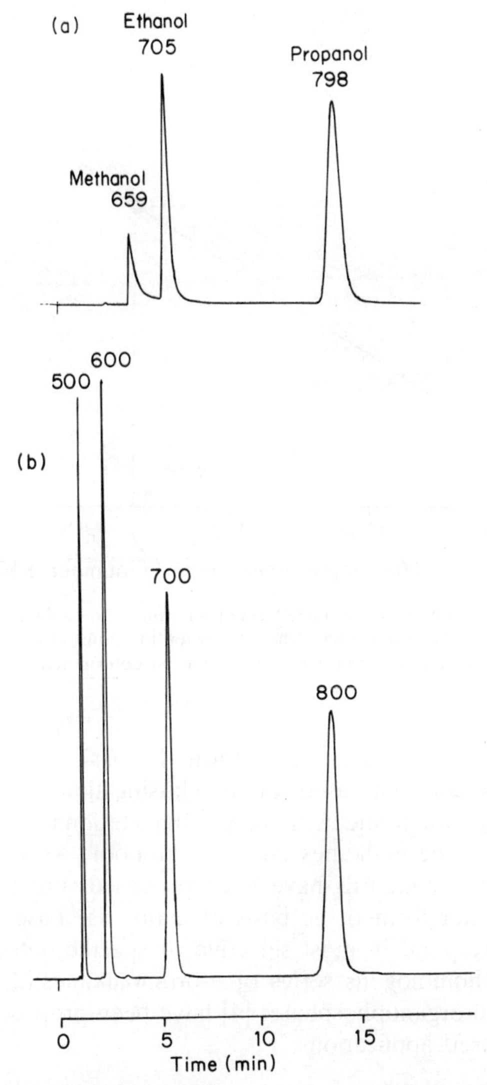

Figure 6.4 Chromatograms of (a) n-alkanols and (b) n-alkanes showing how the retention indices of the alkanols are based on interpolation between the alkane standards. Column 10% polypropyleneglycol column at 45°C.

on different column materials can be identified. It has been shown that with care, retention indices can be recorded in a single laboratory with an accuracy of about ±0.4 units and interlaboratory comparisons of 2–5 units.

As was seen earlier (Section 4.3b), the Kováts retention indices of a set

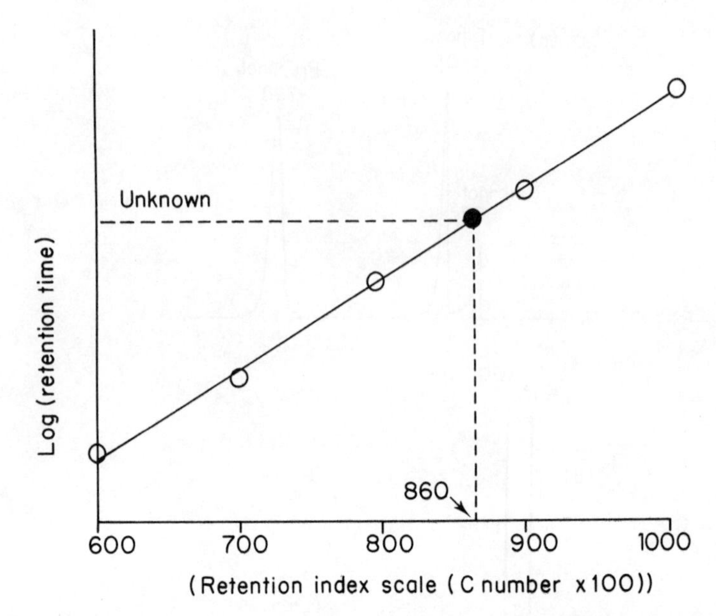

Figure 6.5 Graph of log adjusted retention times on n-alkanes compared to carbon number. The open circles show log retention times of n-alkane standards. The retention of an unknown (full circle) can be determined from the graph as shown.

of standard column test compounds form the basis of the Rohrschneider and McReynolds' constants used for the classification of stationary phases.

Some problems are found with very polar stationary liquid phases and the peak shapes of the n-alkanes can often be poor. As a result, alternative sets of polar index standards have been proposed, but none have gained wide acceptance nor formed the basis of major databases. Because the n-alkanes do not respond in most selective or specific detectors, specialised scales based on homologous series of n-bromoalkanes [2], n-alkyl trichloroacetates [3] and organophosphates [4] have been proposed as alternatives for these specialised applications.

6.3 SAMPLE QUANTIFICATION IN GAS–LIQUID CHROMATOGRAPHY

Probably the majority of GLC separations are carried out as part of a quantitative study, in which the reproducibility of the results is of prime importance. Over the linear response range of the detector, the area under the peak on the recorder or integrator trace is directly proportional to the quantity of analyte injected. However, the response of different compounds will differ depending on their functional groups and structure and the selectivity of the detector. Whereas with a flame ionisation detector, the relative areas under the peaks are an approximate guide to the amount of each compound in a sample mixture, with an electron capture detector there will be very large differences in the areas for the same quantity of different compounds. In both cases, for accurate quantitation it is necessary to calibrate the detector with samples of known size over the range of interest. This will also enable the linearity of the detector to be confirmed. The calibration can either be carried out by comparison with a set of standard solutions of the analyte run separately as an external calibration or by comparison between the peak areas of the analytes and that of a known amount of a reference compound added to the sample in an internal calibration standard.

a. *External calibration*

In the external calibration method, a series of solutions of different concentrations of the analyte are injected, which cover the expected range of analyte concentration in the sample. The peak areas are used to prepare a calibration curve, which can then be used to quantify the presence of the analyte in sample solutions. However, this method is not very satisfactory because of problems with the reproducibility of the injection step.

The injection step is a major cause of variation in GLC because of the difficulty of reproducibly handling small volumes of liquids (0.05–10 μl) and of avoiding small air bubbles in the syringe needle. While the delivery from a cold syringe at room temperature may be highly reproducible, the syringe needle is heated in the injection port of the GLC. As a result different amounts of solution may boil out of the tip onto the column before the injection takes place, depending on the speed of insertion of the needle and the delay between insertion and the plunger being depressed. Some of this solution will be lost by condensation on the cold parts of the injector or go onto the column over a period of time so that no discrete peaks will be observed. The actual volume injected onto the column by the syringe as a sharp band may therefore vary in different injections and the errors can be independent of the nominal sample size. In the preparation of a calibration curve, it is therefore important that each injection has the same nominal volume, which will also be used in the assay. A constant injection routine should also be used so that the amount of solution transferred to the column is reproducible. Different nominal volumes of a single concentration should not be used as these may suffer from systematic errors. Some of the manual handling variations can be avoided by using an automated injection system, which reproducibly times each step and the rate of injection.

These problems are particularly significant with the 1 μl plunger-in-needle syringe, which is frequently used for 0.1 to 1 μl injections. The sample solution is only drawn up into the needle, not into the body of the syringe, and cannot be seen by the operator. If small air bubbles, are present in the needle, they would be undetected and might reduce the effective volume that was injected. If the injection port is hot, the needle containing the sample will be heated during insertion and sample boil-off can easily occur. It is much better to use a more dilute solution and a larger 5 or 10 μl syringe in which the sample can be drawn up into the cold barrel before injection. This also reduces boil-off from the needle during injection and can be combined with a solvent-wash technique (see Appendix 2.1.c).

Gaseous samples cause particular problems in quantitative analysis because changes in the injection pressure or even atmospheric pressure can alter the sample size. Syringe injection is generally unsuitable and a fixed volume loop injector (1–2 ml) is normally used at atmospheric pressure to ensure reproducibility. For maximum accuracy the results should be corrected for changes in the atmospheric pressure from day to day.

b. Internal calibration

The preferred method of calibration is to add a solution of an internal standard of known concentration to both the calibration solutions and analyte samples before injection. Any variations in the volume injected will then affect both the quantity of standard and analyte proportionally but their ratio should remain constant (Figure 6.6). An internal standard can thus serve two roles, primarily to compensate for the actual sample volume injected and secondly as a check on the retention times. Its retention will be sensitive to any changes due to leaking septa, temperature variation, or aging of the column. If the GLC analysis is preceded by an extraction or derivatisation step, the internal standard, if correctly chosen to have comparable properties to the analyte, can be added to the original sample and also used to monitor the reproducibility of the sample preparation stages.

The selection of an internal standard for quantitative analyses should follow the criteria set down earlier (Section 6.2) for standards for qualitative comparisons. In addition the standard compound should be stable in a suitable solvent so that a stock solution can be prepared and used over a period of time. As noted earlier, the internal standard should be absent from the original sample. Check separations should be carried out in the

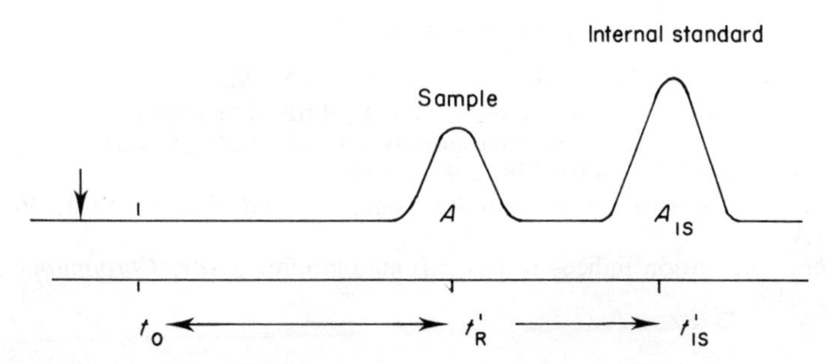

Figure 6.6 Relative peak areas and relative retention times compared to an internal standard reference compound. The relative retention time is t'_R/t'_{IS}, and the relative peak area is A/A_{IS}.

absence of the internal standard to ensure that the sample contains no components that might interfere or co-elute with the standard. If the analytes are being detected with a selective detector, the internal standard should give a similar response and selectivity. It is not normally possible to use internal standards with gaseous samples because of the problems of accurately mixing known volumes of gases.

If precautions are taken it should be possible to obtain quantitative results in GLC with a precision of 1–2% or better relative standard deviation as long as the measurements are not near the limit of determination.

BIBLIOGRAPHY

Sample preparation

W. Dünges, *Pre-chromatographic micromethods*, Alfred Hüthig, Heidelberg, 1986.

W.G. Jennings and A. Rapp, *Sample preparation for gas chromatographic analysis*, Alfred Hüthig, Heidelberg, 1983.

F.W. Karasek, R.E. Clement and J.A. Sweetman, "Preconcentration for trace analysis of organic compounds", *Anal. Chem.*, 1981, **53**, 1050A–1058A.

K.C. Van Horne (Ed.), *Sorbent extraction technology*, Analytichem Int., Harbor City, CA 1985.

Sample identification

R.C. Crippen, *Identification of organic compounds with the aid of gas chromatography*, McGraw-Hill, New York, 1973.

D.A. Leathard and B.C. Shurlock, *Identification techniques in gas chromatography*, John Wiley, Chichester, 1970.

Kováts retention index scale

M.V. Budahegyi, E.R. Lombosi, T.S. Lombosi, S.Y. Mészáros, Sz. Nyiredy, G. Tarján, I. Timár, and J.M. Takács, "Twenty-fifth anniversary of the retention index system in gas–liquid chromatography", *J. Chromatogr., Chromatogr. Rev.*, 1983, **271**, 213–307 (includes 1392 references).

L.S. Ettre, "The Kováts retention index system", *Anal. Chem.*, 1964, **36** (July), 31A–41A.

J.K. Haken, "Retention indices in gas chromatography", *Adv. Chromatogr.*, 1976, **14**, 367–410.

Compilations of retention indices

R.E. Ardrey and A.C. Moffat, "Gas-liquid chromatographic retention indices of 1318 substances of toxicological interest on SE-30 or OV-1 stationary phase", *J. Chromatogr.*, 1981, **220**, 195–252.

Gas-chromatographic retention indices of toxicologically relevant substances on SE-30 or OV-1, 2nd ed., VCH, Weinheim, 1985.

C. Lora-Tamoyo, M.A. Rams and J.M.R. Chacon, "Gas chromatographic data for 187 nitrogen- or phosphorus-containing drugs and metabolites of toxicological interest analysed on methyl silicone capillary columns", *J. Chromatogr.*, 1986, **374**, 73–85.

A. Nakamura, R. Tanaka and T. Kashimoto, "Retention indexes for electron capture gas chromatography on programmed temperature high resolution capillary column", *J. Assoc. Off. Anal. Chem.*, 1984, **67**, 129–132.

K. Pfleger, H. Maurer and A. Weber (Eds.), *Mass spectral and GC data of drugs, poisons and their metabolites*, VCH, Weinheim, 1985.

The Sadtler standard gas chromatography retention index library, Vols. 1–4, Heyden, London, 1986.

A. Yasuhara, M. Morita and K. Fuwa, "Temperature programmed-retention indices of 221 halogenated organic compounds with 1-bromoalkanes as references", *J. Chromatogr.*, 1986, **328**, 35–48.

Quantitative analysis

E. Katz (Ed.), *Quantitative analysis using chromatographic techniques*, John Wiley, Chichester, 1987.

J. Novák and P.A. Leclercq, *Quantitative analysis by gas chromatography 2nd ed.*, Chromatographic Science Series, Vol. 41, Marcel Dekker, New York, 1988.

A. Shatkay, "Effect of concentration on the internal standards method in gas–liquid chromatography", *Anal. Chem.*, 1978, **50**, 1423–1429.

A. Shatkay and S. Flavian, "Unrecognised systematic errors in quantitative analysis by gas–liquid chromatography", *Anal. Chem.*, 1977, **49**, 2222–2228.

REFERENCES

1. E. Kováts, "Gas-chromatographische Charakterisierung organischer Verbindungen", *Helv. Chim. Acta*, 1958, **41**, 1915–1932.

2. F. Pacholec and C.F. Poole, "Evaluation of a calibration marker scheme for open tubular column gas chromatography with on-column injection and electron-capture detection", *J. Chromatogr.*, 1984, **302**, 289–301.
3. T.R. Schwartz, J.D. Petty and E.M. Kaiser, "Preparation of n-alkyl trichloroacetates and their use as retention index standards in gas chromatography", *Anal. Chem.*, 1983, **55**, 1839–1840.
4. J. Enqvist and A. Hesso, "Precision gas chromatography as an analytical tool. Part 1. A new automatic system based on two-channel retention index monitoring", *Kemia*, 1982, **9**, 176.

CHAPTER 7

GAS CHROMATOGRAPHY: SPECIAL TECHNIQUES

7.1 PROBLEM SAMPLES IN GAS–LIQUID CHROMATOGRAPHY

Not all compounds are suitable for direct GLC analysis because they are either involatile or thermally unstable. Many compounds of biological importance, such as carbohydrates and amino acids, are particularly problematic as they are often polar or ionisable and therefore frequently have limited volatility. Although it might seem that in many of these cases, an alternative chromatographic technique such as HPLC could be used, many of these compounds also face different but still troublesome problems in liquid chromatography (see later).

Two main approaches have been adopted for the examination of analytes that do not satisfy the normal criteria of stability and volatility. Either they can be converted by chemical derivatisation into a related compound that is suitable for chromatography (Section 7.2) or they can be degraded under controlled conditions by pyrolysis to give characteristic volatile fragments (see Section 7.3). Derivatisation can also be used to enhance resolution, selectivity, or sensitivity of the analytical technique.

If the matrix of the sample is a solid or contains a high proportion of involatile materials it will usually be necessary to extract the analyte before chromatography. However, these stages are time-consuming, introduce reproducibility problems, and in the case of bulk solid samples, such as rubber or plastics, are impracticable. If the analyte is reasonably volatile, some of these problem matrices can be examined using headspace analysis or vapour trapping (Section 7.4). Other sample mixtures are so complex that a single chromatogram even on an open-tubular column may not resolve all the components. In these cases coupled separation or column switching will often enable increased resolution of at least part of the sample without unduly extending the analysis time (Section 7.5).

7.2 DERIVATISATION REACTIONS IN GAS–LIQUID CHROMATOGRAPHY

Usually derivatisation reactions are carried out to confer volatility or thermal stability on compounds that are too polar or unstable for direct analysis. In addition, derivatisation can be used to enhance separations, particularly of optical isomers, or to increase the sensitivity of specific analytes towards selective detectors.

The reactions used for derivatisation should give a high and ideally quantitative (100%) yield and any excess of the reagent should not interfere with the subsequent chromatographic analysis. The reagents should be simple and cheap and the reaction should be specific for the desired functional group. The reaction conditions should be as mild as possible so that other parts of the analyte molecule are unaffected and no degradation can occur. The most widely used reactions are relatively simple one-step reactions, which can be carried out at room temperature.

Typical compounds that need to be derivatised are those containing hydroxyl, carboxylic acid, or amino functional groups. These include many compounds of great biochemical interest, such as the fatty acids, carbohydrates and amino acids, which are frequently present in low concentrations in complex biological matrices. GLC would be an ideal technique for their analysis by providing both high selectivity and sensitivity but the analytes must first be converted into volatile derivatives.

The field of derivatisation has been well reviewed with considerable detail both in textbooks and specific articles (see Bibliography) and therefore only the principal chemical groups of interest and most commonly used reactions will be discussed here.

a. *Hydroxyl groups*

The low-molecular-weight alkanols are readily analysed by GLC without difficulty but many larger or polyfunctional alcohols, such as the sterols and carbohydrates, are much less volatile and on heating suffer dehydration or decomposition. Three main approaches can be used to derivatise hydroxyl groups, either alkylation, acylation, or silylation.

i. Alkylation to form ethers

This method is not very important because of experimental problems with the reaction (Equation 7.1).

$$ROH + Ag_2O + MeI \xrightarrow{DMF} ROMe + 2AgI + H_2O \qquad (7.1)$$

but the products are very stable and reasonably volatile.

ii. Acylation

Both acetic anhydride and trifluoroacetic anhydride in pyridine or tetrahydrofuran (THF) have been used to convert hydroxyl groups into the corresponding acetates (i.e. Equation 7.2). The trifluoroacetyl derivatives are often used because they also give an enhanced response in the electron capture detector.

$$ROH + (CF_3CO)_2O \xrightarrow{pyridine} ROCOCF_3 \qquad (7.2)$$

Alternative acylation reagents include trifluoroacetylimidazole and perfluorinated reagents, such as heptafluorobutanoic anhydride or pentafluorobenzoyl chloride, which yield the corresponding perfluoro esters.

iii. Silylation

The conversion of the hydroxyl group to the trimethylsilyl (TMS) group (Me_3Si–O–) (Equation 7.3) is the most widely used derivatisation

$$ROH \longrightarrow ROSiMe_3 \qquad (7.3)$$

reaction for alcohols and has been particularly valuable for polyols and carbohydrates, including disaccharides. TMS ethers are readily synthesised and are thermally stable and volatile (Figure 7.1). These properties have also led to their use to prepare volatile derivatives for mass spectrometry. However, TMS ethers are very susceptible to hydrolysis, giving back the free hydroxyl group, and therefore water must be rigorously excluded from the reaction mixture.

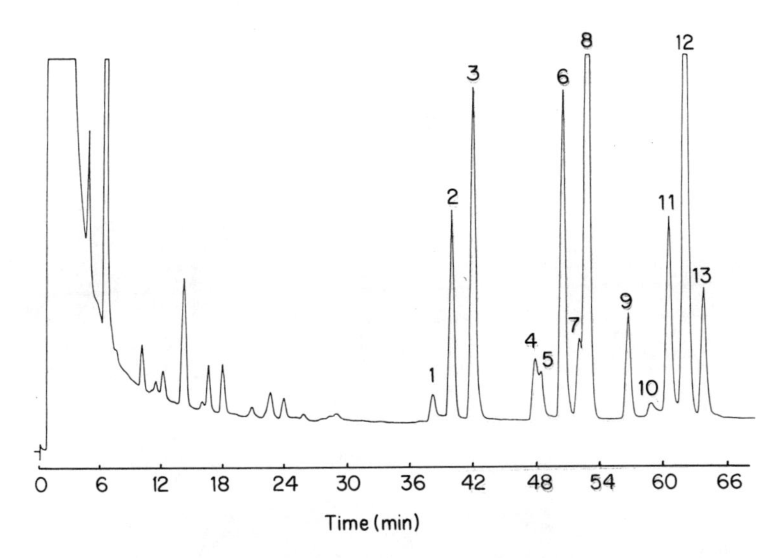

Figure 7.1 Separation of trimethylsilyl ethers of a mixture of carbohydrates on a 10% E-30 column on 100–120 mesh Diatomite CQ. Temperature programmed from 100°C to 230°C at 2°C min^{-1}. Peaks: 1, γ-fucose; 2, α-fucose; 3, β-fucose; 4, α-mannose; 5, γ-galactose; 6, α-galactose; 7, α-glucose; 8, β-galactose and β-mannose; 9, β-glucose; 10, inositol; 11, α-galactosamine; 12, α-glucosamine and β-galactosamine; 13, β-glucosamine. (Reproduced by permission of Philips Analytical from P. J. Ridgeon, *Gas chromatography separations*, 3rd ed., Pye Unicam, Cambridge, 1976, J. p. 18.)

Many of the silylation reagents will also form volatile trimethylsilyl derivatives of carboxylic acids, phenols, amines and thiols, by the replacement of active hydrogens, and the products can often also be used for chromatography.

A number of different reagents can be used to synthesise TMS ethers. One of the original reactions is that of Sweeley and coworkers [1], using trimethylsilyl chloride and hexamethyldisilazane in pyridine (Equation 7.4). The reaction proceeds to completion in 5–10 minutes at room temperature and gives quantitative yields.

$$ClSiMe_3 + Me_3NH.SiNHSiMe_3 + \text{pyridine} + ROH \longrightarrow ROSiMe_3 \qquad (7.4)$$

Alternative reagents that have been widely used include *O,N*-bistrimethylsilylacetamide (BSA) and trimethylsilylimidazole (Tri-Sil Z).

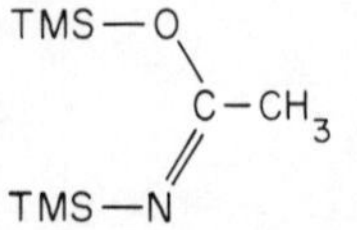

TMS—O
C—CH_3
TMS—N

BSA

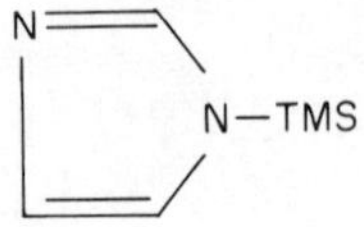

N
N—TMS

Tri-Sil Z

The latter reagent is not as sensitive as the other reagents to the presence of moisture in the sample as long as the reagent is in excess but it will not derivatise amino groups. The acetamide reagent is slower and is unsuitable for some samples, in particular carbohydrates, but is the method of preference for phenols.

A number of related silyl reagents have also been used to give derivatives carrying other alkyl groups. *N*-Methyl-*N*-t-butyldimethylsilyltrifluoro-acetamide will yield t-butyldimethylsilyl ethers. They are particularly useful in GC–mass spectrometry as the loss of the t-butyl group from the molecular ion gives a fragment ion ($M^+ - 87$), which is very characteristic and enables the derivatives to be clearly identified. Partially halogenated methylsilyl reagents (i.e. $ClCH_2Me_2Si–X$) have been used to introduce halogen atoms into derivatives in order to give an increased response in the electron capture detector and enhanced specific sensitivity.

b. Carbohydrates

Carbohydrates are frequently examined as their pertrimethylsilyl ethers, prepared by reaction with either trimethylsilyl chloride/hexamethyldisilazane with or trimethylsilylimidazole. With the former reagent, amino sugars are silylated on both O and N atoms. Both α and β anomeric C-1 TMS derivatives can be formed depending on the reaction conditions and the original composition of the carbohydrate. The anomers are separated on GLC (Figure 7.1). Both mono- and disaccharide derivatives are sufficiently volatile for analysis. Carbohydrates that exist as both pyranose and furanose forms may give additional peaks.

An alternative widely used reaction for the derivatisation of the carbohydrates is the formation of alditol acetates. The carbonyl group of the carbohydrate is first reduced with sodium borohydride to give the alditol, which is then peracylated with acetic anhydride or trifluoroacetic anhydride (Figure 7.2). The products are usually stable and separate well on chromatography (Figure 4.4). However, because two isomers can be formed by the reduction of a ketose, fructose gives two peaks corresponding to glucitol and mannitol. In addition, because of the symmetry of the alditols, two different carbohydrates can often yield the same alditol. Both D-arabinose and D-lyxose give D-arabitol, and although D-glucose and D-gulose yield D- and L-glucitol, respectively, the acyl derivatives of these enantiomers are chromatographically identical on most columns.

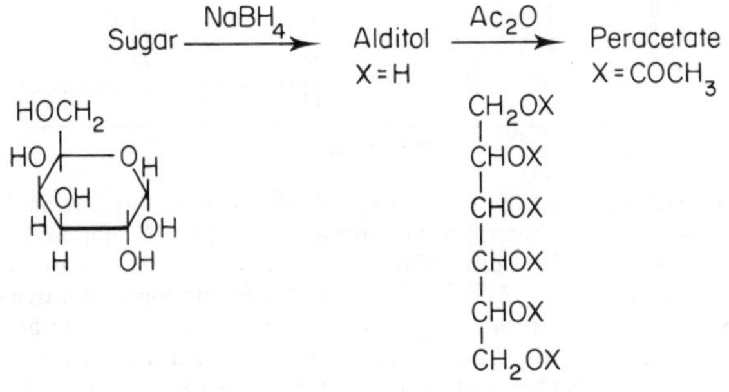

$$\text{Sugar} \xrightarrow{NaBH_4} \underset{X=H}{\text{Alditol}} \xrightarrow{Ac_2O} \underset{X=COCH_3}{\text{Peracetate}}$$

CH_2OX–CHOX–CHOX–CHOX–CHOX–CH_2OX

Figure 7.2 Formation of alditol acetates.

c. *Carboxylic acids*

Although carboxylic acid groups will form trimethylsilyl esters with many silylation reagents, the most important derivatives are alkyl esters. Usually the volatile methyl esters are prepared (Figure 7.3) and even long-chain fatty acids can be examined (Figures 4.7 and 6.3).

The less volatile propyl and butyl esters are often used for low-molecular-weight acids so that the derivative is not lost on working up the reaction mixture. A number of esterification reagents have been used.

i. *Diazomethane*

On a small scale and with caution, diazomethane (CH_2N_2) is chemically an excellent and apparently ideal reagent (Equation 7.5).

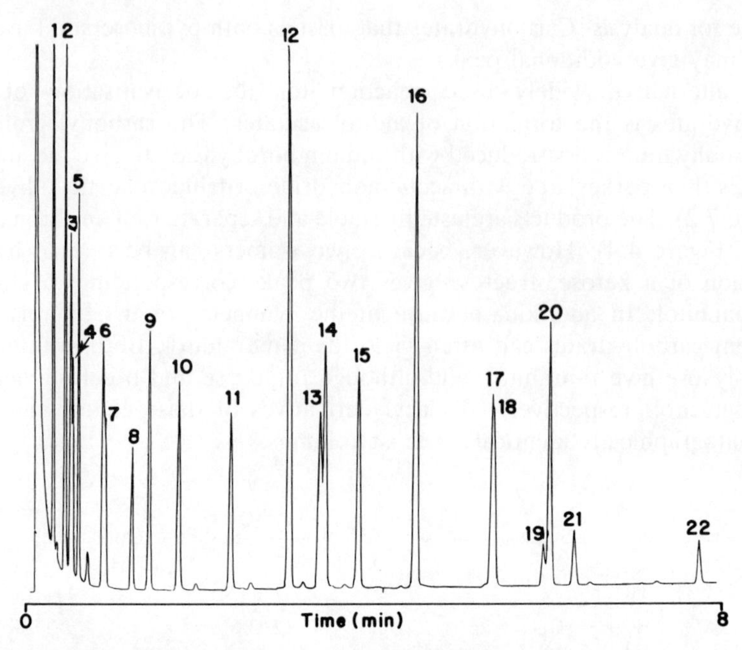

Figure 7.3 Separation of methyl esters of 22 non-volatile organic acids following direct injection onto a thick-film wide-bore CP-Sil 19 CB column 10 m × 0.53 mm. Film thickness 2.0 μm. Carrier gas hydrogen. Column temperature programmed from 80°C to 280°C at 25°C min^{-1}. Flame ionisation detector. Peak identification: 1, lactic acid methyl ester; 2, propyleneglycol; 3, butanediol; 4, β-hydroxybutyric acid methyl ester; 5, methylmalonic acid methyl ester; 6, pyruvic acid ethoxime; 7, succinic acid methyl ester; 8, benzoic acid methyl ester; 9, glutaric acid methyl ester; 10, adipic acid methyl ester; 11, 2-hydroxyphenylacetic acid methyl ester; 12, 4-phenylbutyric acid methyl ester; 13, α-ketoglutaric acid ethoxime; 14, α-ketoadipic acid ethoxime; 15, C_{10} dicarboxylic acid methyl ester; 16, 4-hydroxyphenylacetic acid methyl ester; 17, homovanillic acid methyl ester; 18, hippuric acid methyl ester; 19, 4-hydroxymandelic acid methyl ester; 20, indoleacetic acid methyl ester; 21, 2-hydroxyhippuric acid methyl ester; 22, 5-hydroxyhippuric acid methyl ester. (Reproduced with permission from *Chrompack application note*, 8 GC.)

$$RCOOH + CH_2N_2 \longrightarrow RCOOMe + N_2 \qquad (7.5)$$

The only by-product from the reaction is gaseous nitrogen and the reaction is complete in a few seconds in an ice bath. However, as a neat liquid diazomethane is potentially explosive so it should always be used as a cold ethereal solution. In addition, diazomethane is toxic and is a suspected carcinogen. It is now usually synthesised *in situ* in a closed reaction system but the reagents used are also suspected carcinogens. Despite these hazards it was widely used in the past but should now be avoided if possible.

ii. Acid-catalysed esterification

Because of the concern about the safety of the diazomethane reaction, conventional esterification reactions are preferred. The carboxylic acid is refluxed with an alcohol and a catalytic mineral or Lewis acid (Equation 7.6). The reaction mixture is then diluted with water, neutralised and the ester derivatives are extracted from the aqueous solution for analysis.

$$\mathrm{RCOOH + R'OH \xrightarrow[BF_3\ or\ HCl\ (dry)]{H_2SO_4\ (conc)\ or} RCOOR'} \qquad (7.6)$$

Prepared mixtures of boron trifluoride and methanol are commercially available and this is usually the reagent of choice. Dry hydrochloric acid gas is messy to prepare and concentrated sulphuric acid can degrade some samples. A number of activated esterification reagents have also been developed, including dimethylformamide dialkylacetals (Equation 7.8).

$$\mathrm{Me_2N{-}CH(OR')_2 + RCOOH \longrightarrow RCOOR'} \qquad (7.7)$$

iii. On-column esterification reactions

If carboxylic acids are mixed with tetramethylammonium salts, no reaction takes place at room temperature, but when the mixture is injected into the hot injection port of a gas chromatograph a thermal reaction takes place with the formation of the methyl ester (Equation 7.8).

$$\mathrm{RCOOH + {}^{+}NMe_4{}^{-}OH \xrightarrow{\Delta} RCOOMe} \qquad (7.8)$$

This reaction therefore avoids any work-up steps.

Similar thermal reactions to form methyl esters take place with trimethylanilinium hydroxide. This reagent has also been used to convert the weakly acidic barbiturates into their corresponding *N,N'*-dimethyl derivatives, which give better shaped peaks on GLC analysis.

iv. Derivatisation for thermionic detection

The reaction of carboxylic acids with phosphonate acetals yields phosphorus-containing esters, which will give a specific response in a thermionic ionisation detector (Equation 7.9).

$$RCOOH + HO\text{–}CH_2\text{–}PO(MeO)_2 \xrightarrow{DCC} RCOOCH_2\text{–}PO(OMe)_2 \quad (7.9)$$

(where DCC is dicyclohexylcarbodiimide, a catalyst).

d. Amines

Problems of tailing and poor peak shapes are often observed with analytes containing the amino group. These are primarily caused by their basicity and interactions with active acidic silanol groups on the surface of the support material or the glass tubing of the column. The two most widely used derivatisation reactions are acylation, to form a non-basic amide, and arylation, which can convert a strongly basic aliphatic amino group into a weakly basic aromatic amino group.

i. Acylation

Acylation is usually carried out by reacting the amino group with acetic anhydride or trifluoroacetic anhydride in pyridine (Equation 7.10 where $R' = CH_3$ or CF_3).

$$RNH_2 + (R'CO)_2O \longrightarrow RNHCOR' \quad (7.10)$$

The presence of the trifluoro group also increases the response of the derivative in a electron capture detector.

Trifluoroacetylimidazole has also been used as a reagent. These three acylation reagents will also convert hydroxyl groups to the corresponding acyl derivatives. However, *N*-methylbis(trifluoroacetamide) will give a specific reaction with amino groups.

ii. Arylation

The most commonly used arylation reaction is to treat the amine with fluoro-2, 4-dinitrobenzene (FDNB, Sanger's reagent) (Equation 7.11).

$$RNH_2 + FDNB \longrightarrow RNH\text{–}DNB \quad (7.11)$$

(where DNB is 2, 4-dinitrobenzene). The electron-withdrawing nitro groups also increase the sensitivity of these derivatives in the electron capture detector.

iii. Silylation

Amino groups will react with many of the silylating reagents listed earlier (Section 7.2.a.iii) to give *N*-TMS derivatives.

e. Amino acids

In solution, amino acids are ionised or are zwitterions and therefore present special problems for GLC. In order to obtain a neutral non-ionised derivative, it is usually necessary to react both the amino and carboxylic acid groups, usually to give a *N*-acetyl alkyl ester (Equation 7.12).

$$RCH(NH_3^+)(COO^-) \longrightarrow RCH(NHCOCF_3)(COOPropyl) \quad (7.12)$$

The propyl ester is usually formed as the methyl esters of some amino acids are volatile and they may be lost during the work-up of the synthesis. The acidic group is reacted first, otherwise it interferes with the reaction on the amino group. Because of the range of different amino acids, the separation of the derivatives usually requires temperature programming (Figure 4.10). Amino acids can also be separated in the liquid phase by dedicated ion-exchange chromatographs, but for small numbers of samples GLC analysis has the advantage that the equipment can also be used for other assays.

f. Other functional groups

If possible, chromatographers will avoid reaction steps because of the time required and risk of poor reproducibility, so that few other groups are routinely derivatised to enhance separation. However, reactions can also be used to provide derivatives with improved detection particularly towards selective detectors. For example, ketones can be converted to oximes ($R_2C{=}N{-}O{-}R'$) in which the R' group is halogenated, i.e. pentafluorobenzyl, and therefore gives a selective response in the electron capture detector.

Except for low-molecular-weight and permanent gases, inorganic compounds are not usually suitable for GLC analysis but some neutral organometallic compounds are sufficiently volatile and thermally stable for ready separation. For example transition-metal ions can be analysed after conversion to their volatile hexafluoropentane-2,4-dionate chelates $[(CF_3COCH_2COCF_3)_2M]$.

g. *Chiral separations*

Enzymes and drug receptor sites are usually chiral and the behaviour of the different chiral isomers of biologically active compounds, including drugs, pesticides and their metabolites, can often differ markedly. In extreme cases one isomer may have the desired activity but the other enantiomer will have adverse effects. Drug and pesticide manufacturers are therefore being required to test the activity, toxicology, and metabolism of both enantiomers. As noted earlier (Chapter 4) the resolution of chiral compounds can be carried out by using stationary phases that are themselves chiral (Figure 4.10). However, this method may not always be suitable. An alternative procedure is to convert a pair of chiral isomers into non-equivalent diastereoisomers by derivatisation with a single enantiomer of a chiral reagent (Equation 7.13). Typical reactions include the use of a (−)-menthol as the alcohol in an esterification reaction (Equation 7.14). As the diastereoisomers have different properties, they can be separated using an conventional achiral liquid phase column.

$$\underset{\text{(enantiomers)}}{\text{D-form} + \text{L-form}} \xrightarrow{\text{D-reagent}} \underset{\text{(diastereoisomers)}}{\text{DD} + \text{DL-forms}} \qquad (7.13)$$

$$\text{R}-\underset{\text{OH}}{\underset{|}{\text{CH}^*}}\text{COOH} + (-)\text{-Menthol} \longrightarrow \text{R}-\underset{\text{OH}}{\underset{|}{\text{CH}^*}}\text{COO}-(-)\text{-Menthyl} \qquad (7.14)$$

Because many of the chiral compounds of interest are biologically important, they often contain functional groups, such as acids or hydroxyls, which would need to be derivatised even for non-chiral chromatography. The design of the reagent and hence the derivative is important to obtain significantly different retentions of the diastereoisomers. The separation is usually enhanced if the two chiral sites in the derivative (from the analyte and reagent) are in close proximity and are surrounded by substituents with very different sizes. The derivative should also contain functional groups near to the chiral centres, which can interact with the stationary phase.

h. *Abstraction techniques*

Chemical abstraction reactions can occasionally be used to remove selected components of a mixture before GLC analysis, either by reacting the sample with a reagent solution or by using a pre-column before the analytical column on the gas chromatograph. By comparing the chromatograms of the sample before and after abstraction, useful information can often be obtained about the structure of unidentified components. The main advantage of this technique was that it could often be carried out on a very small scale, but it is now less popular as GC–MS and GC–FTIR have become more widely available as alternative sources of structural information.

Typical reactions would be the addition of 2,4-dinitrophenylhydrazine hydrochloride to the sample, which would react with any components containing carbonyl groups, significantly altering their retentions. Alternatively, on adding 5 Å molecular sieves to the sample solution any n-alkanes will be selectively absorbed and lost from the chromatogram but any branched alkanes would be unaltered.

7.3 SAMPLE PYROLYSIS

Synthetic polymers and biological macromolecules are involatile and therefore cannot normally be examined by GLC. However, if they are pyrolysed under controlled conditions in the injection port of a gas chromatograph, any volatile fragment molecules will be carried onto the column and separated. The chromatographic pattern of the fragments should be characteristic of the original sample and can be used for identification.

Synthetic organic polymers frequently fragment to give mixtures in which the corresponding monomers predominate (Figure 7.4). Comparison of the chromatograms with those of polymers of known composition can be used to identify the component monomers and their proportions.

Because of the greater complexity of biological macromolecules, such as proteins and nucleic acids, the fragmentation patterns on pyrolysis are often very complex. Frequently very few of the fragments can be identified, even if the chromatograph is coupled to a mass spectrometer. However, the chromatograms can be used as fingerprint patterns for comparison purposes. This technique has been developed into a rapid method for the identification of anaerobic bacteria by pyrolysing the intact microorganism and comparing the gas chromatographic pattern with a database file of characteristic chromatographic traces. This method is much faster than the conventional microbiological culturing techniques and enables early appropriate treatment of patients to be started.

Two main methods have been used to pyrolyse the sample in the injection port. The temperatures used for pyrolysis range from 400–1000°C and with both methods heating times of less than 0.5 s can be achieved, ensuring that the fragments enter the column in a sharp band. The most versatile method is the controlled heating of the sample on a filament. An alternative, very

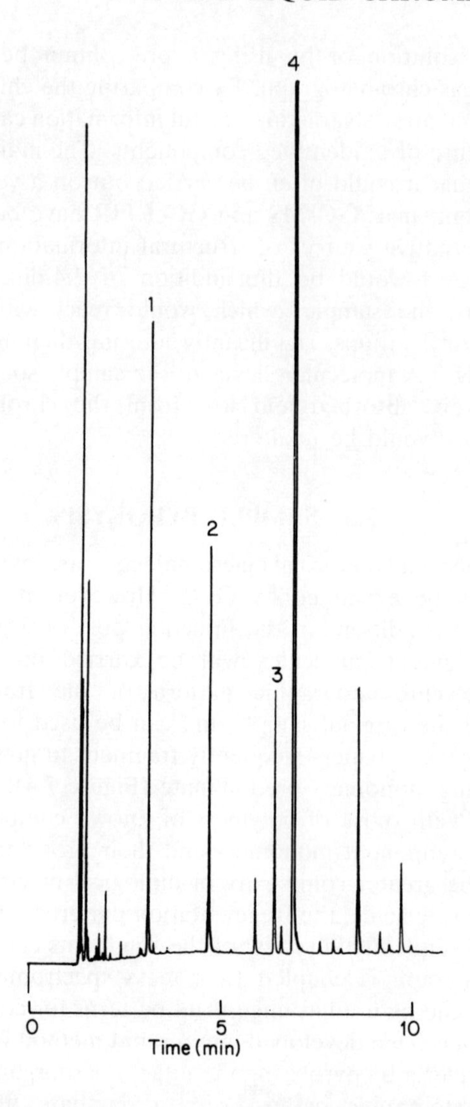

Figure 7.4 Pyrogram of polystyrene. Pyrolysis at 800°C, separation on BP-1 methylsilicone, 25 m × 0.32 mm i.d. Film thickness 0.5 μm. Temperature 40°C for 1 min then 10°C min^{-1} to 130°C. Flame ionisation detector. Under these conditions the secondary pyrolysis products, benzene, toluene and ethylbenzene, are also found as well as the styrene monomer. Peaks: 1, benzene; 2, toluene; 3, ethylbenzene; 4, styrene (monomer). (Reproduced by permission of Scientific Glass Engineering from *SGE applications*, GC4/86.)

simple, technique is the Curie point pyrolyser in which the sample is placed on a metal wire in the carrier gas stream. The wire is heated by an external induction coil. When the temperature of the wire reaches its Curie point, the wire ceases to be magnetic so that induction heating stops. By using wires of different compositions, iron, nickel, etc., different final temperatures can be reached. Other techniques, including pyrolysis of the samples with a laser, have also been reported.

7.4 MATRIX PROBLEMS

If the analytes of interest are present in an involatile matrix, such as drugs in plasma and blood or pesticides in soils and plant material, they must be extracted into a solvent to avoid a build-up of involatile material in the injection port, which would eventually interfere with the separations. The extraction can also enable the sample to be concentrated. However, solid matrices, such as plastics, cannot be efficiently extracted and some emulsions, such as latex samples, would be miscible with an extraction solvent.

a. Headspace analysis

An elegant solution to the problems encountered with these and similar samples is to use headspace analysis. In this technique, the sample is placed in a sealed container. If volatile components are present, the space above the sample (the headspace) will contain a proportion of the analytes, representative of their concentration in the sample. The sample is allowed to come to equilibrium, usually at an elevated temperature, and a sample of the headspace vapour is withdrawn and injected onto a GLC. The injected sample will contain only gaseous components so that no involatile material is transferred to the column. The sampling step can be manual but it is difficult to obtain high reproducibility so that an automatic or semi-automatic system is often used. The vial is held in a thermostated enclosure, pressurised with carrier gas, and then the pressure is released to carry the gas and part of the vapour onto the column.

This technique has found application in a number of areas, including the determination of monomers in plastics used as food packaging, the concentration of ethanol in blood samples from drunken drivers, the analysis of flavours from foods, and solvent levels in the bloodstreams of workers in industrial plants (Figure 7.5). In some of these examples, the alternative method of solvent extraction could not have been used because the extraction solvent would have interfered with the analysis of the volatile analytes.

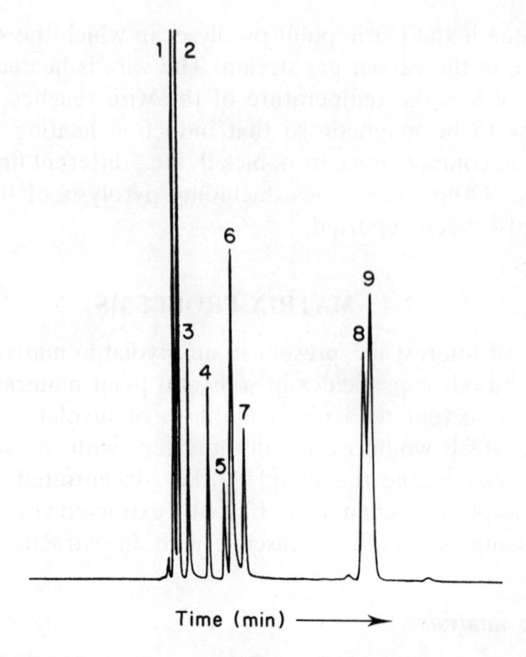

Figure 7.5 Detection of solvents in blood sample by headspace analysis as part of an industrial hygiene study. Sample held at 60°C. Column UCON LB 550, 25 m, at 40°C. Peaks: 1, diethyl ether; 2, hexane; 3, acetone; 4, dichloromethane; 5, methyl ethyl ketone; 6, ethanol; 7, benzene; 8, methyl isobutyl ketone; 9, toluene. (Reproduced by permission of Dani S.p.A.)

Although only part of the analyte is injected, quantitative results can be obtained by using reference samples or using the method of standard additions by spiking samples with known concentrations of the analyte. For problem samples in which matrix effects are difficult to reproduce, such as plasticisers in solid plastics, repeated analysis from the same container can be extrapolated to give the original concentration.

b. Trapping of volatile samples

The sensitivity of static headspace analysis is limited if the analyte concentrations are low or the analyte is not very volatile, because only a small proportion of the vapour is injected onto the column. Increased sensitivity can be obtained using dynamic headspace analysis or vapour trapping. The vapour above a sample is collected by drawing it through a small tube containing a highly absorbent material such as Tenax or Porapak, which traps any organic compounds. The tube is then transferred to the GLC injection port and is rapidly heated in the carrier gas stream so that the volatile compounds are released onto the column. Alternatively, the trapped compounds can be washed out of the tube into a small volume of a volatile solvent, such as carbon disulphide, and injected in solution. Both methods effectively concentrate all the vapour from a large sample volume into a single injection.

An adaptation of this technique is used to monitor chemical and solvent vapours in the workplace or laboratory. A passive adsorption tube traps the vapours that diffuse into it from the surrounding air over a prolonged period. The tube is then capped and transferred to the laboratory. The contents are then thermally desorbed in the inlet of the GLC and analysed. The adsorbent in the tubes is usually Tenax or similar material. These materials do not adsorb water vapour from the air, which might otherwise swamp the tube and deactivate it towards the compounds of interest.

7.5 COLUMN SWITCHING

Many complex samples contain such a wide range of components that a complete analysis could only be achieved by using a very efficient column, a long separation time and a complex temperature program. In practice, usually only part of the chromatogram is of interest and the analysis can often be speeded up by first using a relatively simple low-resolution column to carry out a rough separation. As the compounds of interest are being eluted, the outlet carrier gas stream is switched to a second column, which completes the separation (Figure 7.6). The remaining components from the first column are passed to waste. This heart cutting technique enables a detailed separation to be carried out on just part of the total sample.

A second column switching technique can be used if only the rapidly eluted components of a sample are of interest. Rather than waiting for the later components to be eluted, the flow of carrier gas through the column is reversed. The less volatile components, which have only travelled a short way down the column, are back-flushed out of the column to waste. The column is then ready for the next injection. The reverse elution takes place in the same time as the original separation, whereas complete elution in the original direction might have required an extensive run and temperature programming.

Other column switching techniques can also be employed to increase the resolution of complex samples or to avoid complicated pre-column purification stages.

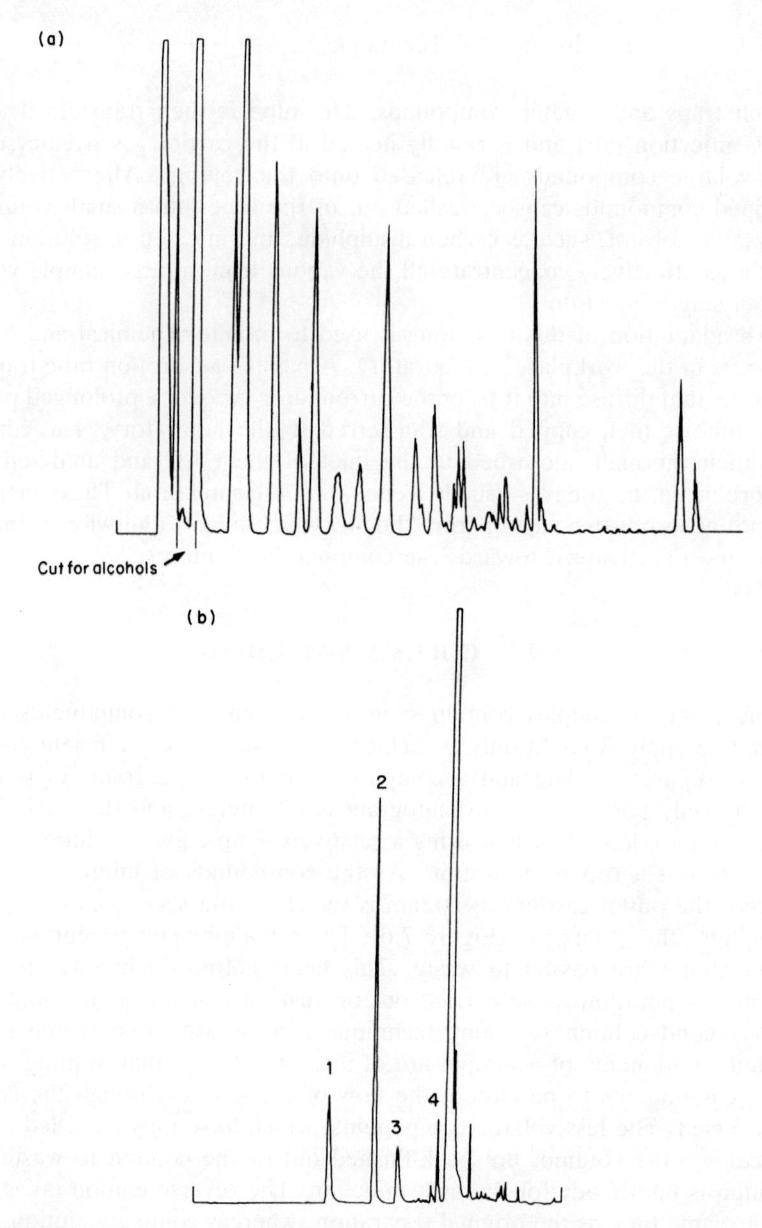

Figure 7.6 The separation and detection of alcohols introduced into petrol as anti-knocking agents using a heart cut technique. (a) Chromatogram of total gasoline fraction on FFAP column 25 m × 0.2 mm i.d. Column temperature 35°C for 4.25 min then to 150°C at 20°C min^{-1}. The section marked is heart cut to a second column. (b) Separation of cut portion on a second column of BP-10 25 m × 0.2 mm i.d. Column temperature 50°C for 4.25 min then to 150°C at 20°C min^{-1}. Peaks: 1, isopropanol; 2, t-butanol; 3, n-propanol; 4, isobutanol. (Reproduced by permission of Philips Analytical.)

BIBLIOGRAPHY

General texts on derivatisation in chromatography

S.Ahuja, "Derivatization for gas and liquid chromatography", in S. Ahuja (Ed.), *Ultratrace analysis of pharmaceuticals and other compounds of interest*, John Wiley, New York, 1986, pp. 19–90.

K. Blau and G.S. King (Eds.), *Handbook of derivatives for chromatography*, Heyden, London, 1977.

R.W. Frei and J.F. Lawrence (Eds.), *Chemical derivatization in analytical chemistry*, Vol. 1, *Chromatography*, Plenum, New York, 1981.

R.W. Frei and J.F. Lawrence (Eds.), *Chemical derivatization in analytical chemistry*, Vol. 2, *Separation and continuous flow techniques*, Plenum, New York, 1982.

D.R. Knapp, *Handbook of analytical derivatization reactions*, John Wiley, New York, 1979.

J.F. Lawrence and R.W. Frei, *Chemical derivatization in liquid chromatography*, J. Chromatogr. Library, Vol. 7, Elsevier, Amsterdam, 1976.

Specialised texts and reviews on derivatisation in gas chromatography

V.G. Berezkin, *Chemical methods in gas chromatography*, J. Chromatogr. Library, Vol. 24, Elsevier, Amsterdam, 1983 (derivatisation and pyrolysis).

W.P.Cochrane, "Application of chemical derivatisation techniques for pesticide analysis", *J. Chromatogr. Sci.*, 1979, **17**, 124–137.

H.B.S. Conacher and B.D. Page, "Derivative formation in the chromatographic analysis of food additives", *J. Chromatogr. Sci.*, 1979, **17**, 188–195.

J. Drozd, "Chemical derivatization in gas chromatography", *J. Chromatogr., Chromatogr. Rev.*, 1975, **113**, 303–356.

J. Drozd, *Chemical derivatization in gas chromatography*, J. Chromatogr. Library, Vol. 19, Elsevier, Amsterdam, 1981.

P. Hušek and K. Macek, "Gas chromatography of amino acids", *J. Chromatogr., Chromatogr. Rev.*, 1979, **113**, 139–230.

W.C. Kossa, J. MacGee, S. Ramachandran and A.J. Webber, "Pyrolytic methylation/gas chromatography: a short review", *J. Chromatogr. Sci.*, 1979, **17**, 177–187.

J.D. Nicholson, "Derivative formation in the quantitative gas chromatographic analysis of pharmaceuticals. A review. Part I", *Analyst*, 1978, **103**, 1–28; "Part II", *ibid.*, 193–222.

A.E. Pierce, *Silylation of organic compounds*, Pierce Chemical Company, Rockford, Il, 1976.

Pierce Chemical Company, *Handbook and general catalogue*, Rockford, Il, 1985–86.

C.F. Poole and A. Zlatkis, "Trialkylsilyl ethers (other than TMS) for gas chromatography and mass spectrometry", *J. Chromatogr. Sci.*, 1979, **17**, 115–123.

C.F. Poole and A. Zlatkis, "Derivatization techniques for the electron capture detector", *Anal. Chem.*, 1980, **52**, 1002A–1016A.

P. Schluz and R. Vîlceanu, "Gas chromatography of acylated α-hydroxy-phosphonate esters. Trace analysis of aliphatic carboxylic acids with the thermionic detector", *J. Chromatogr.*, 1975, **111**, 105–115.

Pyrolysis methods

W.J. Irwin, *Analytical pyrolysis: a comprehensive guide*, Chromatographic Science Series, Vol. 22, Marcel Dekker, New York, 1982.
C.E.R. Jones and C.A. Cramers (Eds.), *Analytical pyrolysis. Third international symposium*, Elsevier, Amsterdam, 1977.
S.A. Liebman and E.J. Levy, "Advances in pyrolysis GC systems: application to modern trace organic analysis", *J. Chromatogr. Sci.*, 1983, **21**, 1–10.
S.A. Liebman and E.J. Levy (Eds.), *Pyrolysis and GC in polymer analysis*, Chromatographic Science Series, Vol. 29, Marcel Dekker, New York, 1985.
R.W. May, E.F. Pearson and D. Scothern, *Pyrolysis-gas chromatography*, Analytical Sciences Monograph, No. 3, Chemical Society, London, 1977.

Headspace analysis

G. Charalambous (Ed.), *Analysis of foods and beverages: headspace techniques*, Academic Press, New York, 1978.
J. Drozd and J. Novák, "Headspace gas analysis by gas chromatography", *J. Chromatogr., Chromatogr. Rev.*, 1979, **165**, 141–165.
H. Hachenberg and A.P. Schmidt, *Gas chromatographic headspace analysis*, Heyden, London, 1977.
B.V. Ioffe and A.G. Vitenberg, *Head-space analysis and related methods in gas chromatography*, John Wiley, New York, 1984.
B. Kolb (Ed.), *Applied headspace gas chromatography*, Heyden, London, 1980.
B. Kolb and P. Pospisil, "A gas chromatographic assay for quantitative analysis of volatiles in solid materials by discontinuous gas extraction", *Chromatographia*, 1977, **10**, 705–711.
A.G. Vitenberg, "Theory of gas chromatographic headspace analysis with pneumatic sampling", *J. Chromatogr. Sci.*, 1984, **22**, 122–124.
T.P. Wampler, W.A. Bowe and E.J. Levy, "Dynamic headspace analyses of residual volatiles in pharmaceuticals", *J. Chromatogr. Sci.*, 1985, **23**, 64–67.

REFERENCES

1. C.C. Sweeley, R. Bentley, M. Makita and W.W. Wells, "Gas–liquid chromatography of trimethylsilyl derivatives of sugars and related substances", *J. Am. Chem. Soc.*, 1963, **85**, 2497–2507.

CHAPTER 8

LIQUID CHROMATOGRAPHY

8.1 LIQUID CHROMATOGRAPHIC METHODS

Separation techniques based on the distribution of the analyte between a liquid mobile phase and either a solid or liquid stationary phase are grouped under the term "liquid chromatography". The stationary phase may bc either an immiscible liquid, a liquid film coated on a solid support, an organic layer chemically bonded to a solid support, an inorganic solid such as silica, or a solid porous organic polymer. The primary practical difference in the operation of liquid chromatography systems, compared to gas chromatography, is that the mobile phase has a much higher density and viscosity and lower diffusion rates (see Table 2.1). Thus for optimum efficiency the particle sizes must be smaller by a factor of about 100. The columns will therefore have a higher resistance to flow and much higher back-pressures but their sample capacities are greatly increased.

The simplest liquid chromatographic technique is column chromatography, which uses fairly large particles of a solid stationary phase with a wide size range (60–500 μm), packed into a vertical tube. The mobile phase moves through the column under gravity to give a separation with only 10–50 theoretical plates. The stationary phase is almost always silica or alumina and each column is discarded after the separation has been completed. The eluents, which are usually organic solvents, are collected in fractions and subsequently analysed by spectroscopy or thin-layer chromatography to determine the components. The separation can be speeded up and made more efficient as "flash chromatography" by using columns packed with uniform 40–63 μm particles and forcing the mobile phase through the column with a low air pressure [1]. In analytical chemistry, these non-instrumental column methods are primarily employed to clean up samples before analysis by the removal of very polar materials or insoluble residues, which might otherwise cause interference. Their principal role has been in synthetic and natural product organic chemistry as a preparative technique for the separation of large (from grams to kilograms) quantities of complex mixtures.

Pure liquid–liquid chromatography using two immiscible liquids, one mobile and one stationary, has not been widely used. If a stationary liquid phase is constrained in a series of narrow tubes, it is possible to pass the mobile liquid phase through the system as droplets moving either upwards or downwards depending on the relative densities of the two phases (Figure 8.1) [2–4]. In this technique of droplet counter-current chromatography (DCC), the components are detected spectrophotometrically as they are eluted. However, DCC is relatively slow, with run times of hours for each sample. The range of samples that can be examined is limited, because there are few totally immiscible combinations of eluents, which will also form discrete droplets. However, as the separation conditions are very mild, DCC has found particular application for the preparative separation of unstable samples, such as plant glycosides, but has not really been used for analytical separations. Other liquid-liquid chromatographic methods have been devised including counter-current chromatography but these are only just reaching commercial availability (Chapter 16).

There are two principal approaches to analytical liquid chromatography. In the first, the sample is separated on a planar stationary phase and the analytes are detected *in situ* without elution. These methods include paper chromatography and thin-layer chromatography (TLC, Chapter 9). However, paper chromatography, in which the mobile phase is carried up or down a sheet of paper by capillary action, suffers from low efficiency and long separation times, often many hours. For almost all purposes, it has now been superseded by TLC or HPLC methods.

In the second approach, the analytes are eluted from a column containing the stationary phase before detection. The concept has led to high-performance liquid chromatography (HPLC, Chapters 10–14), which offers greater selectivity and resolution than traditional column chromatography. In both separation approaches, a wide range of different stationary phase–eluent combinations can be used. These fall into two broad catagories, normal-phase and reversed-phase chromatography, but in some cases the distinction is not clear and often the separation depends on a mixture of the two modes.

(a)

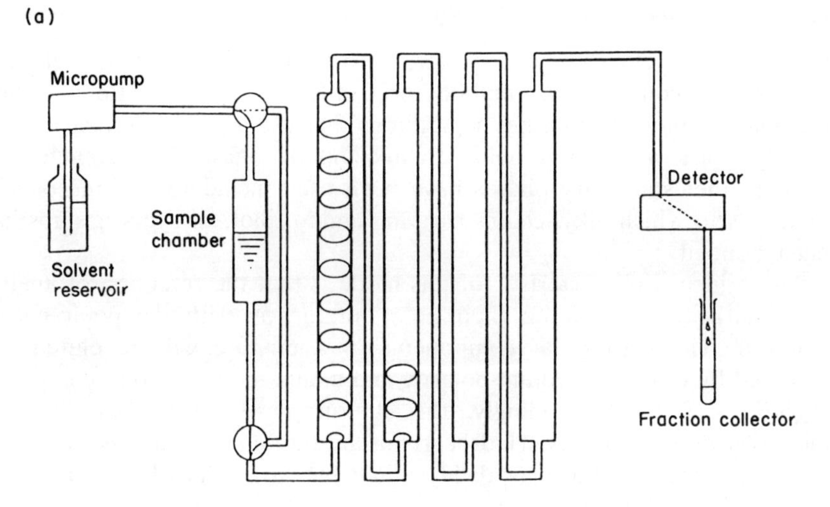

(b)

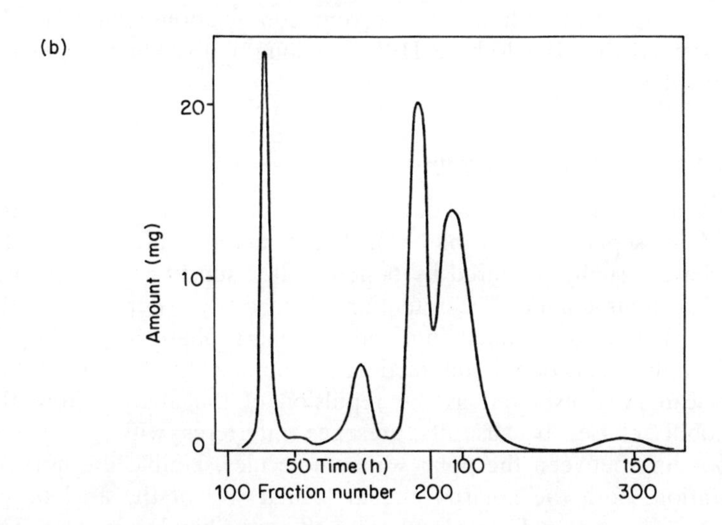

Figure 8.1 (a) Droplet counter-current chromatograph showing droplets of the mobile phase ascending through the stationary liquid phase. (b) Droplet counter-current separation of saponins from *Hovenia dulcis* using 509 tubes (40 cm × 1.65 mm i.d.). The stationary phase is the lower layer from chloroform–methanol–water 35:65:40. The upper phase is used as the ascending mobile phase. (Reproduced by permission of Elsevier Science Publishers from Y. Ogihara *et al.*, *J. Chromatogr.*, 1976, **128**, 219–223.)

a. *Normal-phase chromatography*

Normal-phase chromatography refers to separations carried out using "normal' or conventional methods originally based on traditional column chromatography. It describes separations that use a polar stationary phase, usually silica, and a non-polar organic mobile phase. To a much lesser degree other stationary phases have been used including alumina or silica materials to which alkyl-chains carrying amino, diol, or cyano groups have been bonded.

The principal characteristic of this mode is that the retention of analytes is primarily dependent on their polar interaction with the polar column material. The least polar components of a mixture will be eluted first, followed by those of medium polarity and finally the most polar compounds (Figure 8.2). Correspondingly, the elution power of the mobile phase increases with its polarity. Hydrocarbons are the weakest eluents and elution strength increases through dichloromethane to tetrahydrofuran (THF), acetonitrile and methanol.

As most gravity column and TLC separations are carried out using normal-phase interactions, the methods and solvent combinations can often be readily transferred to normal-phase HPLC as long as the eluent does not interfere with detection.

b. *Reversed-phase chromatography*

In reversed-phase separations, most of the interactions are reversed compared to normal-phase separations (Table 8.1). These methods use a non-polar stationary phase, usually prepared by bonding alkyl substituents to a silica surface, and a polar eluent. The retention of an analyte depends on the degree to which it is partitioned into the stationary phase and is largely determined by the hydrophobic interactions of the analyte with the mobile phase. This can be considered as the repulsion of the analyte from the aqueous mobile phase, because its presence interferes with the stable hydrogen bonding between the polar water molecules. Unlike the normal-phase separation, both the polarity and molecular size of the analyte are important and large non-polar molecules are the most highly retained. The order for similar-sized polar molecules is the reverse of that on a normal-phase column (Figure 8.3).

Because molecular size plays a role, this method is particularly suitable for the separation of homologues and low-polarity compounds (i.e. benzene and toluene in Figure 8.4). As homologues contain the same functional groups, they would have the same polar interactions and are often poorly resolved in normal-phase chromatography.

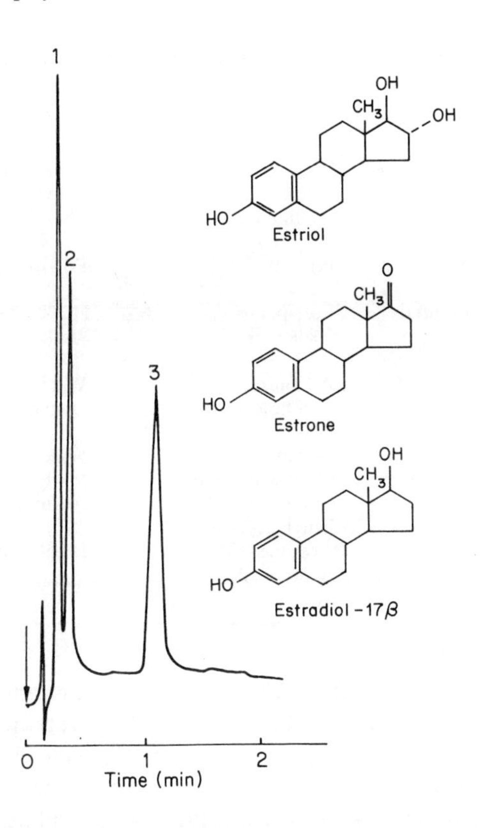

Figure 8.2 Normal-phase separation of estrogens. Peaks: 1, estrone; 2, estradiol; 3, estriol. Column LiChrosorb Si 60, 10 μm, 250 mm × 4.6 mm. Eluent n-heptane–ethanol 95:5. Ultraviolet spectroscopic detection at 280 nm. (Reproduced with permission from E. Merck Darmstadt.)

In the early days of HPLC, non-polar stationary liquid phases, such as β,β′-oxydipropionitrile coated on silica, were used but these tended to be unstable and could be washed out of the column by the eluent. They have largely been replaced by the highly stable alkyl-bonded silicas, particularly octadecylsilyl- (ODS) and phenyl-bonded silicas (see Chapter 12). More recently, homogeneous polymeric column materials, including polystyrene–divinylbenzene (PS–DVB), have also become available. In addition, cyano- and amino-bonded silica columns will behave in a reversed-phase mode when used with an aqueous eluent.

Water is the weakest mobile phase and gives the greatest retentions. The elution strength increases through methanol, acetonitrile and THF. In most

Table 8.1 Comparison of normal- and reversed-phase separation methods.

	Normal phase	Reversed phase
Column packing	Polar	Non-polar
Column material	Silica Alumina	Alkyl-bonded silica PS–DVB
Separation interaction	Polarity	Polarity and size
Most rapidly eluted compounds	Low-polarity alkanes	High-polarity salts
Eluents: weak	Alkanes	Water
strong	Alcohols	Dichloromethane
Elution gradient increasing strength	Hexane Dichloromethane Chloroform Ethyl acetate Acetone Acetonitrile Methanol	Water Methanol Acetonitrile THF Dichloromethane

cases, mixtures prepared from these three solvents and water or a buffer solution are used as eluents. For very strongly retained hydrocarbons or lipids, non-aqueous eluents, such as dichloromethane, have also been used.

For many reasons, primarily because of their high reproducibility and stability, coupled with wide versatility, reversed-phase separations have become the method of choice for most separations by HPLC, but normal-phase separations on silica still dominate TLC.

8.2 RELATED SEPARATION TECHNIQUES

The equipment and technology of HPLC (and to some extent TLC) can also be used to carry out related separation methods, which although often termed "chromatography" are not strictly based on the distribution of an analyte between two phases.

a. Ion-exchange chromatography

In this technique the stationary phase is an inert support, which is coated with chemically bound acidic (RSO_3H or ArOH) or basic (NR_4^+) groups. These groups interact with ionisable analytes of the opposite charge depending on the pH and competing ions in the eluent. Elution can be

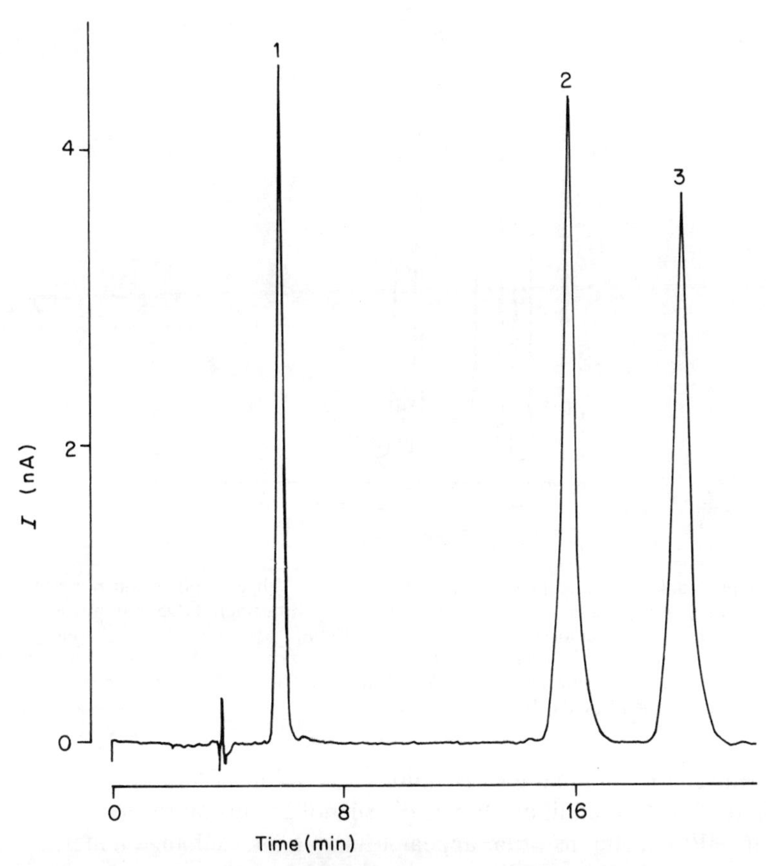

Figure 8.3 Reversed-phase separation of estrogens. Peaks: 1, estriol; 2, estradiol; 3, estrone. Column LiChrosorb RP-18, 5 μm, 250 mm × 4.6 mm. Eluent acetone–water 50:50 with lithium perchlorate and acetic acid at 0.7 ml min^{-1}. Electrochemical detection on a glassy carbon detector at potential of 1.0 V. (Reproduced by permission of Metrohm Ltd.)

achieved by displacement of the analyte from the ionic groups or by the formation of a stronger complex with the analyte in the mobile phase. The method has found application as an analytical technique for the separation of transition-metal ions (Figure 8.5), lanthanides, actinides, nucleic acids (Figure 8.6), amino acids and peptides.

Although a popular method in the early days of HPLC, the separations often gave poor efficiencies. It was difficult to prepare columns with the relatively soft ionic polymers that were then available. More recently, specialised ion-exchange media based on rigid supports have been developed

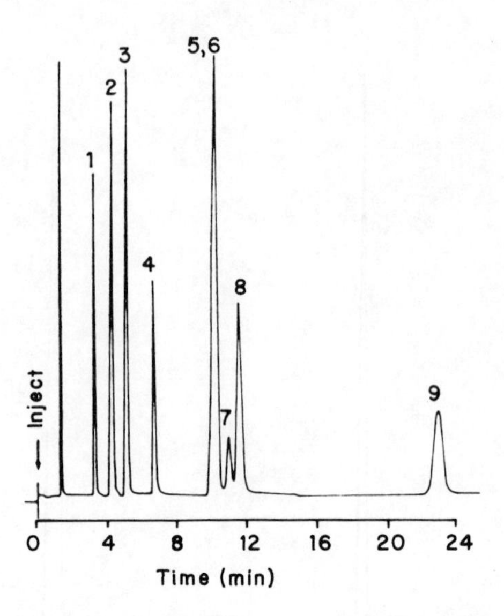

Figure 8.4 Reversed-phase separation of a wide range of polar and non-polar aromatic compounds and homologues. Column Ultrasphere-ODS 150 mm × 4.6 mm. Eluent methanol–water 50:50 at 1.0 ml min^{-1} and 30°C. Peaks: 1, phenol; 2, benzaldehyde; 3, acetophenone; 4, nitrobenzene; 5, methyl benzoate; 6, anisole; 7, fluorobenzene; 8, benzene; 9, toluene. (Reproduced by permission of Beckman Instruments, Inc.)

and are used for the separation of anions and cations in "ion chromatography' (Section 12.5a). In addition, the acidic silanol groups on the surface of silica used in HPLC columns often appear to act as ion-exchange materials. This is often a disadvantage in the separation of basic compounds as it causes peak tailing. However, some separation methods make use of this interaction with appropriate eluents to give good separations of basic drugs (Section 12.5).

b. Size exclusion chromatography

In this analytical technique, analytes are separated on a porous column packing material on the basis of their molecular size. In its original form, gel permeation chromatography, the column was packed with a soft polymeric gel, such as Sephadex, and an aqueous eluent was run through under gravity. The method was widely used to separate biological macromolecules, such as proteins, peptides, etc. The analytes do not interact with the stationary phase and the separation is not dependent on a distribution process as in other chromatographic methods. The smallest molecules in the sample can penetrate all the pores of the gel, whereas medium-sized molecules can only

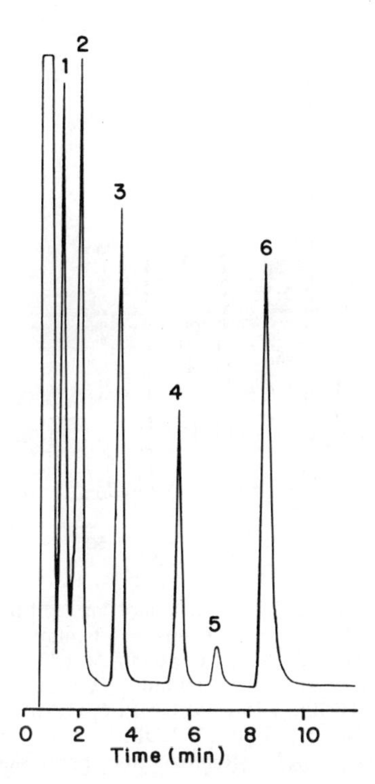

Figure 8.5 Separation of transition-metal ions on a cation ion-exchange column with conductometric detection. Column Bio-Gel TSK IC-Cation SW. Eluent 2.5 mM ethylenediamine–8.5 mM citric acid at 1 ml min^{-1} and 30°C. Peaks: 1, Cu, 5 ppm; 2, Ni, 5 ppm; 3, Co^{2+}, 5 ppm; 4, Fe^{2+}, 5 ppm; 5, Cd^{2+}, 5 ppm; 6, Ca^{2+}, 20 ppm. (Reproduced by permission of BioRad Laboratories from *Biorad catalog*.)

enter some pores and the largest molecules are completely excluded from the gel. Thus the largest molecules are eluted first, as they pass through the smallest interior volume of the column, whereas the smallest molecules, which can penetrate most of the volume of the column, have an effectively larger volume to pass through and are eluted last. All the analytes are thus eluted within one column volume and the separation capacity is limited.

In recent years, rigid polymers and inorganic matrices have been developed that can be used with aqueous or organic solvents and are stable at the flow rates used in HPLC. This has led to more efficient columns and readily enables organic polymers and proteins to be separated and quantified (Figure 8.7) (and see Section 12.6).

Because they have largely been used in the relatively specialised areas of biological and synthetic polymers, size exclusion methods will not be

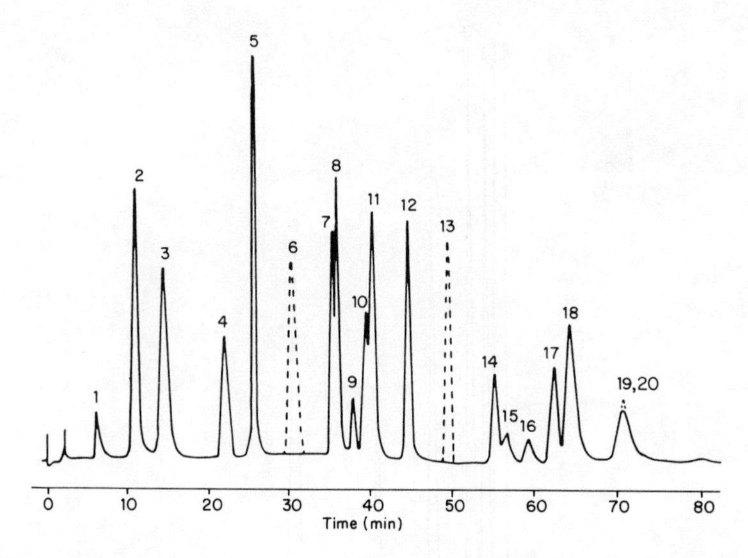

Figure 8.6 Ion-exchange separation of mono-, di– and triphosphate nucleotides on Partisil-10 SAX, 250 mm × 250 mm × 4.6 mm (strong anion exchanger ($-NR_3^+$)). Eluent: linear gradient from 0.007 M KH_2PO_4, pH 4.0, to 0.25 M KH_2PO_4, 0.50 M KCl, pH 4.5, at 1.5 ml min^{-1}. UV detection at 254 nm. Peaks: 1, CMP; 2, AMP; 3, UMP; 4, IMP; 5, GMP; 6, (XMP); 7, dTDP; 8, UDP; 9, CDP; 10, IDP; 11, ADP, 12, GDP; 13, XDP; 14, UTP; 15, dTTP; 16, CTP; 17, ITP; 18, ATP; 19, GTP; 20, XTP. (Reproduced by permission of Whatman.)

discussed in detail. The main methods and applications have been extensively reviewed in recent years (see Bibliography and Appendix 1).

8.3 PREPARATIVE SEPARATIONS

Most analytical instrumental chromatography methods can be scaled-up for the separation and purification of individual compounds. Usually the main change required is an increase in the capacity of the stationary phase, which can be achieved by increasing the cross-sectional area of the column. In TLC, thicker plates (up to 1 mm) or longer glass sheets (20 cm × 1 m) enable samples in the milligram range to be purified. After separation, the bands are detected using a non-destructive technique. The stationary phase is scraped off the plate and the compounds are extracted with a suitable solvent. Some adaptations of TLC, such as radial and centrifugal chromatography, are designed as continuous flow systems so that the bands are eluted from the plates for collection.

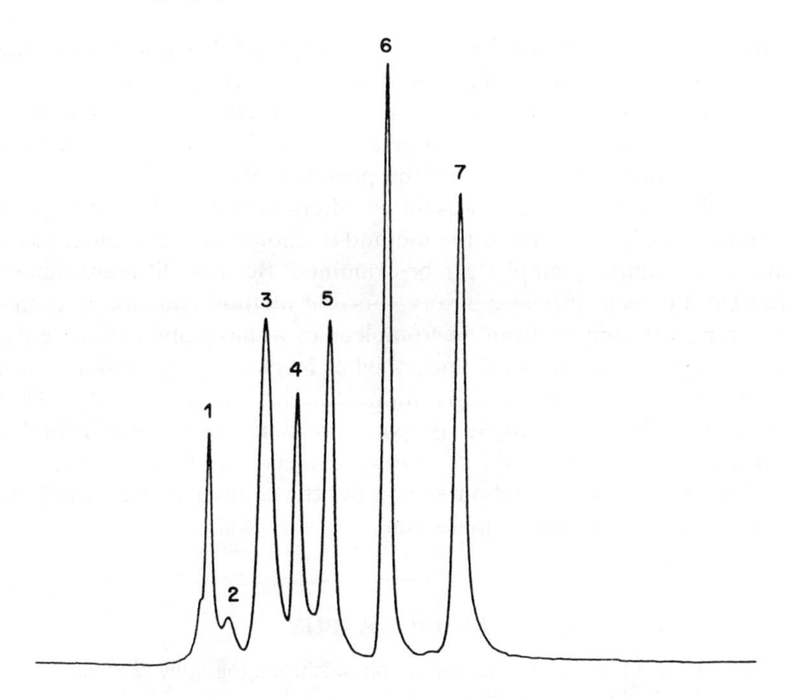

Figure 8.7 Separation of protein standards by gel filtration chromatography on Zorbax Bio Series FP-250 column. Mobile phase 0.2 M K_2HPO_4, pH 7, 0.1% NaN_3. UV detection at 280 nm. Ambient temperature. Peak identification: 1, thyroglobulin; 2, impurity; 3, IgP; 4, bovine serum albumin; 5, ovalbumin; 6, myoglobin; 7, Iysozyme. (Reproduced by permission of E. I. DuPont.)

In HPLC, columns with diameters of 25 mm i.d. and above have been used for preparative separations. A few manufacturers have designed specialised preparative chromatographs with high-capacity eluent pumps for use with columns up to 50 mm diameter. These systems can separate up to 1–2 g per injection, by cycling the sample through the column more than once. Industrial process-scale systems have also been built with large columns, 10–20 cm diameter, which are capable of separating 100–1000 g of sample on each run with a flow rate of 1–8 l min^{-1}.

8.4 SELECTION OF AN ANALYTICAL METHOD

The selection of TLC or HPLC as the most appropriate analytical method for a particular sample depends on a number of factors. Both methods have the advantage that there are very few limitations, other than the requirement that the sample must be soluble in the mobile phase. For qualitative analyses of fairly simple mixtures, TLC is often the preferred method. It uses generally low-cost equipment and can separate multiple samples in one run, so has the advantage of speed and economy. It is ideal for monitoring chemical reactions and testing for the presence of impurities.

However, accurate quantification is often difficult with TLC and for quantitative analyses HPLC is the method of choice. It is also more efficient, so that more complex samples can be examined. Because different separation modes can be used, it is also a very versatile method and can be used with a wide range of samples from macromolecules to inorganic ions. The capital costs are higher than for TLC and a higher level of operator skill is usually needed. As reversed-phase separations can be carried out directly with aqueous samples, the sample preparation time with aqueous/biological samples can be greatly reduced. However, each analysis generally takes 15–30 min, so sample throughput is slow but the equipment can be automated for prolonged or overnight operation.

BIBLIOGRAPHY

For more details of reversed- and normal-phase chromatography see Chapter 12 and for details of TLC see Chapter 9.

General texts

O. Mikeš, *Laboratory handbook of chromatographic and allied methods*, Ellis Horwood, Chichester, 1979.

Preparative chromatography

B.A. Bidlingmeyer (Ed.), *Preparative liquid chromatography*, J. Chromatogr. Library, Vol. 38, Elsevier, Amsterdam, 1987.

K. Hostettmann, M. Hostettmann and A. Marson, *Preparative chromatography: applications in natural product isolation*, Springer Verlag, Berlin, 1986.

K.K. Unger and R. Janzen, *J. Chromatogr.*, 1987, **373**, 227–264.

Ion-exchange chromatography

For more recent studies on the role of ion-exchange chromatography for the separation of anions and cations as ion- chromatography see Chapter 12.

H.F. Walton (Ed.), *Ion-exchange chromatography*. Benchmark papers in analytical chemistry, No. 1, Dowden, Hutchinson and Ross, Stroudsburg, PA, 1976.

Size exclusion chromatography

T. Kremmer and L. Boross, *Gel chromatography: theory, methodology, application*, John Wiley, New York, 1979.

W.W. Yau, J.J. Kirkland and D.D. Bly, *Modern size-exclusion liquid chromatography. Practice of gel permeation and gel filtration chromatography*, John Wiley, Chichester, 1979.

REFERENCES

1. W.C. Still, M. Kahn and A. Mitra, "Rapid chromatographic technique for preparative separations with moderate resolution", *J. Org. Chem.*, 1978, **43**, 2923–2925
2. T. Tanimura, J.J. Pisano, Y. Ito and R.L. Bowman, "Droplet counter-current chromatography", *Science*, 1970, **169**, 54–56.
3. Y. Ogihara, O. Inoue, H. Otsuka, K.-I. Kawai, T. Tanimura and S. Shibata, "Droplet counter-current chromatography for the separation of plant products", *J. Chromatogr.*, 1976, **128**, 218–223.
4. G.T. Marshall and A.D. Kinghorn, "Isolation of phorbol and 4α-phorbol from croton oil by droplet counter-current chromatography", *J. Chromatogr.*, 1981, **206**, 421–424.

CHAPTER 9

THIN-LAYER CHROMATOGRAPHY

9.1 SEPARATIONS ON THIN-LAYER CHROMATOGRAPHY

Thin-layer chromatography is almost entirely a technique for the analysis of organic compounds. Its principal advantages are that it is simple and quick to carry out and expensive capital equipment and extensive practical experience are not required. The stationary phase is a thin layer of a solid spread on a glass, plastic, or metal plate. The samples in solution are applied as spots about 1–2 cm from the bottom edge of the dry plate. After the sample solvent has evaporated, the plate is placed in a closed tank so that the bottom edge is immersed in the mobile phase, which is then drawn up the plate by capillary action (Figure 9.1a). When the solvent front approaches the top of the plate, the plate is withdrawn from the chamber and the eluent is allowed to evaporate. The samples that remain as spots on the dry plate are detected either visually or by indirect visualisation after chemical treatment to give coloured spots. Unlike column chromatographic methods, the analytes are not eluted from the stationary phase before detection and the stationary phase is used only once and is then discarded.

The retentions of the analytes are calculated as the distances they have moved from the origin, compared to the distance moved by the solvent front. The retentions are expressed as R_f (relative to front) values based on the origin being 0 and the solvent front being 1.00 (Figure 9.2). The R_f value of an analyte in a particular eluent is characteristic of that compound but careful control is needed to obtain reproducible results in separate

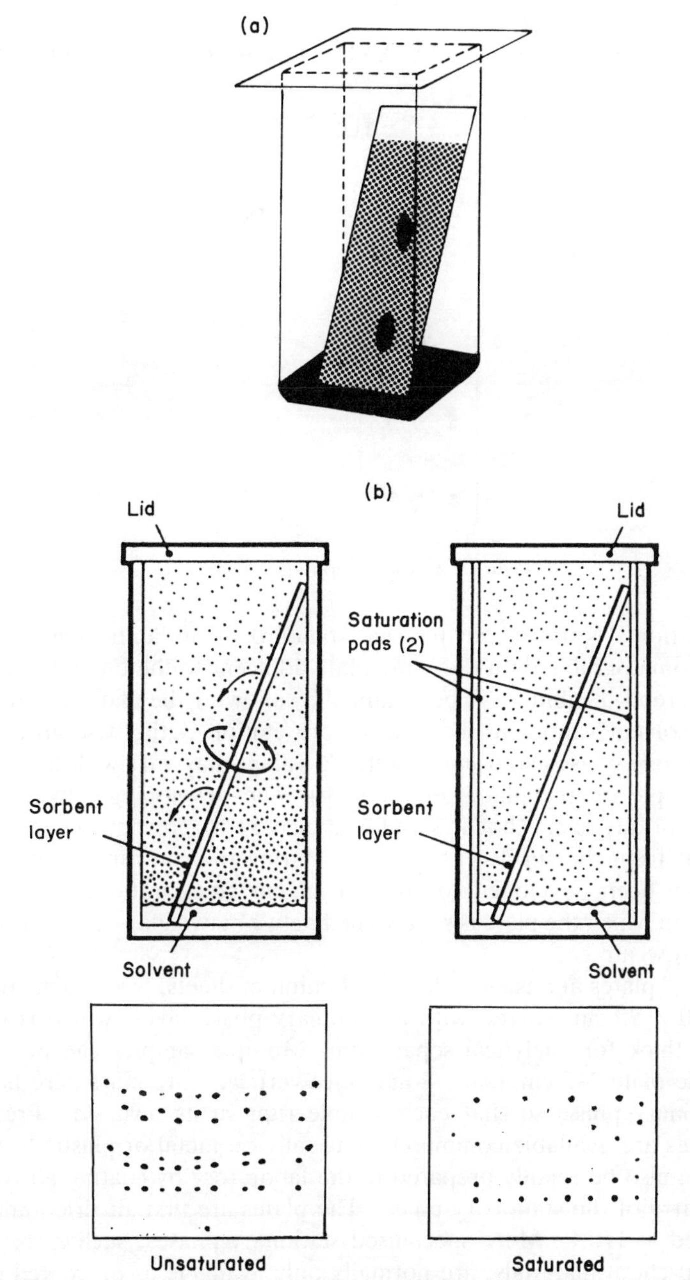

Figure 9.1 (a) TLC system tank and plate showing solvent rising up the plate by capillary action. (b) Effect of partial saturation and complete saturation on the separation on a TLC plate showing irregular R_f values when evaporation has occurred from the surface of the plate. (Reproduced by permission of Analtech. Inc.)

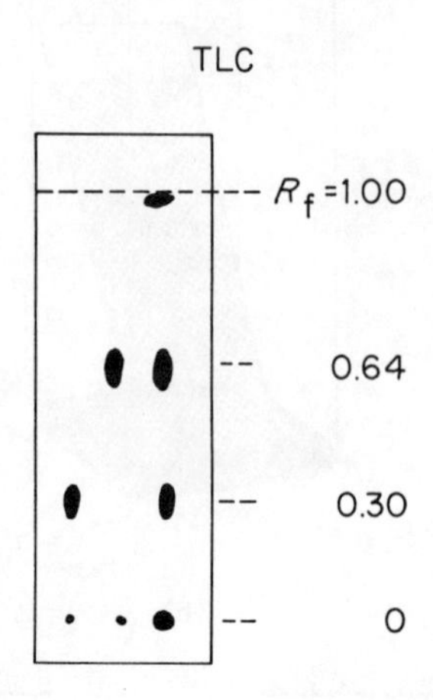

Figure 9.2 Calculation of R_f values.

determinations. Normally an internal standard or authentic sample would be co-chromatographed on the same plate adjacent to the unknown sample. Improved reproducibility can be obtained by allowing the plate to equilibrate with the solvent vapour in the tank before elution. Otherwise an irregular flow of mobile phase can occur and the R_f values may vary with the position of the sample on the plate (Figure 9.1b). The developing tank can be a container of any shape but it should be airtight to prevent evaporation of the eluent from the surface of the plate during the run. Special designs of tanks have been developed to enhance eluent equilibration, including U-chambers in which the plate is sandwiched between two glass sheets separated by a narrow air gap.

The TLC plates are usually glass or aluminium sheets, 5 × 20 cm, 10 × 20 cm, or 20 × 20 cm, coated with a stationary phase layer, which is usually 0.25 mm thick for analytical separations. Multiple samples can be applied to a single plate 1–2 cm apart. Sometimes vertical score marks are made in the stationary phase so that each sample runs in its own slot. Precoated TLC plates are available commercially usually on metal or plastic backings. Plates can also be readily prepared in the laboratory by coating glass plates with a slurry of the stationary phase. The plates are first air dried and then oven dried at 110°C. More specialised stationary phases, such as reversed-phase and chiral materials, are normally only available as precoated plates. A number of specialised designs of plates have also been produced, including plates with a zone of the less active Kieselguhr along the bottom 1–2 cm. Samples placed on this strip are rapidly swept onto the silica layer, where they form a sharp starting band for the separation (Figure 9.3). These plates

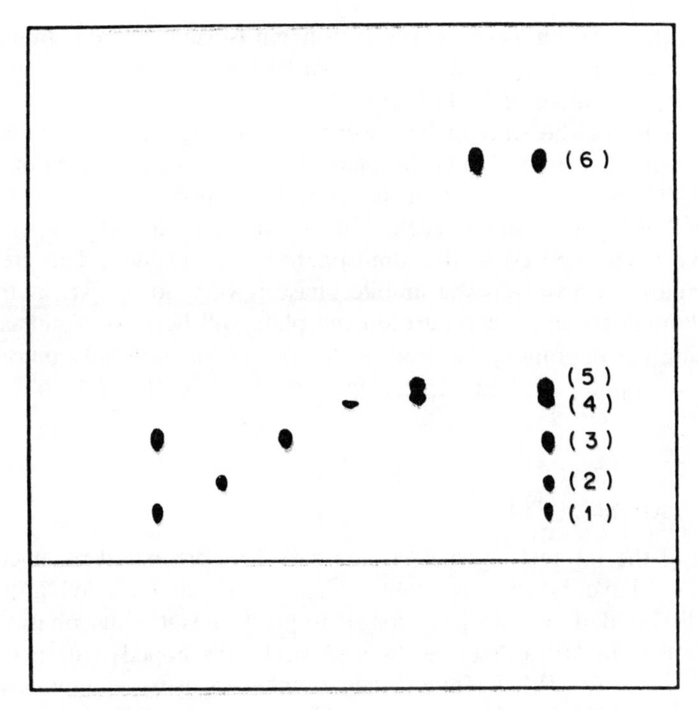

Figure 9.3 Separation of food dyes on silica TLC plate. Eluent butanol–methyl ethyl ketone–water–conc. ammonia 50:38:18:2 v/v. Spots: 1, Yellow No. 5; 2, Red No. 2; 3, Blue No. 1; 4, Red No. 40; 5, Violet No. 1; 6, Red No. 3. (Reproduced by permission of Whatman.)

are also useful for very impure samples of crude plant extracts or food samples as any insoluble residues are left in the concentration zone and do not interfere with the separation.

a. *Adsorption-phase TLC*

Almost all TLC analyses are carried out on silica (10–60 μm) stationary phases using organic solvents as the eluent. The analytes are primarily separated according to their polarity in a normal-phase separation. The silica layer can contain a small amount of gypsum as a binder to increase the mechanical strength (silica gel G) but for some separations this is omitted (silica gel H). In order to aid detection of the samples, an insoluble fluorescent indicator is often incorporated into the plates (silica gel GF_{254}). When the plates are placed under a mercury lamp with a 254 nm filter, the background glows green except where analytes absorb the radiation.

Occasionally a 360 nm active fluorescent agent is used. Some manufacturers also offer a coarser grade of silica (silica P) for the production of thicker (1–2 mm) preparative-scale TLC plates.

The activity of the silica plates is critically dependent on their moisture content. For consistent results the plates should be activated before use at about 110°C for an hour and then stored in desiccator cabinets. Commercial precoated plates may need reactivating by heating to this temperature if they have been exposed to the atmosphere for some time. This treatment is particularly necessary if the mobile phase is very non-polar, as then the effect of small traces of moisture on the plate will be most significant.

Alumina is sometimes also used as an absorption stationary phase but is much less popular and presents some problems in the detection of the analytes.

b. Reversed-phase TLC

Because of the interest in reversed-phase HPLC, precoated reversed-phase TLC plates have become available. C_{18}, C_8, C_2 and phenyl groups are chemically bonded onto a silica support to produce stationary phases similar to those used in HPLC (see Section 12.3). It was hoped that these plates could be used for HPLC method development or as an economic method of duplicating HPLC-type separations. However, in a column the whole of the stationary phase is equilibrated with the mobile phase before the separation commences. In contrast, with TLC the eluent meets a dry stationary phase and many of the separations correspond poorly to HPLC separations. These plates have not gained wide acceptance for this application, although they have generated applications in their own right for which they are preferred to alternative techniques (Figure 9.4).

c. High-performance TLC

The resolution of any chromatographic system is very dependent on the particle size of the stationary phase. This has led to high-performance TLC (HPTLC) plates precoated with a finer particle size silica (5 μm) with a much closer size range, similar to that used in packed HPLC columns. On elution the spots show less lateral spreading and hence the concentration in the spot is higher, which makes the analyte easier to detect and increases the sensitivity. The mobile phase runs more slowly on these plates but the resolution is improved so that a short 10 cm plate is usually used. The high quality control of the preparation of the plates also results in better

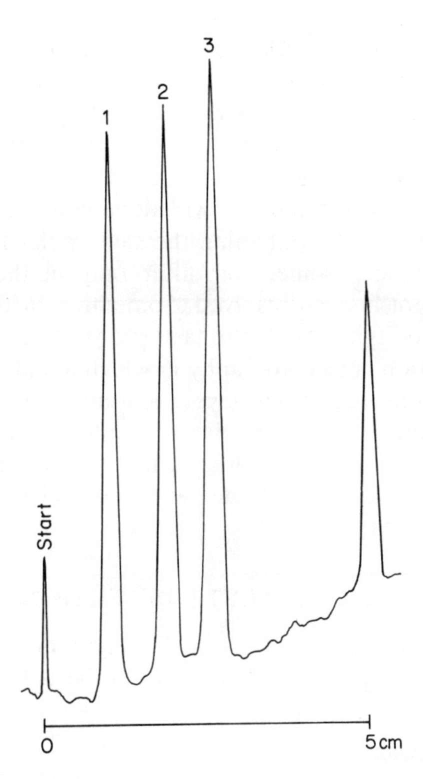

Figure 9.4 Densitometer trace of the separation of steroids on a reversed-phase HPTLC plate. Eluent acetone-water 60:40. Samples: 1, methyltestosterone; 2, Reichstein S; 3, hydrocortisone. (Reproduced by permission of E. Merck, Damstadt.)

reproducibility and these plates have particularly found application for quantitative analyses.

d. Mobile phases

The mobile phase used in TLC depends on the samples under study. Most separations use a polar silica stationary phase and retentions are primarily dependent on a polar–polar interaction between the silica and the sample. The elution power of the mobile phase therefore increases with its polarity, hydrocarbons being weakest, then halogenated compounds, ethers, esters and ketones, and finally alcohols being the strongest eluents. Water is used only rarely as it runs very slowly and tends to cause the stationary phase to lift off the supporting plate. Very frequently mixtures of solvents are used to achieve intermediate polarities. The eluent should be chosen so that the analytes have R_f values between 0.3 and 0.7, as this gives the maximum resolution. For very polar samples that can ionise, such as amines or acids, traces of ammonia or acetic acid are often included in the mobile phase to reduce tailing by suppressing ionisation.

A polar grouping in the analyte will dominate its retention properties. Consequently, compounds containing the same polar functional group, but which are homologues, isomers or differ only in the number of double bonds, will often not be well resolved. Sometimes increased separation can be achieved by modifying the stationary phase. The most commonly used method is argentation chromatography in which about 10–15% silver nitrate is incorporated into the silica layer. Unsaturated compounds will be preferentially retained on these plates as the silver ions form ionic π-complexes with the double bonds. This method has found particular application in the separation of the saturated and unsaturated fatty acids and lipids (Figure 9.5).

9.2 ANALYTE DETECTION

Unlike GLC or HPLC, the analytes on TLC remain on the stationary phase at the end of the analysis and must therefore be detected *in situ*.

a. *Direct examination*

Relatively few organic compounds are coloured, so direct visual examination to detect the analytes is not usually applicable. If a fluorescent indicator has been incorporated into the stationary phase, the plate can be examined under ultraviolet light. The plate should fluoresce green with dark spots where analytes are present which either absorb ultraviolet light or quench fluorescence. These compounds include most aromatic and conjugated organic compounds and some unsaturated compounds. However, the estimation is visual and unless matching standards have also been chromatographed it is difficult to quantify the components. In addition, naturally fluorescent compounds such as riboflavin or polynuclear aromatic hydrocarbons may give blue or orange fluorescent spots. For these samples it may be preferable to use a fluorescence-indicator-free TLC plate so that the analyte emission is not obscured.

b. *General detection reagents*

If the analytes cannot be visualised directly, the TLC plate will have to be treated chemically to produce a coloured or fluorescent spot, by spraying

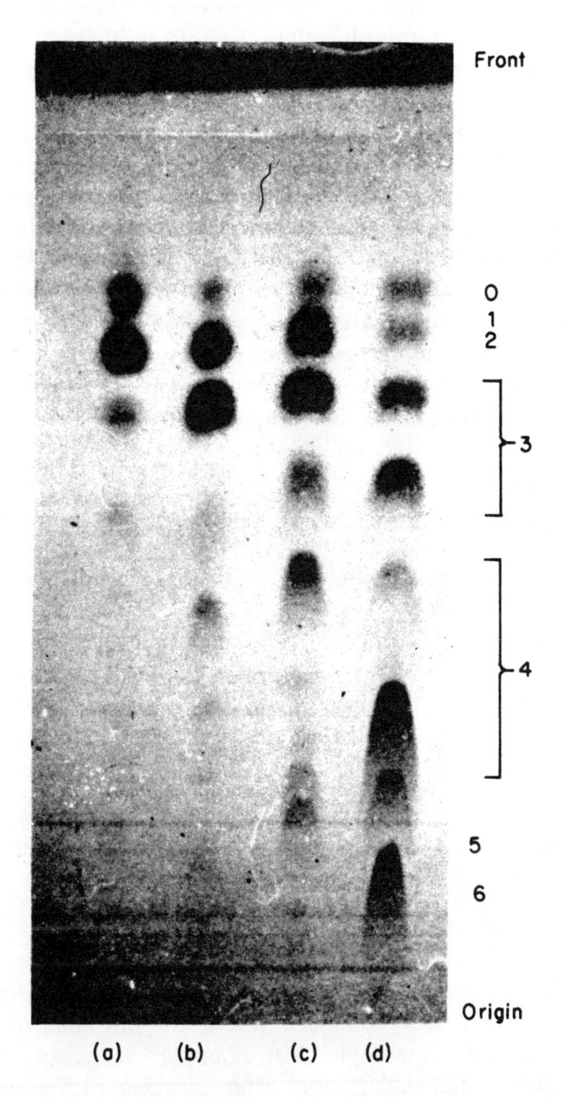

Figure 9.5 Argentation TLC of natural triglycerides on silica gel impregnated with 5% silver nitrate: (a) palm oil; (b) olive oil; (c) groundnut oil; (d) cottonseed oil. Eluent isopropanol–chloroform 1.5:98.5 v/v. Spots were located by spraying with sulphuric acid and charring. The numbers represent the total number of double bonds in each triglyceride molecule. (Reproduced by permission of Chapman and Hall from Gurr and James, *Lipid biochemistry*, p. 101.)

with either a general or specific reagent. Many of these reagents react with the analytes and this method cannot therefore be used for preparative separations, unless only a representative strip along the edge of the plate is treated.

A common non-destructive method is to place the dried plate in a tank containing iodine vapour. Almost all organic compounds are stained to produce brown spots on a pale brown background. Different compounds may behave in different ways and some saturated alkanes do not respond. This method has the advantage that the staining is reversible. The spots can be marked and the iodine allowed to evaporate. The spots can then be scraped off the plate to isolate the analytes or a further reagent can be added. Particular care should be taken to dry the plate before treatment with iodine vapour if acetone is a component of the mobile phase as lacrymatory iodoacetones may be formed.

Another general method of detection is to spray the plate with concentrated sulphuric acid and then to heat it in an oven or under an infrared lamp. Almost all organic compounds will be charred to give stable clearly defined brown spots on a white background (see Figure 9.5). Sulphuric acid will not work as a detection method with alumina TLC plates because they neutralise the reagent.

An alternative general spray reagent is a solution of 5% anisaldehyde in sulphuric acid, which on heating often gives characteristically coloured spots with different chemical groupings. The reagent should be freshly prepared as on prolonged storage decomposition occurs and the colour responses may differ.

c. *Selective detection reagents*

For many analytical purposes a specific reagent that responds to only one class or type of analyte is very useful. Typical reagents form coloured derivatives with the analytes of interest, such as the reaction of 2,4-dinitrophenylhydrazine hydrochloride with ketones and aldehydes to give orange spots. Alternatively one can use a reagent that responds to the acidity or basicity of the sample. For example, acidic analytes appear as yellow spots on a green background after spraying with a slightly alkaline solution of bromocresol green.

Specialised reagents have also been developed for particular groups of compounds (Table 9.1). Basic analytes, including alkaloids and many drugs, will give characteristic purple spots after spraying with Dragendorff's reagent (bismuth subnitrate and potassium iodide) and morphine alkaloids give

Table 9.1 Detection methods in TLC.

Method	Application	Response (spots)
General reagents		
Iodine vapour	All organic compounds	Brown
Conc. sulphuric acid	All organic compounds	Black char
Sulphuric acid–anisaldehyde	Many organic compounds	Coloured spots
Selective reagents		
Dragendorff's reagent (bismuth subnitrate/KI)	Alkaloids and bases	Coloured
Iodoplatinic acid	Bases	Purple-black
Fuchsin	Aldehydes	Magenta
Silver nitrate–base	Aldehydes	Brown
2,4-Dinitrophenylhydrazone	Carbonyl compounds	Orange
Bromocresol green	acids	Yellow on green
Ninhydrin	Primary amines	Purple
	Secondary amines	Brown
Fluorescamine	Primary amines	Fluorescent spots
Mercuric sulphate diphenylcarbazone	Barbiturates	Violet

characteristic colours with iodoplatinic acid. More details are given in the texts listed in the Bibliography.

d. Instrumental detection

In order to improve the precision of detection, instrumental methods can be employed to measure the intensity of the spots, but this often requires costly equipment, which reduces the economic advantages of TLC compared with HPLC. Most methods are based on the use of transmission or reflectance densitometry (Figure 9.6). Transmission methods are hampered by light scattering in the stationary phase and, unless great care is taken, hand-made plates are often not sufficiently reproducible to enable accurate compensation to be made. Methods based on reflectance are easier to use as the scattering of the light beam is largely independent of the plate thickness. Spectroscopic detection can be used either directly or after the analytes have been revealed by a spray reagent. Both ultraviolet–visible spectroscopic and fluorescence spectroscopic methods are used.

A widely used application of direct fluorescence detection is the monitoring of the highly carcinogenic aflatoxins. These are formed on stored grain products by the mould *Aspergillus flavens* (Figure 9.7). In this case the

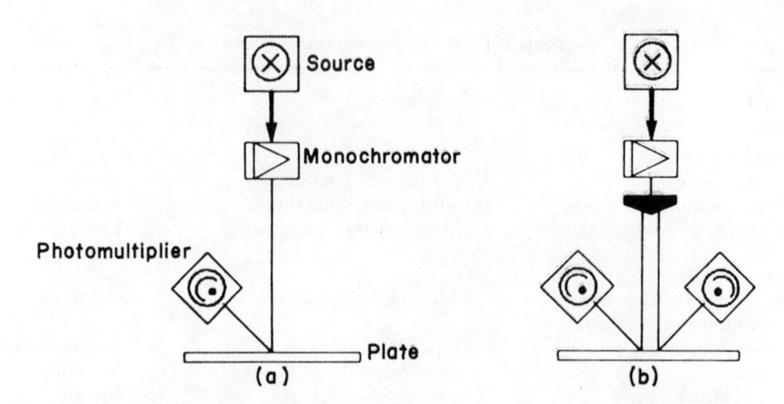

Figure 9.6 Different optical configurations of a densitometer for reflectance spectroscopic scanning of a TLC plate. (Reproduced by permission of Aster Publishing Corp. (*LC Liq. Chromatogr. HPLC Mag,* **1**, *28 (1983).)*

ability to run multiple samples on a single TLC plate enables screening to be carried out more rapidly than with HPLC.

Radioactive analytes can be detected either by scanning with a Geiger counter or by autoradiography. These methods are particularly useful to monitor metabolites from labelled drugs, as the sensitivity can be readily increased by increasing the detection time. Unlike HPLC, all the active components of the sample can be located, even those which remain on the origin but would have been lost on the top of a column.

9.3 SAMPLES AND APPLICATIONS

TLC can be used to separate a wide range of compounds with very few restrictions (for applications see Appendix 1). The only compounds that cannot be examined are those which are insoluble or are so volatile that they would be lost on drying the plate. TLC has found its widest application in organic chemistry, primarily as a qualitative method for the monitoring of organic syntheses or for small-scale preparative separations.

As an analytical technique, it is primarily useful as a screening method for multiple samples and for limit tests where the presence or absence of the analyte is the main criterion. These include drug screening in forensic studies and the test described earlier for aflatoxins. As a test method for sample homogeneity, TLC has the advantage compared to HPLC or GLC that impurities which are very strongly retained and remain on the origin can be detected, whereas with the column instrumental methods these compounds would remain undetected on the column.

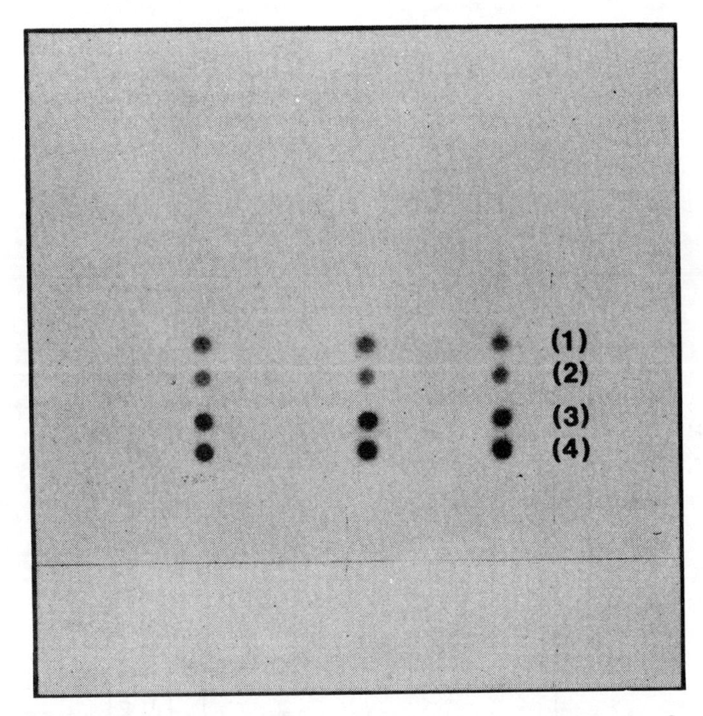

Figure 9.7 Separation and detection of aflatoxins from *Aspergillus flavens*. Whatman LHP-K 10 cm × 10 cm plate. Samples applied to preadsorbent strip. Mobile phase chloroform–diethyl ether 70:30 v/v. Detection by fluorescence under long-wavelength UV light. Sensitivity at least 5 ng. Samples: 1, aflatoxin, B_1, $R_r = 0.47$; 2, aflatoxin B_2, $R_r = 0.39$; 3, aflatoxin G_1, $R_r = 0.31$; 4, aflatoxin G_2, $R_r = 0.24$. (Reproduced with permission from *Whatman chromatography folio*, No. 2.)

a. *Qualitative analysis and identification*

There are many variables that affect the retention of an analyte on TLC. Unless care is taken, the activity of the plates can differ because of changes in the moisture content. The composition of the eluent in the tank may change with time as the more volatile constituents are lost by evaporation. Consequently, R_f values are only poorly reproducible, and for identification studies a reference sample should be chromatographed on the same plate or the retentions should be compared to reference standards. The discrimination power of TLC is limited and related compounds particularly if they contain the same polar group, may be indistinguishable, so that care must be taken to avoid misidentification.

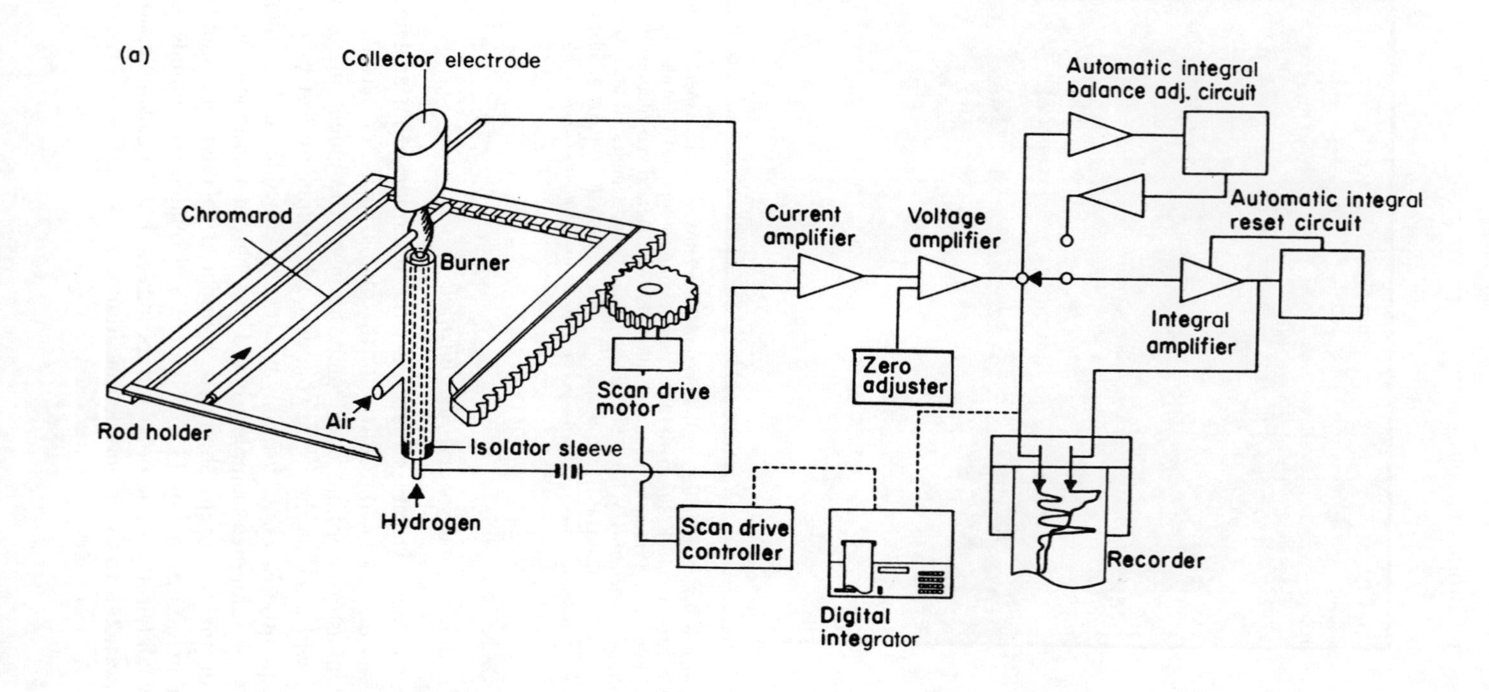
(a)
Collector electrode
Chromarod
Burner
Rod holder
Air
Isolator sleeve
Hydrogen
Scan drive motor
Scan drive controller
Current amplifier
Voltage amplifier
Zero adjuster
Digital integrator
Automatic integral balance adj. circuit
Automatic integral reset circuit
Integral amplifier
Recorder

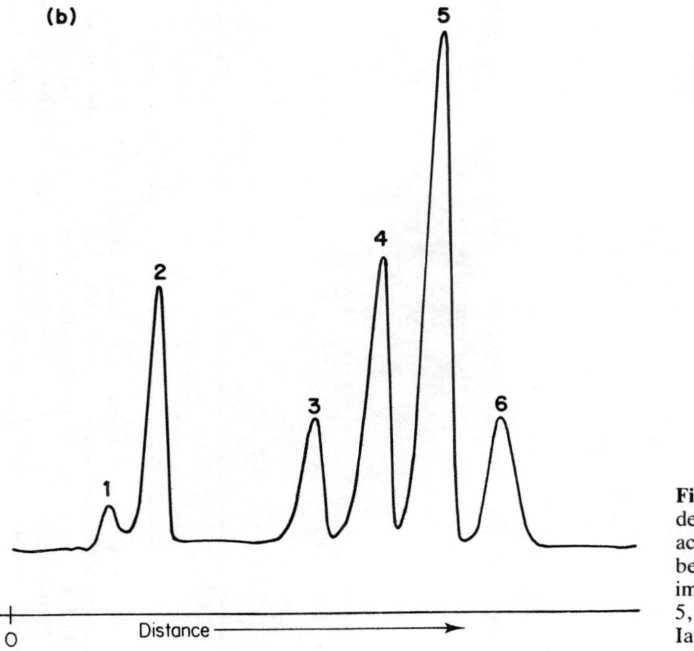

Figure 9.8 (a) Chromarod chromatograph system showing detection with flame ionisation detector. (b) Separation of fatty acids and glycerides on Chromarod system. Mobile phase benzene–chloroform–formic acid 70:30:2. Samples: 1, glycerol and impurities; 2, monoglycerides; 3, 1,2-diglycerides; 4, 1,3-diglycerides; 5, fatty acids; 6, triglycerides. (Reproduced by permission of Iatroscan Laboratories, Inc., Tokyo, Japan.)

b. Quantification

For quantitative determinations a set of calibration samples should be chromatographed on the same plate. A comparison is then made of the relative sizes and intensities of the spots either directly or after treatment with a spray reagent, but it may be difficult to obtain a uniform distribution of the reagent. However, unless instrumental scanning is used, these methods are rather subjective and the precision is poor.

9.4 NEW IDEAS AND TECHNIQUES

In attempting to make TLC a more reproducible technique for both quantitative and qualitative analysis, new developments are continuously being reported. One concept has been the use of a pressurised TLC system (overpressure chromatography) to reduce evaporation of the solvent from the surface of the plate [1, 2]. Other workers have used radial plates in which the capacity increases as the run progresses, or reversed radial plates which concentrate the sample.

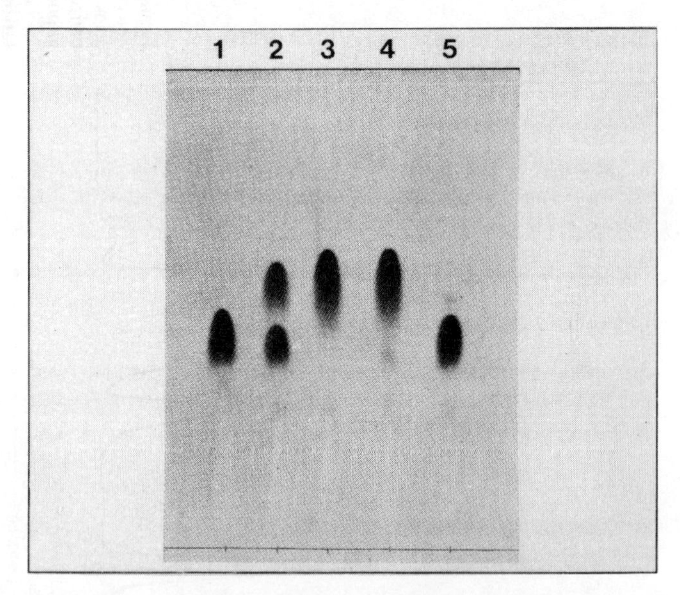

Figure 9.9 Chiral separation of D- and L-Dopa on TLC Chiralplate. Eluent methanol–water–acetonitrile 50:50:30 v/v. Detection by spraying with 0.1% ninhydrin. Samples: 1, L-Dopa; 2, DL-Dopa; 3, D-Dopa; 4, 3% L-Dopa in D-Dopa; 5, 3% D-Dopa in L-Dopa. (Reproduced by permission of Macherey-Nagel & Co. GmbH from Macherey-Nagel, *Chiralplate separation of D,L-amino acids*.)

To improve reproducibility, automatic sample applicators have been developed to deliver exact volumes as precisely defined spots on the plate. Other workers have provided methods that try to retain the open surface nature of the TLC system but place it in an instrumental context. A particularly interesting method is the Chromarod [3]. Rather than using a flat plate, the stationary phase is coated onto a glass rod. The sample is applied a short distance from the end and the rod developed with eluent by capillary action. The solvent is allowed to evaporate. The rod is then placed so that it is in the flame of a modified flame ionisation detector. The detector is moved along the rod and any organic compounds are burnt off and detected. The output is recorded as a response against distance along the rod. The flame also cleans and reactivates the rods so they can be re-used (Figure 9.8).

New stationary phases are continually being introduced following similar developments in HPLC packing materials. As well as alkyl-bonded reversed-phase silicas for reversed-phase chromatography, more recently cyano-bonded silica and ion-exchange materials have become available. A chiral TLC plate has also been introduced coated with silica impregnated with a copper–amino acid complex. This can form an outer complex with suitable chiral analytes giving good resolution of the D- and L-enantiomers (Figure 9.9) [4].

BIBLIOGRAPHY

General techniques and instrumentation

V.G. Berezkin and A.S. Bochkov, *Quantitative instrumental methods in TLC*, Alfred Hüthig, Heidelberg, 1986.

I.M. Böhrer, "Evaluation systems in quantitative thin-layer chromatography", *Topics Current Chem.*, 1984, **126**, 95–118.

D.C. Fenimore and C.M. Davies, "High performance thin-layer chromatography", *Anal. Chem.*, 1981, **53**, 252A–266A.

B. Fried and J. Sherma, *Thin-layer chromatography. Techniques and applications*, 2nd ed., Chromatographic Science Series, Vol. 35, Marcel Dekker, New York, 1986.

J. Gasparič and J. Churáček, *Laboratory handbook of paper and thin-layer chromatography*, Ellis Horwood, Chichester, 1978.

F. Geiss, *Fundamentals of thin layer chromatography*, Alfred Hüthig, Heidelberg, 1987.

H. Halpaap and K.-F. Krebs, "Thin-layer chromatographic and high-performance thin-layer chromatographic ready-for-use preparations with concentrating zones", *J. Chromatogr.*, 1977, **142**, 823–853.

R. Hamilton and S. Hamilton, *Thin layer chromatography*, John Wiley, Chichester, 1987.

P. Jenks and P. Wall, *Thin layer chromatography; a laboratory introduction*, BDH Chemicals Ltd, Poole, undated.
R.E. Kaiser (Ed.), *Planar chromatography,* Vol. 1, Alfred Hüthig, Heidelberg, 1986.
E. Stahl (Ed.), *Thin-layer chromatography: a laboratory handbook*, 2nd ed., Springer Verlag, Berlin, 1969.
J.C. Touchstone and D. Rogers (Eds.), *Practice of thin layer chromatography*, 2nd ed., John Wiley, New York, 1983.
L.R. Treiber (Ed.), *Quantitative thin-layer chromatography and its industrial applications*, Chromatographic Science Series, Vol. 36, Marcel Dekker, New York, 1986.
A. Zlatkis and R.E. Kaiser (Eds.), *HPTLC high performance thin-layer chromatography*, J. Chromatogr. Library, Vol. 9, Elsevier, Amsterdam, 1977.

REFERENCES

1. Z. Witkiewicz and J. Bladek, "Overpressured thin-layer chromatography", *J. Chromatogr., Chromatogr. Rev.*, 1986, **373**, 111–140.
2. H.E. Hauck and W. Jost, "Investigation and results obtained with overpressured thin-layer chromatography", *J. Chromatogr.*, 1983, **262**, 113–120.
3. C.C. Parrish and R.G. Ackman, "Chromarod separation for the analysis of marine lipid classes by Iatroscan thin-layer chromatography with flame-ionization detection", *J. Chromatogr.*, 1983, **262**, 103–112.
4. U.A.Th. Brinkman and D. Kamminga, "Rapid separation of enantiomers by thin-layer chromatography on a chiral stationary phase", *J. Chromatogr.*, 1985, **330**, 375–378.

CHAPTER 10

HIGH-PERFORMANCE LIQUID CHROMATOGRAPHY

10.1 MODES OF OPERATION OF HIGH-PERFORMANCE LIQUID CHROMATOGRAPHY

Instrumental liquid chromatography is characterised by the use of on-line detection systems, controlled eluent flow rates and defined eluent composition. The aim is to increase, compared to column or thin-layer chromatography, the efficiency of separation and the reproducibility and accuracy of retentions and peak areas in order to achieve quantitative and qualitative separations. Because a wide range of separation modes can be used, high-performance liquid chromatography (HPLC) is an extremely versatile technique and can be used to determine virtually any non-gaseous analyte, as long as it is soluble in an organic or inorganic solvent. It is this versatility that has made it one of the most important techniques available to the analyst and is the reason for its rapid growth since the late 1970s.

One of the greatest advantages of HPLC is that the separation can be carried out by making use of a wide range of different interactions between the analyte and the chromatographic system. The two principal modes are normal-phase and reversed-phase liquid chromatography but the same instrumentation, with different column packing and eluents, can also be used for separations based on size exclusion and ion-exchange chromatography. The dominance of reversed-phase chromatography has been shown in a survey of HPLC separations reported in research journals (Table 10.1).

Although the descriptions, normal-phase and reversed-phase (Chapter 8), are widely used and are useful guides to the operation of a particular HPLC separation, they are approximations and in reality the situation is more complex. More than one mechanism may be in operation at the same time,

Table 10.1 Usage of HPLC modes. Based on a survey of the chromatography literature 1982–83 according to the type of column used.

Separation mode		Proportions (%)
Hydrocarbonaceous bonded columns		
Reversed-phase separations		72
C_{18}	54	
C_8	10	
C_2	3	
Phenyl	2	
Others	3	
Other column materials		
Amino/cyano		6
Silica		10
Size exclusion chromatography		6
Ion-exchange chromatography		6

(Reprinted with permission from *Anal. Chem.*, 1984, **56**, 304R. Copyright (1984) American Chemical Society.)

or a single column may be used for different primary modes by changing the solvent. Most silica columns, even after being coated with an alkyl-bonded phase, will still possess some free acidic silanol groups on their surface and will interact with basic analytes by a mixture of ion-exchange and reversed-phase chromatography. The separation of ionisable analytes may thus be effected by parameters, such as pH and ionic strength, that have little effect on non-ionised analytes. Because the silica matrix of the stationary phase has a definite pore size, high-molecular-weight analytes may suffer exclusion from some pores and a form of size exclusion chromatography can be superimposed on the separation.

Much of the discussion and simple theory of bonded-phase separations regards the column as a liquid–liquid partition system and this adequately describes much of the analyte behaviour. However, the bonded phase is rigidly held and has a defined depth, and thus a more detailed treatment of the mechanism of retention must also include consideration of the layering of the stationary and mobile phases. On the other hand, silica columns are described as adsorption chromatography but some mobile-phase constituents can coat the surface and make it behave in a similar manner to a liquid phase.

The selection of the appropriate method for a particular analysis is therefore complex and the experience of the operator can be an important factor. A more detailed discussion and guidance in the choice of the separation mode, stationary phase and mobile phase, for a particular sample and matrix, are given in Chapter 12. Generally ion-exchange separations and size exclusion systems will not be discussed in detail, as both are relatively specialised areas primarily used for the separation of cations/anions and polymeric compounds, respectively.

10.2 INSTRUMENTATION FOR LIQUID CHROMATOGRAPHY

The requirements for HPLC are relatively simple: a solvent reservoir, a pump, an injector, a column and a detector (Figure 10.1). Although HPLC instruments can be purchased as complete integrated systems, many manufacturers offer individual components and in many laboratories the pump, detector and column will frequently come from different sources. This modular approach has been very successful and gives the operator great flexibility to combine the best pump with a particular detector or column. Most of the components are made of 316 stainless steel or Teflon to prevent damage by the corrosive attack of solvents or buffers.

The main influence on the design of HPLC systems is the mobile phase. Because viscosities are much higher than for gases and the diffusion rates

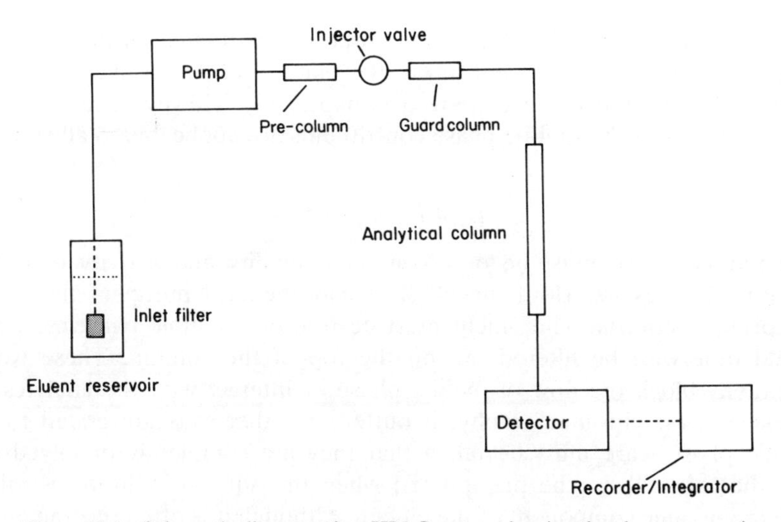

Figure 10.1 Schematic diagram of an HPLC system showing the eluent reservoir, plump, guard-column, injector valve, pre-column, analytical column, detector and recorder/integrator.

in liquids are much lower than in gases, longitudinal band spreading is less significant than in GLC but resistance to mass transfer effects in the mobile phase can cause band broadening. In order to obtain separations with a high efficiency, the stationary phase must therefore have a much smaller particle size, 3–10 μm, to reduce the effective depth of the mobile phase. A high pressure is consequently required to pump the relatively viscous mobile phase through the column. Mixing in the mobile phase is slow and great care must be taken to avoid dead volumes in the connecting tubing, injector, and detector cells, which would cause eddying and band spreading.

Many separations are carried out using isocratic elution, i.e. constant mobile-phase strength. If the sample contains components with very different retentions, then programmed elution can be used, in which the eluent strength is increased during the separation. Although, this technique is possible with normal-phase separations, it is rarely used as re-equilibrating the column to repeat the run is time-consuming. In reversed-phase chromatography with a bonded stationary phase, equilibration is rapid and programmed elution, in which the proportion of the stronger organic phase is increased compared to the weaker aqueous phase, is commonly used for method development and in routine separations (Figure 10.2). In ion-exchange chromatography ionic strength and pH gradients can also be used.

a. *Eluent*

The composition of the eluent being used in a particular chromatographic separation depends very much on the nature of the sample and the mode of separation being employed (Chapter 12). However, many of the requirements for the mobile phase constituents are applicable to all systems.

i. Eluent purity

The mobile phase must be of a consistent quality and activity to obtain reproducible results. Both the physical and chemical purity of the eluent are primary criteria. The eluent must be free of insoluble particles, which would otherwise be filtered out on the top of the column. These would eventually block the flow of mobile phase or interact with the analytes. In reversed-phase chromatography, if buffers or other salts are added to the mobile phase, care must be taken that they are completely dissolved and that the salt will not be precipitated when the aqueous solution is mixed with the organic component of the eluent. Although it is often recommended that eluents should be filtered through a millipore filter before use, this is often omitted unless problems are experienced. However, a 2 or 5 μm

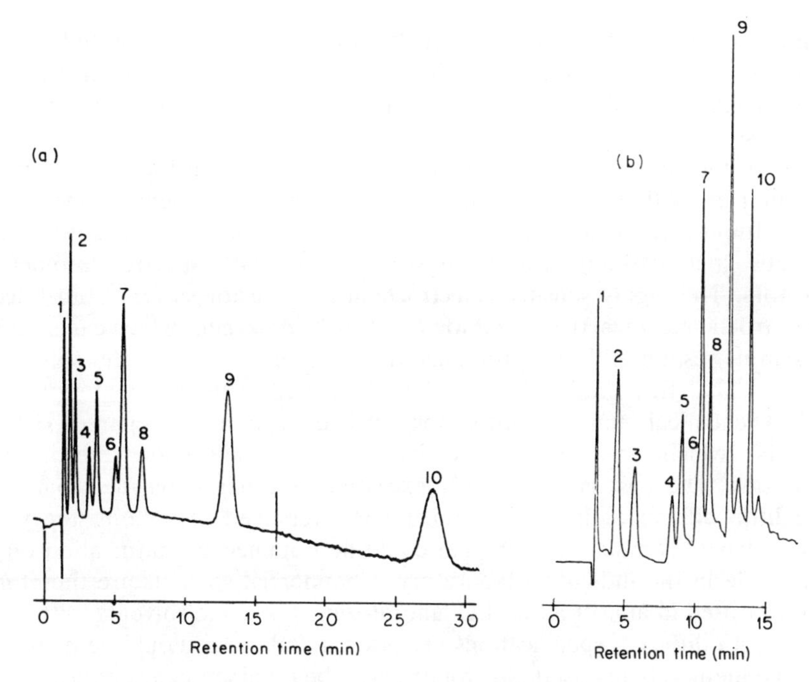

Figure 10.2 Comparison of isocratic and gradient elution of chlorinated hydrocarbons. Column Permaphase ODS 100 cm × 0.21 cm. Ultraviolet spectroscopic detection at 254 nm. Eluent: (a) isocratic methanol–water 50:50. (b) programmed gradient elution from methanol–water 40:60 to methanol 100% at 8% min^{-1}; Peaks: 1, benzene; 2, monochlorobenzene; 3, *o*-dichlorobenzene; 4, 1,2,3-trichlorobenzene; 5, 1,3,5-trichlorobenzene; 6, 1,2,4-trichlorobenzene; 7, 1,2,3,4-tetrachlorobenzene; 8, 1,2,4,5-tetrachlorobenzene; 9, pentachlorobenzene; 10, hexachlorobenzene. (Reproduced by permission of Wiley Interscience from Snyder and Kirkland. *Introduction to mechem liquid chromatography*, © 1979.)

sintered metal filter is usually fitted to the inlet end of the tube in the eluent reservoir to trap any dust or other particles.

Chemical impurities in the mobile phase are particularly a problem in normal-phase chromatography, as traces of polar compounds, particularly water, can have a major effect on the separation. They deactivate the surface of the silica column material and this can markedly alter the retention properties and order of elution. Even very small quantities can have a significant and long-lasting effect, if the nominal eluent has a low polarity. Impurities are generally a less significant problem in reversed-phase chromatography but very non-polar compounds in the eluent can be absorbed onto the stationary phase and alter the selectivity and retention properties of the column if they are present in large amounts.

In a similar way, the properties of the column can also be altered by the solvent used to dissolve the sample if this is stronger than the eluent. Ideally, samples should be prepared in the mobile phase or a solvent with a weaker elution stength.

Even if they do not alter the retention of analytes, impurities in the eluent may interfere with detection, by giving a high background signal or increased noise. Impurities with chromophores, such as plasticisers or ketonic or aromatic hydrocarbons, can seriously interfere with spectrophotometric detectors. The highly sensitive electrochemical detector suffers from traces of electrochemically active metal ions or dissolved oxygen, which can swamp the sample response. These problems can often be traced to buffer solutions and only analytical-grade reagents should be used in their preparation.

Most chemical manufacturers now provide specially prepared HPLC solvents, which are filtered and have specified ultraviolet absorption properties. The cost is not usually excessive as other impurities that will have little effect on the separation are not removed. The consistency of these solvents is much greater than could be obtained by purification on a small scale in the individual laboratory. Commercial solvents are therefore routinely used in almost all HPLC laboratories. For some solvents, different grades with different specifications are offered and these should be matched with the method being used. Acetonitrile can be obtained in a routine HPLC grade, which can be used with detection at 254 nm, but for shorter wavelengths the more highly purified and expensive far-UV grade may be needed. Some solvents are unstable to oxidation on exposure to air. Once bottles of tetrahydrofuran or dioxane are opened, auto-oxidation occurs and the absorption background will increase. Normal reagent-grade chloroform should not be used for HPLC as this can contain 1–2% ethanol as a stabiliser, which will seriously alter normal-phase retentions.

Water is a common constituent of most reversed-phase eluents. Although purified HPLC-grade water is commercially available, many laboratories prepare their own by distillation but care must be taken to avoid contamination. The water should be distilled and stored in glass containers, as significant levels of plasticisers (dialkyl phthalates) can be extracted from the widely used plastic carboys. These relatively non-polar plasticisers will be concentrated on the column until the elution strength is increased. They will then be eluted as anomalous peaks, which can be particularly troublesome in gradient elution (Figure 10.3). A blank gradient run should therefore always be carried out to check for contamination. Deionised water should normally be avoided as it can contain organic impurities from the ion exchange resins, but laboratory-scale systems are available that combine

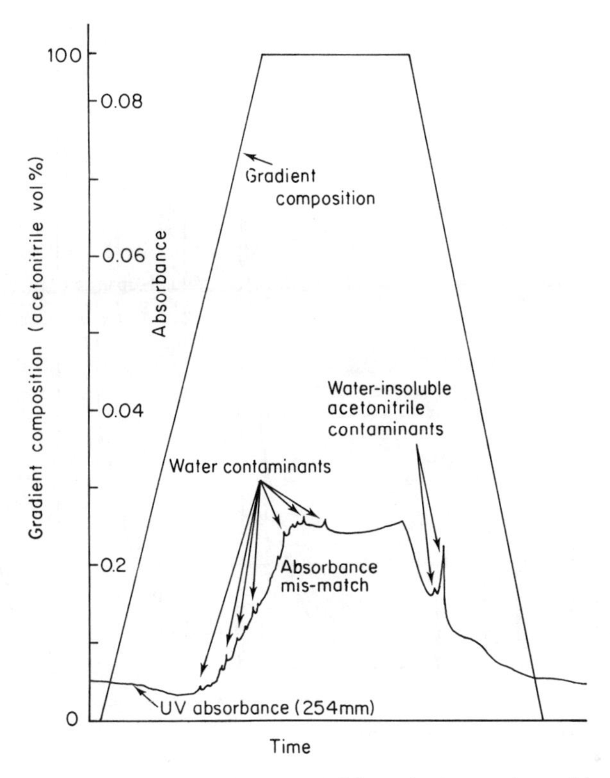

Figure 10.3 Chromatogram of water stored in a plastic container with gradient elution showing the residual materials, which can build up on the column and give background peaks. (Reproduced by permission of International Scientific Communications Inc. from *International Laboratory*.)

ion-exchange with an activated charcoal filter to remove organic constituents and give HPLC-grade water.

ii. Eluent degassing

All liquid mobile phases on standing become saturated with air. This is not in itself a problem during the chromatographic separation. However, the detector is at atmospheric pressure and bubbles of the dissolved air may come out of solution in the detector cell, which will result in spikes on the recorder trace and changes in the background signal (Figure 10.4). These effects can be reduced by good cell design, such as placing the exit at the top so that bubbles are removed as soon as they are formed. Connecting a

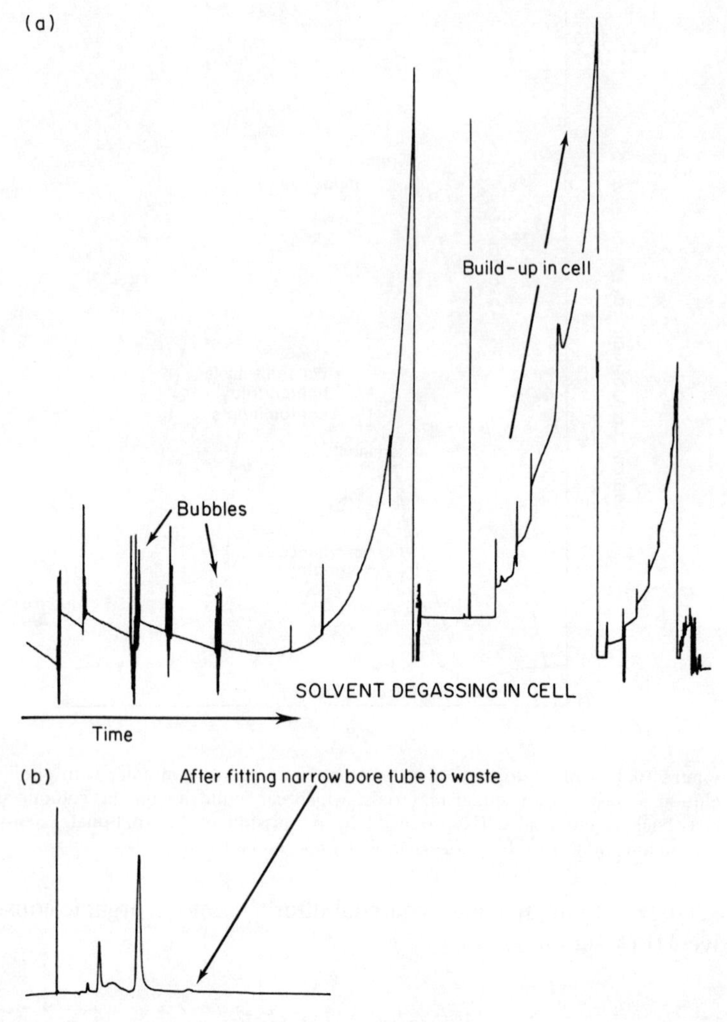

Figure 10.4 (a) Effect on recorder trace of degassing in a UV detector cell. (b) Effect of fitting narrow-bore exit tubing to suppress degassing by increasing pressure in cell.

length of narrow-bore tubing (1 m × 0.15 mm) on the exit of the detector will retain a small positive pressure in the cell and will reduce degassing (Figure 10.4b). However, the flow resistance cannot be too large or the cell may leak. If degassing of the eluent occurs in the pump head, the bubbles will collect in the check valves causing irregular eluent flows and causing problems with reproducibility.

The problem of degassing is particularly severe with aqueous–organic eluent combinations for reversed-phase chromatography as the solubility of air in the mixed eluent is usually less than in the individual components. A newly prepared eluent will thus be supersaturated with air and bubble formation will be inevitable. This problem can be readily seen on mixing water and methanol: the mixture warms up and gas is evolved. In the early days of HPLC this problem received a lot of attention, often unnecessarily, and some of the methods proposed such as using an ultrasonic bath have been found to be unsatisfactory, because once the treatment stops the degassed eluent will rapidly re-equilibrate with the atmosphere and become saturated again.

For isocratic elution, it is often sufficient to apply a low vacuum from a water pump to the eluent after it has been prepared. This will remove any excess of dissolved gas and any remaining air will stay in solution. As long as the solvent reservoir is positioned higher than the pump, so that the eluent flows to the pump head under gravity, and a small back-pressure is maintained in the detector cell, this simple method will normally be sufficient for routine work.

However, more precautions need to be taken if the eluent mixture is prepared on-line by mixing two or more solvents, for either isocratic or programmed elution (Section 10.2.b.v). A supersaturated solution may then be formed at the point of mixing. In these cases, the individual solvents will need to be continuously degassed. The most common method is to bubble helium through the solvents from a sintered filter at a low but continuous flow rate so that it also blankets the surface of the solvent. The helium flushes (or sparges) the other gases out of the solvent and itself has a very low solubility. In some commercial systems, this technique may be combined with heating the solvent as this also reduces the gas solubility. Alternatively, commercial degassing units, based on a permeable membrane to remove any dissolved gas, are available, but these are not in widespread use.

In programmed elution, the problems of degassing can be greatly reduced if, instead of mixing the pure solvents (i.e. methanol and water), premixed solutions representing the limits of the programme are prepared for each pump (i.e. methanol–water 80:20 and methanol–water 20:80), as these solutions will already contain reduced concentrations of air.

As well as the problems of degassing, air must also be removed from the eluent to reduce background absorbance when ultra-high-sensitivity spectrophotometric detection is being used at very short wavelengths (less than 210 nm). A particular problem arises with electrochemical detectors in the reductive mode, as all the oxygen must be removed from the eluent. Otherwise it will also be reduced and would give a high background signal.

b. Pumps

The principal difference between open column chromatography and HPLC is that the latter uses a stationary phase with a much smaller particle size to give vastly increased efficiency. As a consequence, the packing material resists the eluent flow and the mobile phase has to be pumped through the column under pressure. For analytical separations with 3–10 μm particle size columns and flow rates of 0.1–10 ml min^{-1}, pressures of up to 6000 psi (400 bar) are needed. The flow rate must also be highly reproducible to give reliable retention times. These requirements can be satisfied by a number of different designs but most systems use either reciprocating or diaphragm pumps.

The materials used in the pumps and fittings must be resistant to commonly used eluents. Metal parts are usually made from 316 grade stainless steel. However, this material is susceptible to attack by halide ions in the eluent, although some protection can be provided by passivating the metal surfaces with nitric acid. High concentrations of sodium chloride or other halide salts should therefore be avoided if possible. The beads and seats in the check valves controlling the flow of eluent through the pump are usually made from ruby or sapphire. Some manufacturers offer metal-free pumps with wholly ceramic or plastic parts for work with inorganic ions and proteins but it is claimed that under normal conditions the dissolution of metal ions from the surfaces in the chromatograph is minimal compared to the background levels of metal ions in buffer salts.

i. Reciprocating pumps

These are the most widely used pumps and form the basis of almost all low- to medium-cost chromatographs. The pumping action is achieved by a cam-driven reciprocating piston driven by a constant-speed motor. In its simplest form, a single-headed pump is used but this has the disadvantage that the flow stops as the pumping chamber is being filled. The resulting flow fluctuations can appear as baseline noise on the recorder trace, particularly with electrochemical or low-wavelength spectroscopic detectors. An improved design uses an irregular cam to give an even outward flow for most of the cycle with a rapid refill stroke (Figure 10.5). Much of the remaining irregularities can be eliminated by using a pulse damper, usually based on a bellows system, which smoothes out any small changes in flow rate. This combination is very useful for routine separations. Much faster-stroke frequency pumps try to reduce these pulsations but this approach seems to suffer from cavitation and excessive degassing in the pump head.

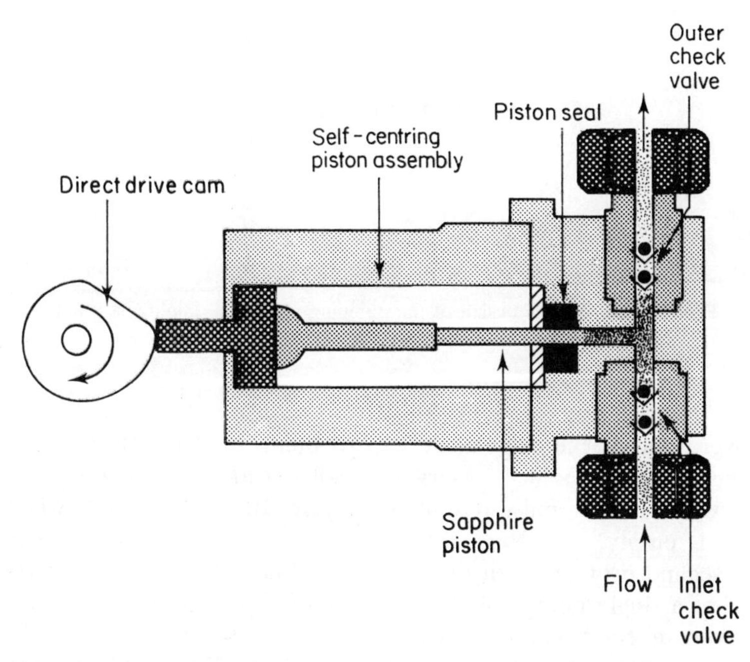

Figure 10.5 Single-head reciprocating pump. (Reproduced by permission of Gilson Medical Electronics, Inc.)

A more uniform flow can be obtained by using a dual-headed pump with two pistons out-of-phase, so that one is filling as the other is pumping. With pulse damping this can give a very reproducible flow, ideal for most analytical applications. An alternative system uses a main piston plus a smaller compensating piston, which acts during the main refill stroke. All the reciprocating pumps have proved to be very reliable and normally the only part susceptible to wear is the easily replaced seal around the piston.

ii. Diaphragm pumps

In these pumps a diaphragm is rapidly pulsed to pump the solvent through the pump head (Figure 10.6). The rate of pulsation is sufficiently high that there are effectively no observable irregularities in the flow rate. Pumps of this type are generally more expensive and are usually only found in fully integrated systems. They are well suited to programmed elution as they have only a small internal volume, which is rapidly flushed out as the eluent composition changes.

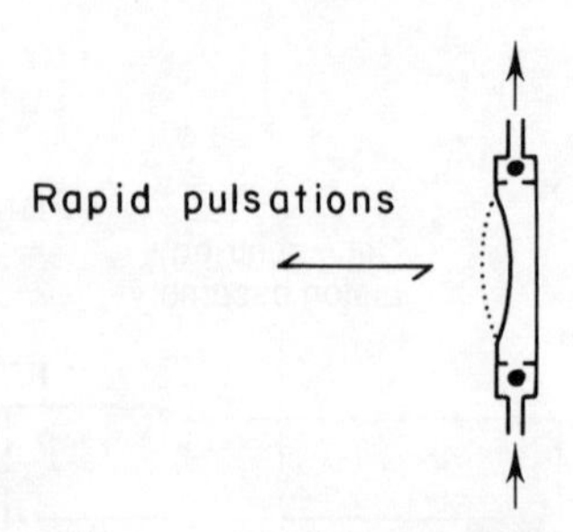

Figure 10.6 Typical design of the pumping head of a diaphragm pump.

iii. Syringe pumps

These were one of the original designs of pump used in HPLC. A screw-driven piston acts on a large reservoir of solvent like a giant syringe, which gives a very smooth pulse-free flow (Figure 10.7). However, when the reservoir is empty, the slow refill cycle causes a major disruption in the flow. A second problem with this pump was that the large internal volume means that a single pump could not be used for programmed elution. Two pumps were needed whose relative pumping rates could be altered, thus increasing the cost. These pumps have largely been displaced by the simpler and cheaper reciprocating pumps and they are not now in widespread use.

In the last few years, two forms of syringe pump have been revived. Firstly, small versions have been used for microbore columns, which require very low flow rates, typically 1 to 100 μl min^{-1}. A single fill of the piston will thus last for a prolonged period. A larger version of the syringe pump is also being used in some automated systems for routine isocratic separations,

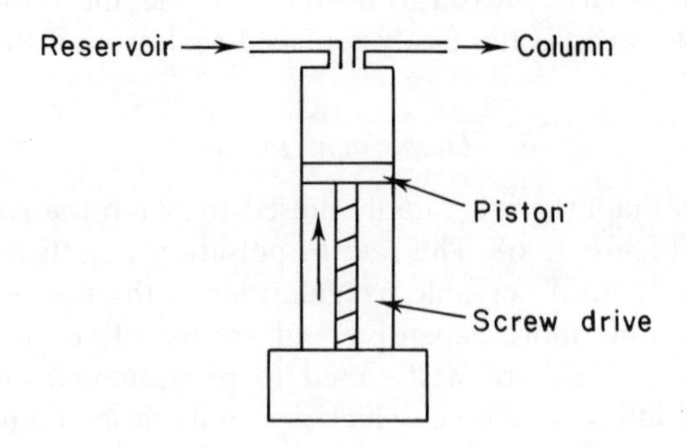

Figure 10.7 Design of a syringe pump used for isocratic elution in HPLC.

as the refill stroke can be automatically arranged to coincide with the end of each separation run.

iv. Pressure amplification pumps

Although rarely used for analytical chromatography, pressure amplification pumps, in which high pressures are achieved by using a single piston with a large area on the low-pressure side and small area on the high-pressure side (Figure 10.8), are often found in chromatography laboratories as part of column packing systems. The ratio of the areas defines the amplification, typically between 1:15 and 1:122, so that moderate pressures applied from a gas cylinder can deliver quite high final pressures. Because they deliver a constant-pressure output rather than constant flow, they are liable to give irreproducible retentions but can provide a high flow rate at high pressure and are thus ideal for column packing. They also have no electrical connections and thus are safe to operate in the presence of large volumes of solvents.

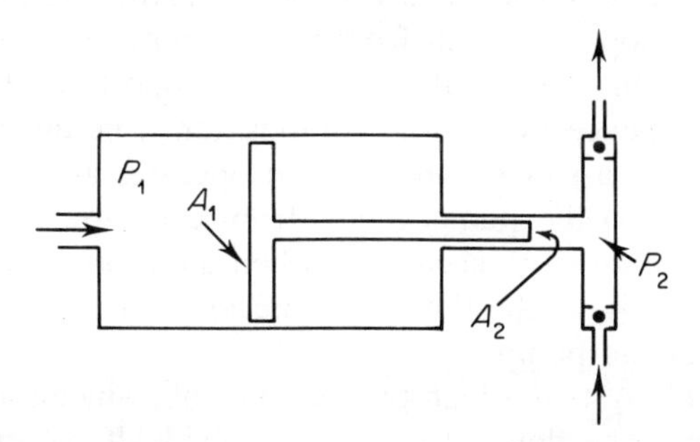

Figure 10.8 Pressure amplifier pump. The pressure amplification is the ratio of the areas of the ends of the pistons, $P_2/P_1 = A_1/A_2$.

v. Programmed elution

As noted earlier, eluent programming enables a wide range of analytes to be eluted during a single run (see Figure 10.2), but because the equilibration times on silica columns are very slow, its application is effectively limited to reversed-phase chromatography. Two methods can be used to alter the composition of the eluent (Figure 10.9). Usually two solvents or mixtures are used but some controllers enable more complex mixtures to be prepared. Most programmes use a linear change from one solvent to another expressed at percentage change per minute (% min^{-1}). Concave or convex changes

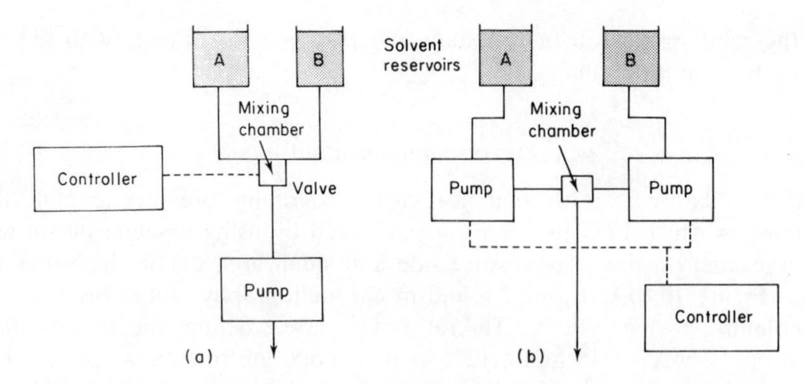

Figure 10.9 Flow diagrams for (a) low-pressure mixing and (b) high-pressure mixing systems for eluent programming in HPLC.

are also possible or the changes can be programmed in steps to enable an initial slow increase and then a more rapid change to flush any residual compounds from the column.

The cheaper method is to use low-pressure mixing. The eluent is prepared by switching the flow of solvent from two or more reservoirs using a valve. Changing the timing of the valve alters the proportion of the solvent from each source and hence the overall composition. The eluent goes through a mixing chamber to a single pump, which operates at a constant flow rate. Because the flow to the pump comes from each source in turn, many of the controllers do not work accurately when the proportion from one source is greater than 90%, as the flow from the minor source may coincide with the fill stroke of the pump.

A more reliable system is high-pressure mixing, which uses separate pumps for each solvent. The flow rates can be individually controlled and pumps must therefore be more sophisticated (and hence more expensive) than those used in low-pressure mixing. A more sophisticated controller is also required as it must be capable of controlling the pump flow rates. The solvents are blended after the pumps to give the eluent. A smaller mixing chamber can be used as each flow is continuous. Often for simplicity, single solvents (rather than mixtures) are used with each pump.

For both types of programming it is important to degas the solvents in the reservoirs to avoid out-gassing in the pumps or in the detector. For low-pressure mixing partially premixed solvents can be a great help.

c. *Sample injection systems*

In the early days of HPLC, a syringe injection system similar to that in GLC was often used but it became impracticable when column pressures increased with the use of smaller packing materials. To avoid the problem of injecting into a high-pressure system, stop-flow systems were devised, which interrupted the eluent flow during injection.

Nowadays, virtually all HPLC systems use valve injectors, which separate the sample introduction from the high-pressure eluent system. In the *LOAD* position (Figure 10.10a) the sample is injected into the fixed volume loop at atmospheric pressure. Meanwhile the eluent is passed directly from the pump to the column at high pressure. When the valve is rotated to the *INJECT* position (Figure 10.10b), the loop becomes part of the eluent flow system and the sample is flushed onto the column. The filling tubes to the loop can then be flushed. The basic six-port design has a fixed loop, typically 10 or 20 μl, which gives high reproducibility as the volume injected is mechanically fixed by the dimensions of the loop. Because of the wall effects, a relatively large sample volume (5–10 loop volumes) must be flushed through the valve, to ensure that all the eluent has been washed out of the loop and that it is completely filled with sample.

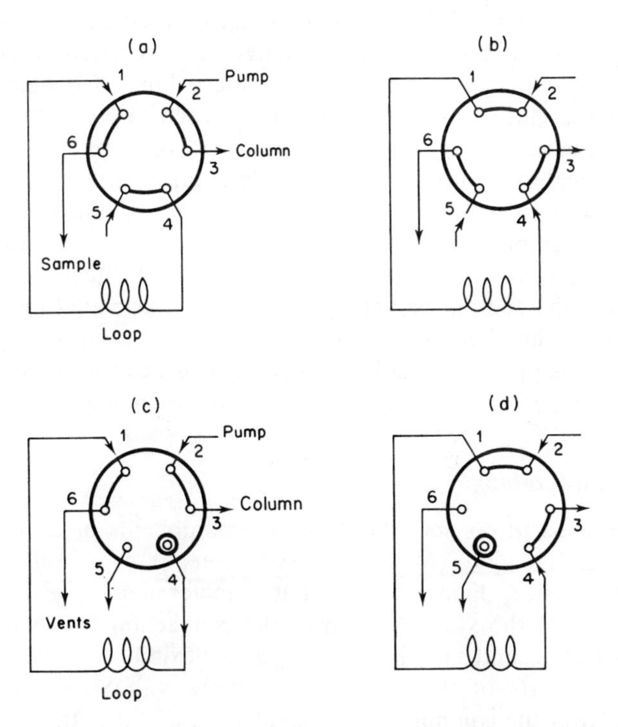

Figure 10.10 Six-port injection valve: (a) *LOAD* position; (b) *INJECT* position. Seven-port injection valve for use with syringe partial loop injection: (c) *LOAD* position; (d) *INJECT* position. (Reproduced by permission of Rheodyne, Inc.)

An alternative and very popular design is the seven-port valve in which a variable volume of sample can be injected directly into the sample loop with a syringe (Figure 10.10c and d). The precision in this case is limited by the reading of the syringe. The sample volume should not exceed 50% of the loop volume (typically 10 μl for a 20 μl loop). The syringe should not be removed from the fill port until after the valve has been rotated to the *INJECT* position, to ensure that no leakage can occur from the loop. These valves have the advantage that only the volume needed for the analysis is required and the injection volume can be readily altered.

Typical precisions of injection for a 10 μl sample are 0.5% relative standard deviation, if a completely filled loop is used, and 0.8% for syringe injection partially filling a 20 μl loop.

Normally for an analytical column (4.6–5 mm i.d.) the sample volume should not exceed 20 μl or band broadening can occur. Smaller-diameter columns only permit correspondingly smaller volumes of sample. With the introduction of microbore columns (1 mm i.d.) micro-injectors have become available with sub-microlitre internal fixed loops (0.5 μl).

In selected cases much larger samples can be used as long as the sample solvent is such a weak eluent for the analyte that the compounds of interest will effectively be trapped on the top of the column. Changing the eluent will then cause elution. This method can be used to concentrate non-polar pollutants such as hydrocarbons from aqueous solutions.

In the normal valve injector, the sample is mixed with the eluent stream and is spread across the diameter of the column on injection. As a result parts of the sample can suffer spreading because of wall effects in the column. There have been theoretical arguments that injection of the sample centrally into the column bed can give a higher efficiency (infinite-diameter effect) and this has been demonstrated in practice using syringe injection. However, it has proved difficult to implement this curtain flow arrangement with a valve-type injector and no systems are commercially available.

d. Connecting tubing

The tubing used to connect the different components of an HPLC system should be made of stainless steel or Teflon, usually 1/16th inch outside diameter. The latter material can be used for low-pressure parts of the systems, such as the connection between the column and the detector, and has the advantage of flexibility. Connections in the high-pressure parts of the system are usually with Swagelock or similar couplings. After the column, Teflon ferrules are useful as they can be readily replaced when the system is altered. Any couplings used in the system should be designed for HPLC to avoid significant internal dead volumes and suitable fittings are now readily available.

To reduce the dead volume of the system, all the tubing through which the sample passes, between the injector and the detector, should have a miminal internal diameter, typically 0.15 mm, and should be kept as short as possible. With microbore columns this becomes particularly important as the sample peaks are smaller and any spreading is more significant. Consequently the injector and detector cell are often connected directly to the column.

e. *Columns*

As the applications of HPLC have increased, columns of different types and sizes have become available, ranging from preparative columns with sizes up to 5 cm × 50 cm or even larger, and down to wall-coated capillary columns. A much more limited range of columns, packed with 3, 5, or 10 μm particles of stationary-phase, are routinely used in analytical chemistry. The nature of the stationary phase materials will be discussed in more detail in Chapter 12.

i. Analytical columns

Most routine HPLC separations are carried out using columns with an internal diameter of 4.6 or 5 mm, an external diameter of 1/4 inch and eluent flow rates of 1 or 2 ml min^{-1}. The columns are normally 10, 15, or 25 cm long, the shorter columns usually being used only with 3 and 5 μm packing materials. Very short 3 cm "high-speed" columns, packed with 3 μm packings, have been used for rapid separations. Most of the columns are made from stainless steel and can be either prepacked or packed in the laboratory (see later). Prepacked column materials are also available as glass- or metal-walled cartridges, which can be mounted in metal holders with built-in end fittings.

In order to increase the reproducibility and lifetime of columns, some packing materials are available in flexible-walled plastic cartridges, which can be radially compressed in specially designed column units. Any loss or settling of the stationary phase is immediately taken up by the compression. Similar advantages are claimed for columns that can be tightened axially.

ii. Microbore columns

Over the last few years, great interest has been expressed in microbore columns with internal diameters of 1–2 mm and 25–50 cm long, which promised higher efficiency and greater sensitivity for limited sample sizes. The overall flow rates are much reduced and the cross-section of the columns is much smaller than conventional analytical columns. As a result the volume of the peaks is much smaller. These columns therefore impose strict limitations on the maximum volume of the detector cell and the sample injection volume, otherwise the system efficiency can be lost. Unfortunately, the limitations of small sample capacity, the dead volume of HPLC systems and the slow response time of detectors and recorders have meant that the advantages predicted by theory have frequently not been obtained in practice.

The small internal diameter of these columns means that only low flow rates of 10–100 μl min^{-1} are used. This has advantages when the column is coupled to a mass spectrometric detector as the volume of solvent vapour passing into the spectrometer source is reduced and a high vacuum can be maintained. The low flow rates may also enable exotic eluent solvents to be used.

Particularly high column efficiencies can be obtained by coupling microbore columns together to give total column lengths of 500 or 1000 mm (Figure 10.11) and up to one million theoretical plates have been achieved [1]. However, the separation times taken to achieve these efficiencies are typically several hours so that these columns are not practicable for routine analytical use.

For most purposes microbore columns appear to offer few advantages over analytical columns, except reduced solvent consumption, but impose stringent operational requirements on the chromatograph. However, they have an important role when the absolute sample size is limited because

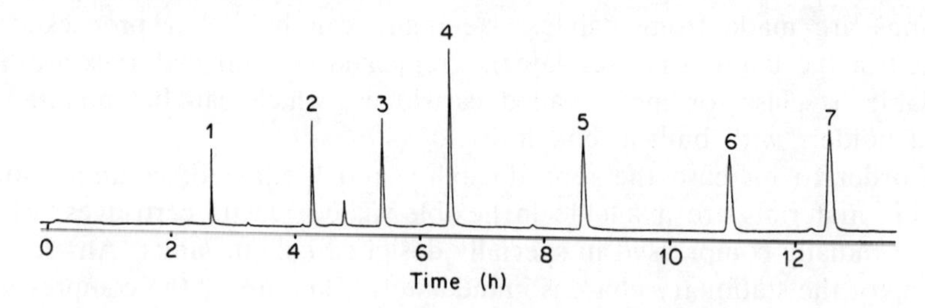

Figure 10.11 High-efficiency separation of aromatic hydrocarbons on linked microbore columns. Column MBC-ODS, 3 m × 1 mm i.d. Eluent acetonitrile–water 65:35 at 25 μl min^{-1}. Efficiency 130 000 plates. Peaks: 1, benzene; 2, naphthalene; 3, biphenyl; 4, acenaphthene; 5, anthracene; 6, fluoranthene; 7, pyrene. (Reproduced by permission of Shimadzu.)

they reduce the dilution of the sample in the detector and enable greater sensitivity to be obtained.

iii. Capillary columns

As yet narrow open-tubular columns with wall-coated stationary phases or very narrow columns packed with bonded-phase materials are only of research interest in liquid chromatography. High efficiencies comparable to open-tubular columns in GLC cannot be reached with open-tubular columns in HPLC unless the internal diameters are in the range of 5–20 μm because of the much lower diffusion rates in liquids. These columns have very low sample capacities and the detection of the analyte is a major problem. As a consequence almost all the current work has used either electrochemical detection or fluorescence detection using a laser excitation source. However, both methods, although highly selective and sensitive, have only limited applicability. The sample injection volume must be correspondingly very small and as yet no commercial equipment is available.

iv. Pre- and guard columns

Short disposable guard columns, containing the stationary phase, are sometimes placed between the injection valve and analytical column to trap any insoluble materials or very highly retained compounds, so that these do not build up on the main column (see Figure 10.1). Often these guard columns can be hand-packed with coarser material than the analytical column or with pellicular beads without significantly detracting from the overall separation efficiency.

A pre-column (or scavenger column) packed with large-particle silica is often placed between the pump and the injector as a sacrificial column. This will saturate the mobile phase with silica and thus protect the analytical column from dissolution, a common cause of column deterioration and early failure. These columns are particularly needed if high pH buffers (pH 8.5–9.0), are being used, as under these conditions column attack is high. Silica columns should not in any case be used above pH 9.0 because of their poor stability.

With both columns it is essential that they are regularly inspected and replaced as needed, because once they are depleted any deleterious effects will pass onto the analytical column *v. Column ovens*

The majority of HPLC separations are carried out at ambient temperature. However, retention times are temperature dependent (Figure 10.12). In order to obtain reproducible qualitative studies, it is necessary to thermostat the column and pre-column. Some separations, particularly of high-molecular-weight analytes, such as steroids and cyclosporins, are improved by raising the column temperature, as the efficiency of the separation increases (Figure 10.12). This is because diffusion rates increase with temperature and thus the resistance to mass transfer in the mobile and stationary phases is reduced (see Chapter 2, equation 2.17). Temperature control is also particularly important for ion-pair separations (see Chapter 12). The temperature of a column can be controlled with either a water jacket, electrical oven block, or air blown oven. Unlike GLC, programing the temperature during the separation is not used in HPLC, and care should be taken that the system has fully equilibrated at the analytical flow rate

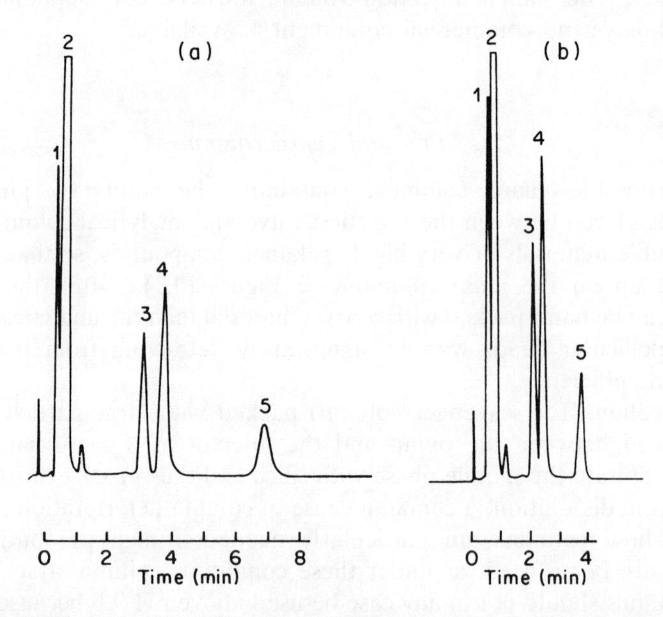

Figure 10.12 Effect of temperature dependence on efficiency of separation and analyte retention time in HPLC. Column RP 18 cartridge column 100 mm × 4.6 mm. Eluent methanol–water 60:40 at 1 ml min^{-1}. Ultraviolet spectrophotometric detection at 254 nm. (a) Ambient temperature; (b) 60°C, showing shorter retentions and sharper peaks. Peaks: 1, *o*-nitrophenol; 2, *p*-nitrophenol; 3, benzophenone; 4, ethyl salicylate; 5, phenyl salicylate. (Reproduced by permission of Philips Analytical.)

before measurements are made, as the resistance to flow in the column also produces an internal temperature gradient.

vi. Packing columns

Although many users purchase prepacked HPLC columns, analytical columns can be readily packed in the laboratory with considerable cost saving, but until a reasonable amount of practice has been obtained, the results may not be reliable. Most column packing materials are available loose in small quantities, although a few manufacturers only make their materials available in a prepacked form.

Most commercially available column packing systems use a pressure amplification pump to pump a thin slurry of the column packing material (2 g in 25 ml), suspended in a suitable solvent such as methanol, rapidly into the column at high pressure and flow rates. Details of the procedures to be used for particular packing materials are usually supplied by the manufacturers. Pre-column and guard columns can be packed in the same way.

A test mixture (for example Figure 10.13) should be used to check the efficiency and selectivity of any column, whether home or commercially

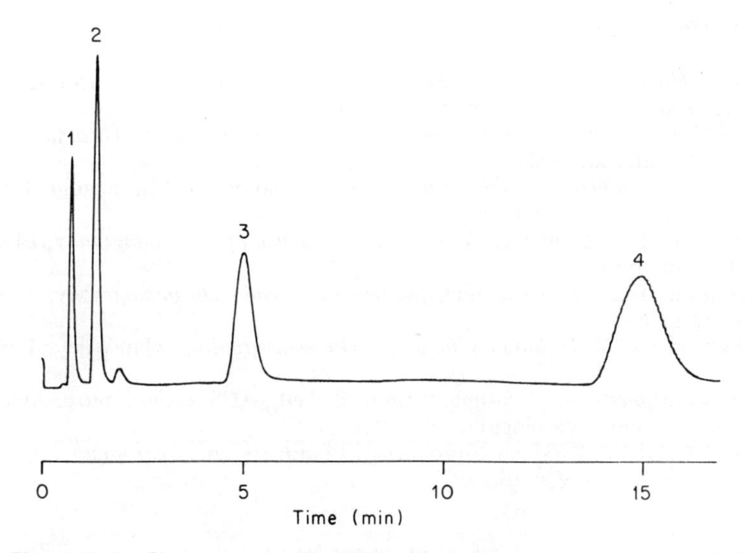

Figure 10.13 Chromatogram of typical general column test mixture for an alkyl-bonded silica column. Peaks: 1, benzamide; 2, acetophenone; 3, benzophenone; 4, biphenyl. Separation on ODS-Hypersil 100 mm × 5 mm. Eluent 60:40 methanol–water at 1.0 ml min^{-1}. Detection at 254 nm.

packed, as a guide to the reproducibility of the packing process. The test should be repeated at regular intervals during the life of the column to test reproducibility. Normally the test mixture will be designed for the eluent conditions in use in the assay to avoid changing the eluent to carry out the test.

BIBLIOGRAPHY

General texts on HPLC

Many also include detailed sections on instrumentation and columns.

A. Braithwaite and F.J. Smith, *Chromatographic methods*, Chapman and Hall, London, 1985.
P.A. Bristow, *Liquid chromatography in practice*, HETP, Wilmslow, 1976.
H. Engelhardt (Ed.), *Practice of high performance liquid chromatography*, Springer Verlag, Berlin, 1986.
M.T. Gilbert, *High performance liquid chromatography*, Wright, Bristol, 1987.
R.J. Hamilton and P.A. Sewell, *Introduction to high performance liquid chromatography*, 2nd ed., Chapman and Hall, London, 1982.
E. Heftmann, *Chromatography. Fundamentals and applications of chromatographic and electrophoretic techniques*. Part A, *Fundamentals and techniques,* Part B. *Applications*, J. Chromatogr. Library, Vol. 22, Elsevier, Amsterdam, 1983.
J.H. Knox (Ed.), *High performance liquid chromatography*, Edinburgh University Press, Edinburgh, 1978.
C.K. Lim (Ed.), *HPLC of small molecules—a practical approach*, IRL Press, Oxford, 1986.
S. Lindsey, *High performance liquid chromatography*, John Wiley, Chichester, 1987.
V. Meyer, *Practical HPLC*, John Wiley, Chichester, 1988.
I. Molnar (Ed.), *Practical aspects of modern high performance liquid chromatography*, de Gruyter, Berlin, 1983.
N.A. Parris, *Instrumental liquid chromatography*, 2nd ed., J. Chromatogr. Library, Vol. 27, Elsevier, Amsterdam, 1984.
C.F. Poole and S.A. Schuette, *Contemporary practice of chromatography*, Elsevier, Amsterdam, 1984.
C.F. Simpson (Ed.), *Practical high performance liquid chromatography*, Heyden, London, 1976.
C.F. Simpson (Ed.), *Techniques in liquid chromatography*, John Wiley, London, 1982.
L.R. Snyder, *Modern liquid chromatography*, 2nd ed., ACS Audio Course, American Chemical Society, Washington, DC, 1981.
L.R. Snyder and J.J. Kirkland, *Introduction to modern liquid chromatography*, 2nd ed., John Wiley, New York, 1979.

Equipment for HPLC

Achieving accuracy and precision with Rheodyne sample injectors, Rheodyne Technical Note, No. 5, Rheodyne, Cotati, 1983.

HPLC sample injectors: design characteristics and accuracy/precision capabilities, Rheodyne Technical Note, No. 1, Rheodyne, Berkeley, 1979.
J.F.K. Huber (Ed.), *Instrumentation for high-performance liquid chromatography*, J. Chromatogr. Library, Vol. 13, Elsevier, Amsterdam, 1978.
H.M. McNair, "Equipment for HPLC—VI", *J. Chromatogr. Sci.*, 1984, **22**, 521–535.

Microbore columns

Includes both narrow-bore 1–2 mm and open-tubular capillary columns

P. Kucera (Ed.), *Microcolumn high-performance liquid chromatography*, J. Chromatogr. Library, Vol. 28, Elsevier, Amsterdam, 1984.
M.V. Novotny and D. Ishii (Eds.), *Microcolumn separations. Columns, instrumentation and ancillary techniques*, J. Chromatogr. Library, Vol. 30, Elsevier, Amsterdam, 1985.
R.P.W. Scott (Ed.), *Small bore liquid chromatography columns: their properties and uses*, John Wiley, New York, 1984.

REFERENCE

1. H.G. Menet, P.C. Gareil and R.H. Rosset, "Experimental achievement of one million theoretical plates with microbore liquid chromatographic columns", *Anal. Chem.*, 1984, **56**, 1770–1773.

CHAPTER 11

DETECTION IN HIGH-PERFORMANCE LIQUID CHROMATOGRAPHY

11.1 GENERAL CRITERIA

The main factors that govern the requirements of detection in HPLC are much the same as those in GLC (Chapter 5): the need for sensitivity, reproducibility, selectivity, stability, economic capital and operating costs, and a wide linear range.

Whereas in GLC, it is usually easy to differentiate between the organic analytes and the inert inorganic carrier gas, in HPLC both the analyte and mobile phase can be organic compounds. As a result, all the main detectors in current use are based on a selective response for the analyte compared to a low or negligible response for the eluent. Invariably, this also means that different analytes will respond to different extents. As yet HPLC lacks a universal detector that can effectively detect all analytes but will ignore the mobile phase.

Almost every conceivable method for the measurement of a physical property of an organic or inorganic compound in solution has been applied to the detection of analytes in HPLC eluents. Most of the methods are only of academic interest and have not been used outside the research laboratory. However, five types of detector are in widespread use, based on ultraviolet–visible spectrophotometry, refractive index determination, fluorescence spectrophotometry, conductivity and electrochemical analysis

Table 11.1 Usage of different HPLC detectors. Proportion of operators who use each kind of detector. The total comes to more than 100% because many operators use more than one type of detector. Based on a user survey in 1985.

Detector type	Usage (%)
Ultraviolet spectroscopy	
Fixed-wavelength	53
Variable-wavelength	78
Diode array	18
Refractive index	37
Fluorescence spectroscopy	31
Electrochemical	21
Conductivity	15
Other detectors—each	<10

(Reproduced by permission of Aster Publishing Corp. (*LC.GC Magazine*, 1986 **4**, 526.)

(Table 11.1). In addition, HPLC systems can be coupled to infrared and mass spectrometers, but both techniques are expensive and, as yet, not as successful as in GLC. A few detectors based on other detection mechanisms, such as flame ionisation or vapour-phase nephelometry, are commercially available, but generally these methods have achieved limited acceptance and are used in relatively few applications.

Additional new detectors are continually being proposed but, to be successful, they must demonstrate that they can either satisfy a particular demand or, compared to existing detectors, can offer a significantly greater sensitivity or specificity. These factors usually mean that it is difficult for a new detector to gain wide acceptance. Even most of the new detectors produced commercially fade away after an initial popularity, usually because of low sensitivity, poor reproducibility, or poor reliability, and cease to be manufactured.

Because of the relatively low flow rates in HPLC, it is particularly important that the volume of the detector cell is appropriate for the column diameter and sample size and will not add to the peak broadening. The total volume of a well resolved rapidly eluted peak on an HPLC column is typically 100 μl for a 4.6 mm column but can be only about 5 μl for a 1–2 mm microbore column. As the efficiency achieved on the column must not be lost by mixing in the detector, the cell must be considerably smaller than these peak volumes. Thus the cell volume should be no more than 8 μl for the widely used 4.6 mm columns or 1 μl for a microbore column. The recent trend to short 3 cm columns (fast LC) with faster flow rates produces additional problems as the peaks are often eluted very rapidly. Care must be taken that the response times of the detector, amplifier and recorder are very short or electronic band broadening will occur. These demands often mean that equipment that was satisfactory for 4.6 mm columns needs to be modified before it can be used with microbore or fast LC columns. Older detectors may incorporate a 1–2 s damping to reduce background noise, particularly on higher sensitivities, which cannot be reduced by the operator.

The materials of the detector must be chemically inert to the analytes, common solvents and buffers and must be compatible with the analytical system being used. A frequent problem is that optical cells must be capable of withstanding internal pressures of 2–3 bar without leaking but still transmit all wavelengths of light between 200 and 700 nm. The design of the shape of the flow cell was the subject of much early study. The most important features are that the eluent must flow through the cell without dead volumes or spaces that can trap air bubbles. Consequently, Z-shaped and double L-shaped cells are used, with the outlet being placed at the top of the cell (Figure 11.1). The widely used 8 μl flow cell for routine analytical separations is based on a 1 mm diameter tube with a 10 mm path length. To obtain smaller cell volumes, the path length has to be reduced, which with spectrophotometric detectors also reduces the sensitivity.

The selection of a detector for the analysis of a particular sample will depend very much on the structure of the analyte and the interferences that may occur from other components of the sample or the eluent. The choice may also restrict the eluent that may be used in the separation. Because of

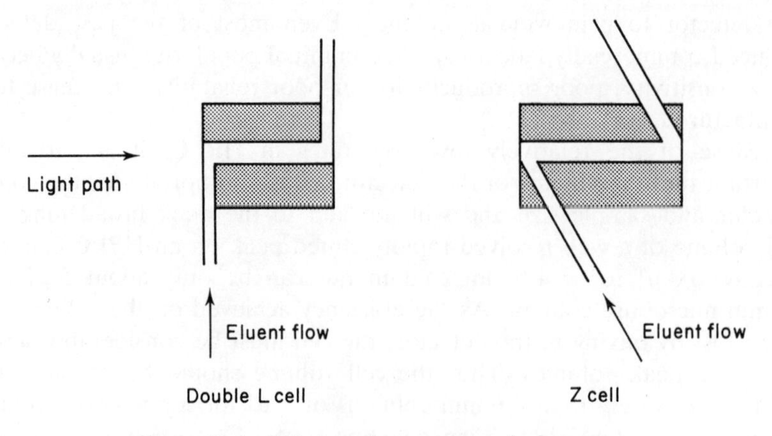

Figure 11.1 Cell shape for spectroscopic detectors.

the limited range of detectors, many analytes cannot be detected directly or only give a weak response. In some cases, these compounds can be derivatised by either pre- or post-column reactions to introduce detectable chromophoric groups (Chapter 14).

11.2 SPECTROPHOTOMETIC DETECTORS

Spectrophotometric detection provides the most popular group of detectors used in HPLC and can be regarded as the standard method for most assays.

a. Ultraviolet–visible spectrophotometric detectors

A high proportion of all organic compounds can be detected by their absorption of ultraviolet or visible light. Compounds with chromophores include all conjugated and aromatic compounds, some simpler unsaturated organic compounds, such as ketones and esters, and a number of inorganic ions and complexes.

The sensitivity and response, when using a spectrophotometric detector, depend primarily on the extinction coefficient of the analyte at the wavelength being used for detection. This will differ markedly for different compounds depending on their chromophores. For example the sensitivity can be quite high for conjugated aromatic compounds but unconjugated carbonyl groups can only be detected at short wavelengths with a weak absorbance and low sensitivity.

The advantage of the ultraviolet detector is that it responds to most compounds as a pseudo–universal detector. However, by careful selection of the wavelength used for detection it can become quite selective and the response of a particular analyte can sometimes be enhanced at the expense of possible interferences. Generally, at shorter wavelengths most compounds will respond, but greater selectivity can be obtained at longer wavelengths as relatively fewer compounds will respond (Figure 11.2). As the response of all analytes will vary with the wavelength of detection, the analytical separations and any calibrations must clearly be carried out at the same wavelength. As long as the light being used for detection is monochromatic, the magnitude of the absorption signal will be proportional to the concentration of the analyte according to the Beer–Lambert law

$$\text{absorbance} = \epsilon \times \text{cell path length} \times \text{molar concentration}$$

and directly dependent on the molar extinction coefficient ϵ at the wavelength of detection.

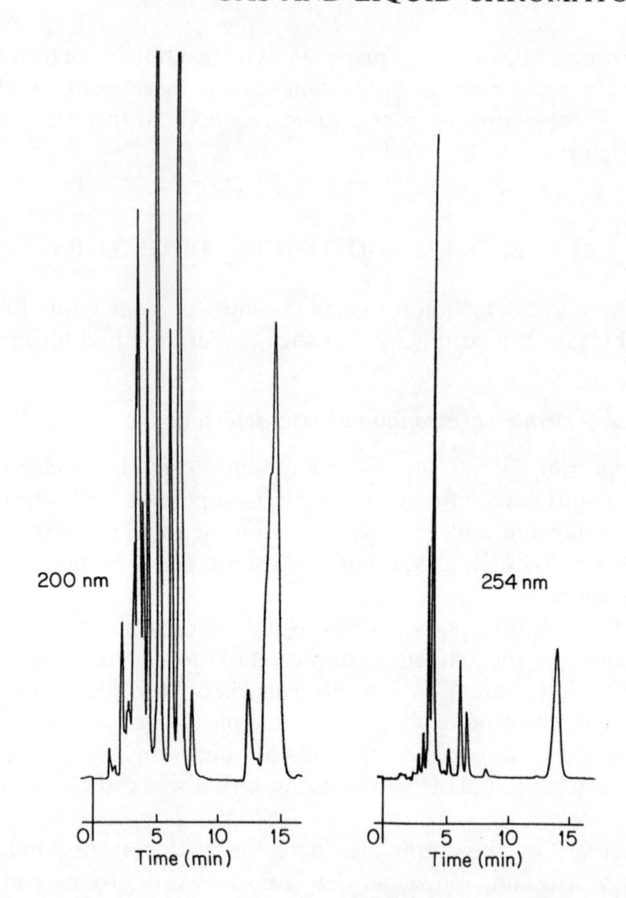

Figure 11.2 Examination of peppermint oil extract at two wavelengths, 200 and 254 nm, showing the greater selectivity of longer wavelengths. Column ODS Spherisorb 10 μm, 250 mm × 4.6 mm. Eluent acetonitrile–water 75:25. (Reproduced by permission of Milton Roy (UK) Ltd.)

Different designs of spectroscopic detectors have been used for HPLC. All are based on well established ultraviolet–visible spectrophotometer designs and therefore can be manufactured relatively cheaply. As they contain few moving parts, they are robust and usually have a good reliability record. The detectors can be divided into three groups with increasing versatility and complexity: fixed-wavelength, variable-wavelength and diode array spectrometers.

i. Fixed-wavelength detectors

Simple filter colorimeters are rarely used as detectors for HPLC because the bandpass of the filter (20–50 nm) means that the response is often non-linear. However, a filter can be used to isolate a single emission line from a line source such as a mercury lamp, effectively providing monochromatic light (Figure 11.3). The most popular version of this detector uses the strong line at 254 nm from the mercury lamp, which is conveniently a wavelength of significant absorption for most conjugated and aromatic compounds. A shorter-wavelength line at 214 nm can be obtained by using a zinc lamp and a filter. The advantage with these detectors is that the optical path is very simple and no monochromator is needed (Figure 11.4).

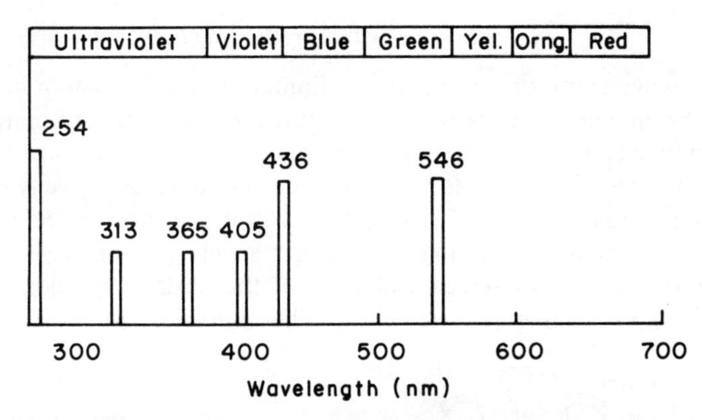

Figure 11.3 Emission line spectrum from mercury lamp. The 254 nm line can be readily isolated with a simple filter. (Reproduced by permission of Waters.)

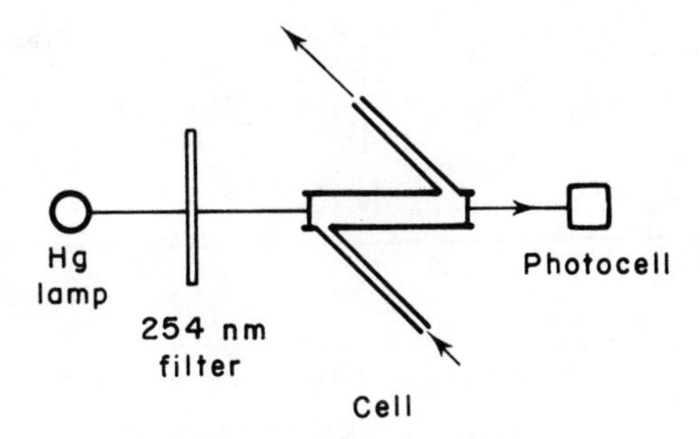

Figure 11.4 Fixed-wavelength detector. Wavelength 254 nm from a mercury lamp and filter.

Some early instruments used the 254 nm line to excite phosphors in order to generate other wavelengths, such as 280 nm, which is of interest for biological samples. Nowadays, a variable-wavelength detector would probably be used as a more readily selected source. The popularity of the 254 nm detector can be seen by the frequent adoption of this wavelength as the "standard" detection wavelength when an adjustable spectrophotometric detector is being used.

In order to compensate for background signals and variations in the lamp output, most fixed-wavelength instruments use a reference beam and a second solvent cell. In practice, this reference cell is frequently left empty as an air blank. If a reference solution is used, unless it is continually pumped through the cell, it will heat up and air bubbles will form in the light path.

ii. Variable-wavelength detectors

The fixed-wavelength detector is rather limited because it cannot be adjusted to give the maximum sensitivity for a particular analyte, by matching the detection wavelength to the wavelength of maximum absorption of the analyte. It is also not possible to use longer and more selective wavelengths for detection (as Figure 11.2). For this reason the most popular general-purpose detector is the variable-wavelength spectrophotometer, in which a grating monochromator selects radiation of the desired wavelength from a continuum light source (Figure 11.5). The source is usually a deuterium

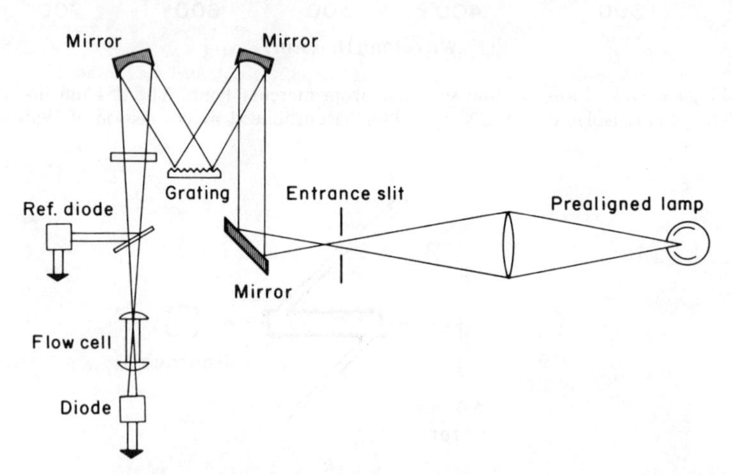

Figure 11.5 Variable-wavelength detector, showing deuterium lamp, optical path, reference photodiode and monochromator. (Reproduced by permission of Pharmacia LKB Biotechnology.)

lamp for the ultraviolet region (190–350 or 370 nm) or a tungsten lamp for the much less popular visible region (350–700 nm) of the spectrum. Unlike a conventional scanning spectrometer, most detectors do not have a built-in change-over between the two scales. Normally little use is made of the visible region, because even the relatively few compounds that are coloured usually also absorb strongly in the ultraviolet region of the spectrum. Consequently, most variable–wavelength detectors are essentially designed for the ultraviolet range but may be modified for use in the visible region by replacing and realigning the lamp.

Conventional ultraviolet–visible spectrophotometers, even if fitted with small-volume flow cells, are normally unsuitable for HPLC because the light beam is not sufficiently focused through the small-diameter window (1 mm) of the flow cell. Consequently only a small proportion of the normal light intensity will reach the detector and the detector will be very noisy.

iii. Diode array spectrophotometric detectors

The most recent development in spectrophotometric detection has been the introduction of instruments based on the use of a photodiode array detector. By using "reverse optics', all the light from a deuterium lamp is passed through the sample cell onto the monochromator, which spreads out the beam into a spectrum. This falls across an array of 230–350 photodiodes mounted on a silicon chip (Figure 11.6). These can all be read virtually

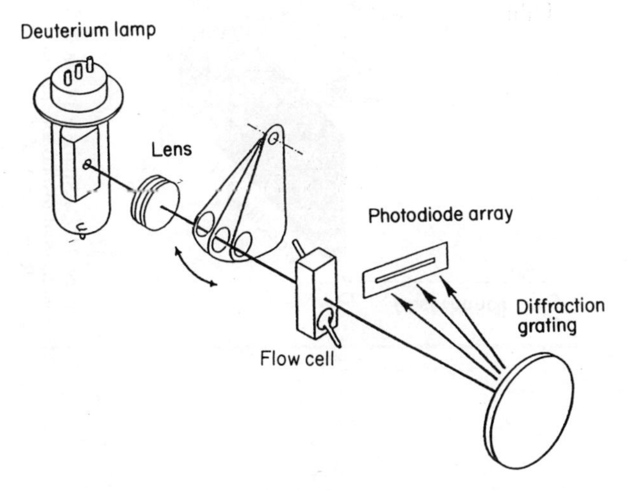

Figure 11.6 Optical path in a photodiode array spectrometer. (Copyright © Hewlett–Packard Co. (1979). Reproduced with permission.)

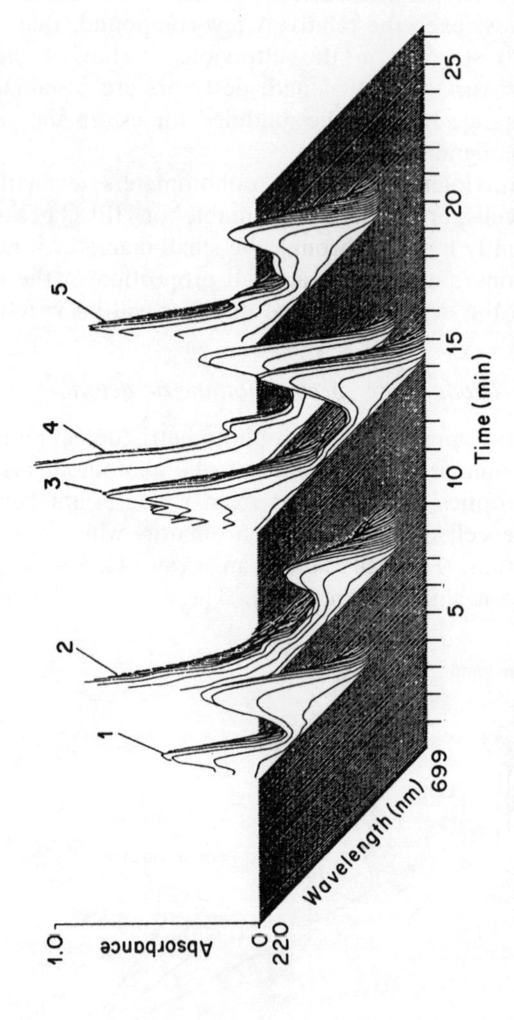

Figure 11.7 Analysis of food dyes using a diode array spectrometer, demonstrating the three-dimensional display of spectra with time and the large amount of information provided. Dyes: 1, Y-4; 2, R-2; 3, B-2; 4, R-102; 5, Y-5. (Reproduced by permission of Shimadzu.)

simultaneously by a microcomputer to provide the full absorption spectrum from 200 to 700 nm every 0.1 s. This detector can thus provide a continuous measurement of the full spectral absorbance of the eluent with time. The chromatogram can be displayed either as the absorbance at a specific wavelength or at a summed group of wavelengths, or as the changes in the full spectra against time using a three-dimensional plot (Figure 11.7).

Because this detector records the full spectrum of the eluent, it can provide much more information about the composition of the sample than is possible with monochromatic detection. This aspect has found particular application in method development as it can enable the components of a mixture to be identified. Therefore any changes in the order of elution on altering the eluent can be monitored. In more routine studies, it can be used for quality control, as it is possible to examine the spectrum at different positions throughout a peak and thus determine the spectral homogeneity (Figure 11.8). Differences in the spectrum between the front and tailing edges of a peak would indicate the presence of an unresolved impurity. In a simpler mode, the ratio of the absorptions at two wavelengths can be monitored as a check on peak purity. The ratio can also be used as a characteristic property for analyte identification, although these values can also be obtained using two variable-wavelength detectors in series.

The full three-dimensional plot can be useful to examine chromatograms in metabolic studies, as peaks with the same chromophore can be readily identified and all the components with chromophores will be detected, whereas at a fixed wavelength some components might have a low extinction coefficient and be undetected.

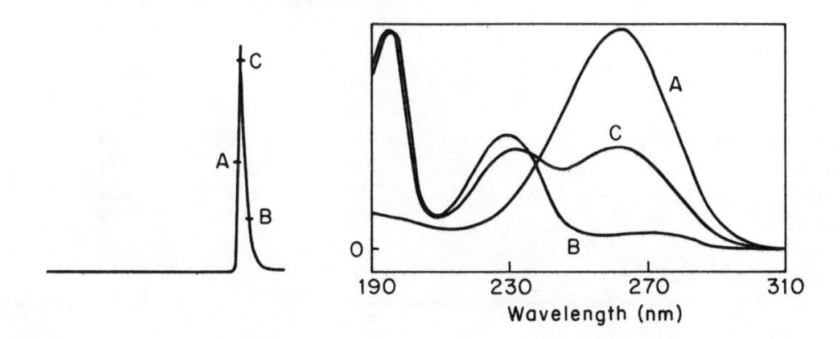

Figure 11.8 Determination of spectral purity of a peak by scanning the spectrum using a diode array spectrometer: A, on the up-slope; B, at the peak maximum; and C, on the down-slope. The front and back of the peak have different spectra, demonstrating that the peak contains two unresolved components. (Reproduced by permission of International Scientific Communications Inc. from *Inter. Laboratory.)*

Because they are very versatile and are supported by sophisticated computer software for the interpretation of the results, diode array detectors are relatively expensive (3–6 times more costly than a variable-wavelength detector). For most applications the simpler monochromatic detectors are just as useful and are often more sensitive.

iv. Eluent restrictions with ultraviolet detection

A major constraint in the use of ultraviolet spectrophotometric detection is that the eluent must not significantly absorb light at the wavelength of detection. If the eluent absorbance is greater than 1.0, the remaining light reaching the detector will be too weak and the background signal will become excessively noisy. It will therefore be difficult to make accurate measurements of the changes caused by the sample. Fortunately, the most popular solvents used for reversed-phase chromatography, namely water, methanol and tetrahydrofuran, all show little absorbance above 210 nm (Table 11.2 and Figure 11.9). Acetonitrile can be more difficult, and hence more costly, to purify and is usually sold in both a general HPLC grade and a specially purified far-UV HPLC grade for work lower than 230 nm. On standing in the air, the absorbance background of tetrahydrofuran increases significantly and opened bottles should be flushed with nitrogen,

Table 11.2 Spectral cut-off wavelength and refractive index values of solvents used as eluents in HPLC.

Solvent	Cut-off wavelength, OD > 1.0	Refractive index at 20°C
Acetone	330	1.359
Acetonitrile	230	1.344
Acetonitrile (far-UV grade)	<200	1.344
Chloroform	245	1.444
Cyclohexane	<200	1.427
Ethyl acetate	255	1.372
Dichloromethane	233	1.424
Diethyl ether	205	1.352
Dioxane	218	1.422
Hexane	<200	1.372
Methanol	205	1.328
Propan-2-ol	205	1.377
Tetrahydrofuran	215	1.407
Toluene	286	1.496
2,2,4-Trimethylpentane (iso-octane)	<200	1.391
Water	<200	1.333

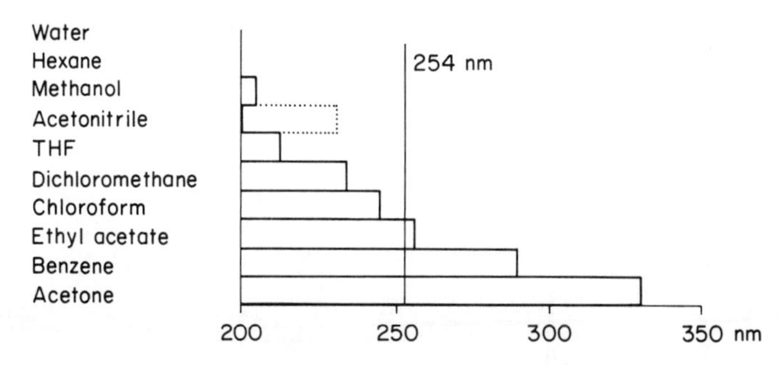

Figure 11.9 Wavelengths at which absorbance is greater than 1.0 for different eluents commonly used in normal- and reversed-phase HPLC. In these areas, ultraviolet spectrophotometric detectors will be unable to operate.

if they are to be stored for prolonged periods. It is not normally necessary to use the very expensive spectroscopic-grade solvents for HPLC. Care must also be taken with any additives to the mobile phase, such as buffer salts or ion-pair reagents, and analytical-reagent grade chemicals should always be used. Phosphate, acetate, or citrate buffers cause few problems but phthalate buffers are unsuitable. As none of the commonly used reversed-phase eluents absorbs UV light, the baseline is unaffected in progammmed elution or if the pH or ionic strength of the mobile phase is altered.

The solvents used for normal-phase chromatography cause more problems. The need for the eluent to be compatible with the detection wavelength may limit the choice of mobile phase for the separation. This often means that eluents used in normal-phase TLC cannot be directly transferred to HPLC systems. Aromatic hydrocarbons and compounds with double bonds, such as acetone or ethyl acetate, cannot be used over a wide wavelength range (Table 11.2 and Figure 11.9). Although the chlorinated solvents, chloroform and dichloromethane, may be used at 254 nm, they show significant absorbance at lower wavelengths. Hydrocarbons, hexane and iso-octane, and alcohols, such as propanol and ethanol, are therefore popular in mixtures with tetrahydrofuran or di-isopropyl ether.

The sample for analysis should usually be prepared as a solution in the mobile phase. Even if a sample solvent has no absorbance, as its relatively high concentration is eluted, the change in the refractive index of the eluent will bend the light beam passing through the detector flow cell. This produces a characteristic differential "refractive index" peak (Figure 11.10). Some manufacturers claim to have reduced this effect with improved cell design. This peak is often used erroneously as an indication of the column void volume but the solvent may itself have been retained by the column.

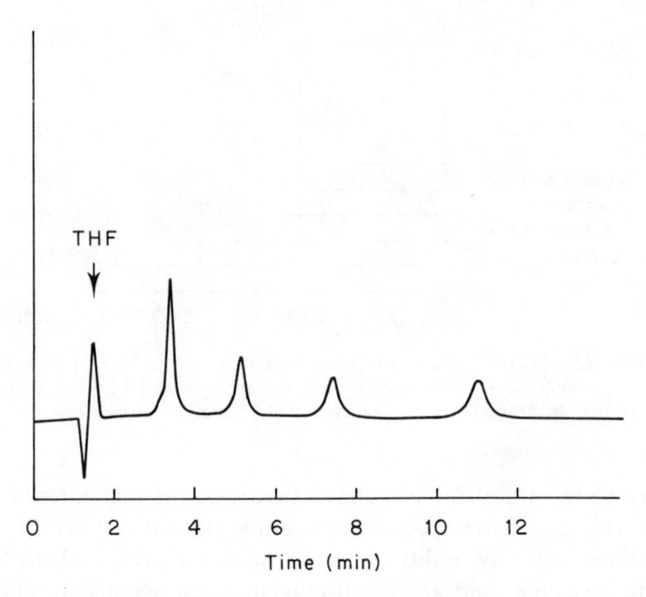

Figure 11.10 Refractive index peak caused by difference between the THF used as the sample solvent for a mixture of alkylphenones (acetophenore–valerophenene) and the mobile phase of THF–water 35:65. Column ODS-Hypersil 5 μm, 100 × 5 mm.

b. Fluorescence detectors

In order to detect an analyte by fluorescence spectrophotometry, it must contain a group that possesses fluorescence properties and the detector must be set to the correct excitation and emission wavelengths for this group. Relatively few compounds are inherently fluorescent but often analytes, such as aliphatic amines or acids, can be converted into fluorescent derivatives. Because of the spectroscopic and chemical specificity, fluorescence detection is very selective. As fluorescence is an emission technique in which the background signal in the absence of a fluorophore is virtually zero, it is also a very sensitive method, often up to 100–1000 fold greater than for the same compound in absorbance spectrophotometry.

The high selectivity means that the design of a fluorescence HPLC detector can be very simple and sophisticated optics are not necessary (Figure 11.11). The excitation and emission wavelengths can be selected using narrow bandpass filters, although dual or single monochromator fluorimeters are also used. The flow cell is typically larger, up to 25 μl, than in ultraviolet spectroscopic detection but the effective volume is similar because only part

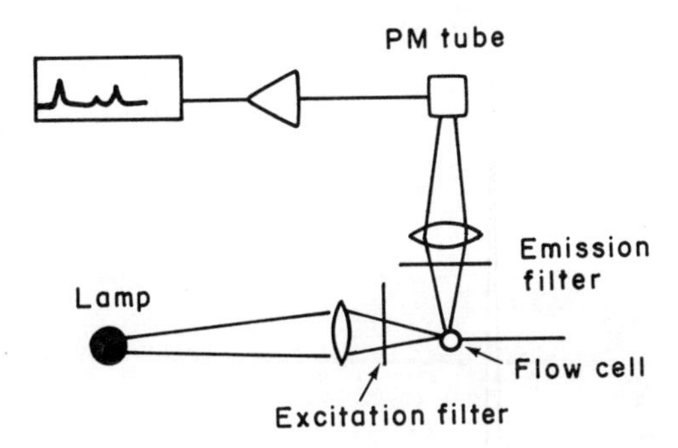

Figure 11.11 Schematic diagram of a simple dual-filter fluorimeter. Excitation using a xenon lamp. Filters used to select the wavelength in both the excitation and emission beams.

of the cell is illuminated with the excitation beam. Special flow cells have also been designed for HPLC in which the emitted light over an arc of 180° is collected with a mirror and focused on the detector, rather than just detecting the light emitted at right angles to the excitation beam. Because they are so selective, fluorescence detectors are often used in sequence with an ultraviolet detector acting as a "universal" detector.

As none of the commonly used HPLC solvents contains a fluorophore, the choice of the eluents for use with the fluorescence detector is not restricted, except that the fluorescence of some analytes may be quenched by the components of the eluent.

Fluorescence detection has found particular application in the detection of polynuclear aromatic hydrocarbons (PAHs) in waste water, the determination of riboflavin in vitamin tablets, the monitoring of aflatoxins in stored cereal products, and the confirmation of the presence of LSD (lysergic acid diethylamide) in forensic samples (Figure 11.12). In each case, the added specificity of the fluorescent detection compared to the ultraviolet detector aids confirmation of the identification. For complex mixtures such as environmental samples, the specific wavelengths required for the fluorescence of the different PAHs may mean that not all the analytes can be detected using a single pair of wavelength settings and repeated runs or a spectrometer that can alter the wavelengths during the separation will be needed. Some compounds that are normally considered to be fluorescent cannot be readily detected following HPLC separations, because they require conditions of pH or solvents that cannot be used with reversed-phase columns. For example, quinine, often used as a standard in fluorescence spectroscopy, only emits strongly if the solvent pH is less than 2.0.

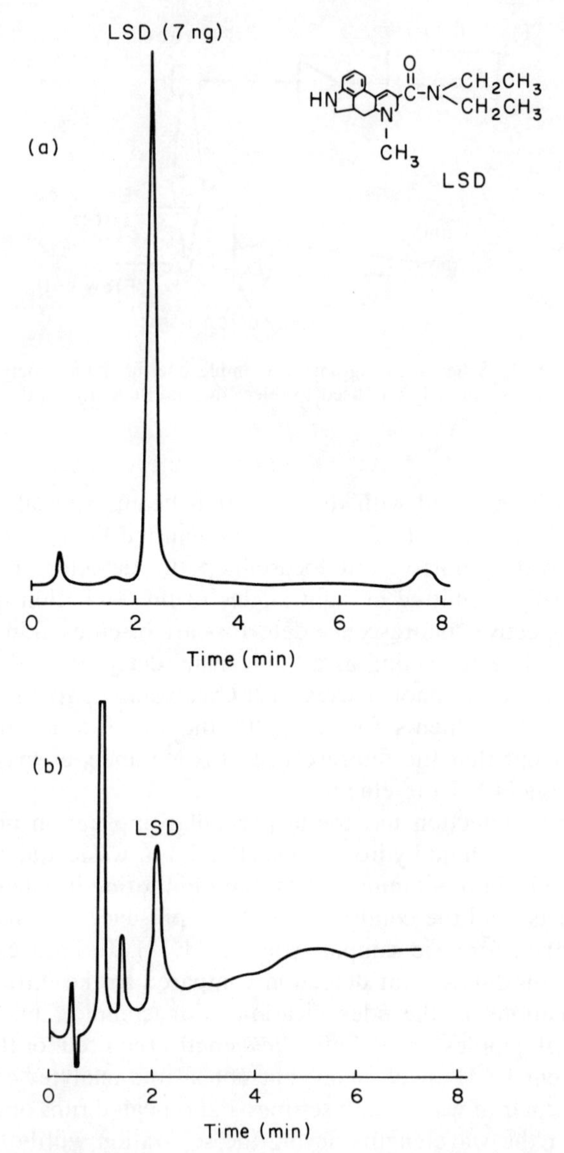

Figure 11.12 Example of the selectivity of fluorescence detection and its application to forensic analysis for the separation of the drug LDS. (a) Fluorescence detection: excitation wavelength 330 nm filter and emission all wavelengths >430 nm. (b) Ultraviolet absorbance detection at 254 nm. Column Micropak MCH-10, 250 mm × 2.1 mm. Eluent acetonitrile–aqueous 0.1 M ammonium carbonate 50:50 at 1.0 ml min^{-1}. (Reproduced by permission of Varian Associates, from *LC at work*, No. 32.)

11.3. ELECTROANALYTICAL DETECTORS

Two main electroanalytical methods have been used for the detection of analytes based either on changes in the conductivity of the eluent or on the electrochemical reduction or oxidation of the analyte.

a. Conductometric detectors

Analytes that are ionised and carry a charge can be detected by measuring the changes in the conductivity of the eluent between two electrodes. To obtain high sensitivity, the background conductivity of the eluent should be low and therefore only weak buffer concentrations can be used. Conductometric detection has therefore been mainly used to detect inorganic anions and cations (Figure 8.5) and organic acids (Figure 11.13) on elution from weak ion-exchange columns. Its principal application has been as the basis of ion chromatography (see section 12.5).

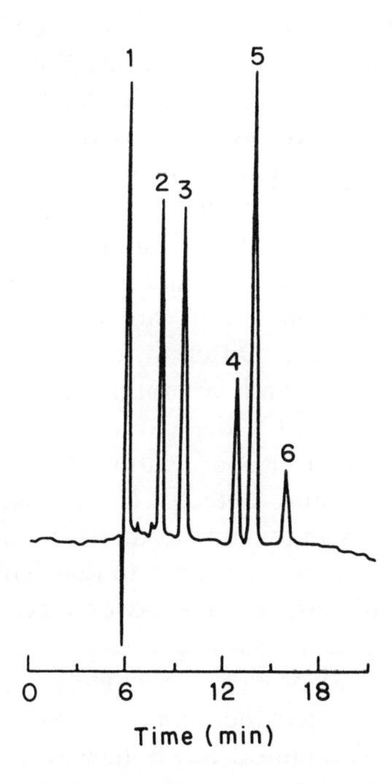

Figure 11.13 Detection of organic acids with a conductivity detector on elution from a weak ion-exchange column. Eluent 0.001 M sulphuric acid. Peaks: 1, oxalic acid; 2, maleic acid; 3, malic acid; 4, succinic acid; 5, formic acid; 6, acetic acid. (Reproduced by permission of Wescan Instruments.)

b. Amperometric and coulometric detectors

Some compounds can be reduced or oxidised in solution by the addition or removal of electrons at an electrode surface. These electrochemical reactions will take place when a positive or negative potential is applied to the solution, which is greater than the characteristic half-wave potential for the analyte. This concept is the basis of polarography, one of the earliest methods applied to the determination of metal ions in dilute solutions. A dropping mercury electrode is used and the current flowing through the cell is monitored as the applied potential is increased. In this amperometric mode, only a small proportion of the analyte reacts and the current that flows through the solution is proportional to the concentration of the analyte. Alternatively, if a fixed potential is applied to the solution, the total number of coulombs (current × time) needed to react fully with the sample can be used to determine the number of moles of the analyte. The same concepts can also be applied, usually using glassy carbon electrodes, to organic compounds, such as phenols and aryl amines, that can be electrochemically oxidised, and to aromatic nitro and azo compounds that can be reduced. Non-reactive compounds give no response. Electrochemical detection is therefore highly selective and even a reactive analyte will not give a response unless the appropriate potential is being used.

Rather than scanning the applied potential, in an amperometric detector for HPLC the potential is usually held constant just above the half-wave potential of the analyte (Figure 11.14), and the current flowing across the cell is continuously monitored. This method provides a selective and very sensitive means of detection, often about 10^4 better than ultraviolet spectrophotometric detection. Different designs of detector have been marketed or developed in research laboratories. They are known collectively as LC electrochemical (LCEC) detectors or sometimes electrochemical detectors (ECD), but this latter abbreviation is best avoided as it can cause confusion with electron capture detector (ECD) used in GLC. Most of the commercial detectors work in the amperometric mode because in the flowing eluent the analyte does not have a chance to react completely in the detector cell. Some detectors with larger surface-area electrodes claim coulometric efficiency.

The eluent is buffered to provide a relatively high concentration of ions in the solution. This enables the mobile phase to act as a supporting electrolyte for any electrochemical reaction, which ensures that the current being measured is limited only by the concentration of the analyte and not by the conductivity of the eluent. LCEC detection cannot therefore be used with non-aqueous eluents and is effectively restricted to reversed-phase

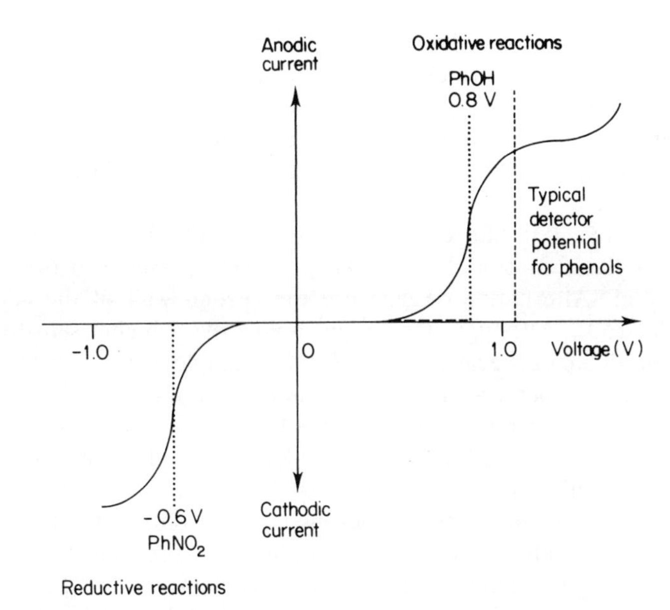

Figure 11.14 Polarographic trace detection showing polarographic waves for typical oxidation analyte and reductive analyte.

separations. The buffer also controls the ionisation of analytes such as phenols and amines, and can be used to optimise their response.

Although LCEC detection has been known for some time, its application has been fairly limited, partly because as a general analytical technique the polarography of organic compounds has not been widely accepted outside a limited group of adherents. A wider use of LCEC detection followed the development of simple and reliable glassy carbon or metal electrodes, which replaced the dropping mercury electrode.

All the commercially available LCEC detectors contain three electrodes. The reaction takes place at the working electrode, usually made from glassy carbon. Gold and platinum electrodes can also be used and have been found to be preferable for some samples, particularly sulphur-containing drugs. The working electrode is held at a fixed potential above a silver/silver chloride or calomel reference electrode. The third counter (or auxiliary) electrode is used to maintain the potential at a constant value as the analyte is eluted. This electrode is necessary because, as the electrochemical reaction occurs and a current flows through the cell, the internal resistance of the detector cell alters.

In the wall-jet type of detector (Figure 11.15a), the eluent enters the detector cell as a jet, which is directed at the glassy carbon working electrode. This design gives a very small working volume, less than 1–2 μl, and is suitable for 4.6 mm and microbore columns. The counter-electrode is built into the tip of the jet and the reference electrode is placed downstream in the waste tube.

The alternative thin-film detector (Figure 11.15b) has a relatively large-surface-area glassy carbon electrode. The element passes over the electrode as a thin film. Alternative designs use the second wall of the cell as the counter-electrode. Although this design should give a higher sensitivity than the wall-jet detector because more of the sample can react, the larger working area also increases the background signal and both designs have a similar signal-to-noise ratio. This design is less susceptible to flow pulsations caused by the pump, which can be a problem with wall-jet electrodes if the eluent has a significant background signal.

These two designs are both amperometric and only a small proportion (2–10%) of the analyte is oxidised or reduced in the detector. The third commercial design of detector is essentially coulometric but again it is the current that is monitored. In this detector the eluent flows through a working electrode made of a block of porous glassy carbon with a large surface area (Figure 11.15c). This gives a much longer contact time for the analyte, so that the electrochemical reaction is essentially complete and the current depends on the total concentration of the analyte. The response is higher than from the amperometric detectors but, because the electrode surface is also larger, higher background currents are observed.

A number of modifications of these three basic designs have been made by different manufacturers. A particularly interesting development has been detectors with dual cells, which have the capability for simultaneous or sequential measurements at two different potentials on the same eluent. If, at the first electrode, a high proportion of the analyte is reacted, by using a complementary potential at the second electrode the reverse electrochemical reaction can be monitored, giving additional selectivity. For example, if a aromatic nitro group is reduced at the first electrode the resulting aromatic amine can be oxidatively monitored at the second electrode without the interferences frequently encountered in reduction reactions (see below).

Almost all the applications of the LCEC detector have been for the analysis of oxidisable groups, particularly phenols and aromatic amines, which have $E_{1/2}$ potentials about +0.4 to +1.0 V compared to an Ag/AgCl reference electrode (Table 11.3). A major application has been for the routine determination of the catecholamines. These phenolic compounds are of clinical importance but are normally present in blood and tissues at levels too low for detection by spectrophotometric detectors (Figure 11.16). Another clinical application has been in the monitoring of estrogen steroids in pregnant women as a guide to the health of the foetus.

Oxidative reactions can be carried out over the range 0 to about +1.2 V. At higher potentials (depending on the pH), the aqueous eluent is oxidised to hydrogen and oxygen and the background signal increases sharply.

Although reductive electrochemical reactions can be monitored, the potential required for the reaction of the nitro group (−0.6 V) is more negative than the reduction potential of oxygen (−0.4 V) on the glassy carbon surface. Air and oxygen must therefore be rigorously excluded from the eluent and the sample solution by flushing with nitrogen. With practice and care, high sensitivity can be obtained and this method has found particular application in the detection of explosive residues and some drug groups.

Much of the potential of electrochemical detection has yet to be developed, and relatively few applications have been demonstrated in the literature. The detectors have also suffered from poor reproducibility because of surface contamination and deactivation of the electrode by the eluent or analyte reaction products. Careful selection of eluent and buffer components is therefore required and analytical reagent-grade chemicals should be used for the preparation of buffers and eluents. However, the very high sensitivity and selectivity of LCEC detectors make them ideal for trace analysis. A probable area of growth, which would expand the scope of application of LCEC detection, is the use of electrochemically active derivatives of otherwise unreactive compounds, although as yet only a limited number of derivatisation reactions have been explored.

11.4 REFRACTIVE INDEX DETECTORS

As any analyte whose refractive index is different from that of the mobile phase is eluted from a column, the refractive index of the eluent will alter. These changes can be monitored using a refractive index detector to provide potentially a nearly universal method of detection. However, because a bulk property of the eluent is being measured, it will be an inherently less sensitive method than with more specific detectors. A comparison of the refractive index values of typical eluent components and analytes (Table 11.2) confirms that the addition of traces of a typical organic compound to a water/methanol eluent will have only a small effect on the overall refractive index. For different analytes, their responses will depend on the difference between their refractive index values and that of the eluent. Unfortunately, as these differences can be both positive or negative, the peaks can also be

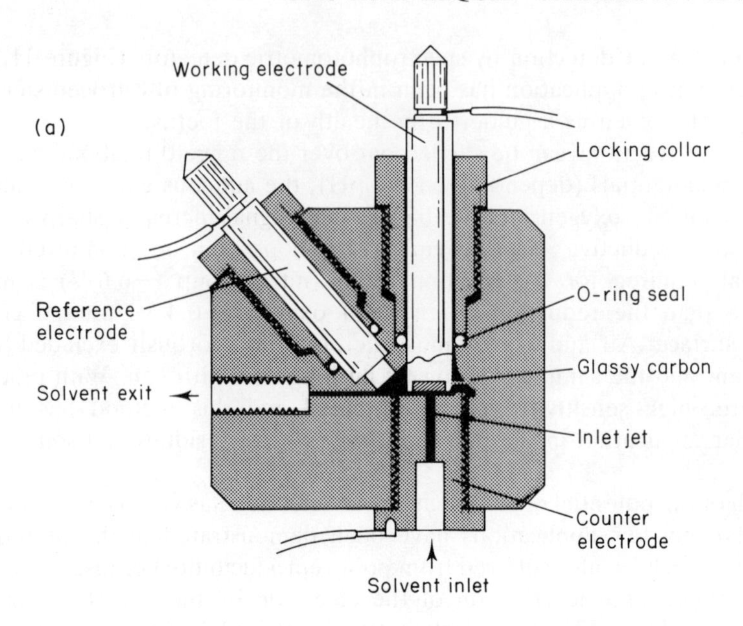
(a)
Working electrode
Locking collar
Reference electrode
O-ring seal
Glassy carbon
Solvent exit
Inlet jet
Counter electrode
Solvent inlet

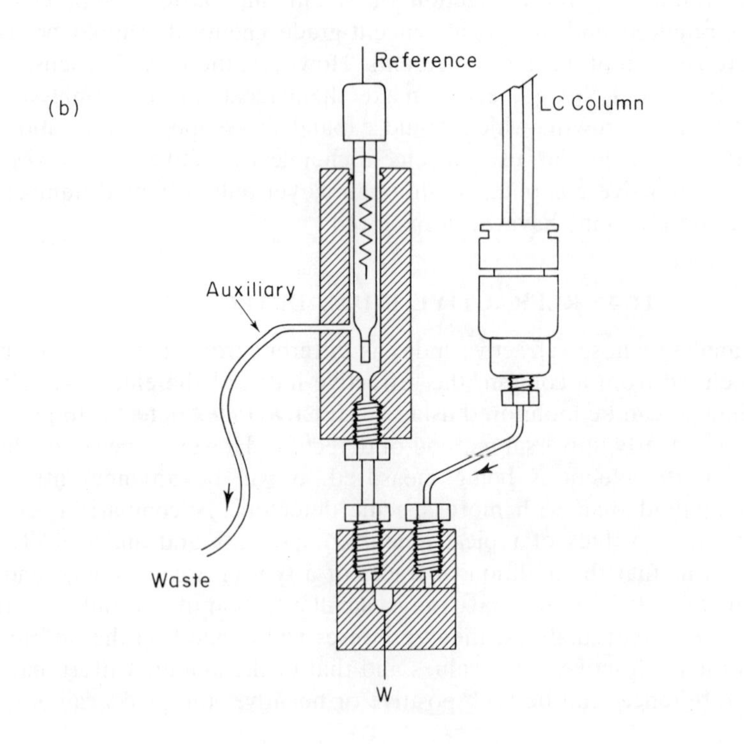
(b)
Reference
LC Column
Auxiliary
Waste
W

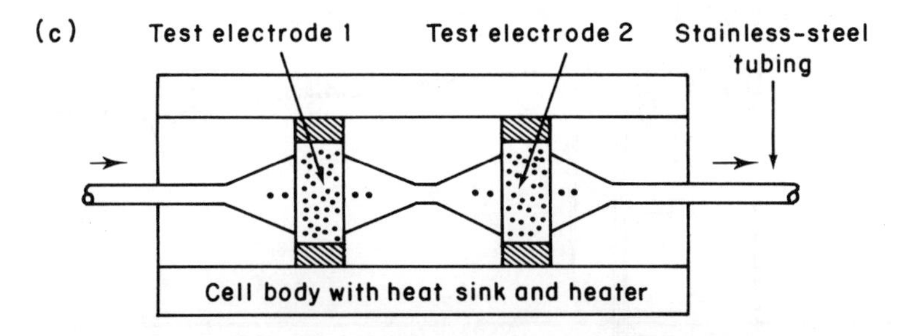

Figure 11.15 Detectors used in LCEC detection with glassy carbon electrodes: (a) (opposite) wall-jet detector cell; (Reproduced by permission of EDT Research). (b) (opposite) thin-film detector (w, working electrode); Reproduced by permission of BioAnalytical Systems). (c) dual porous carbon detector cells. (Reproduced with permission Coulochem).

Table 11.3 Electrochemically active compounds and groups. Half-wave potentials $E_{1/2}$ are approximate and depend on solvent pH and composition.

Type of compound	Examples	$E_{1/2}$ (V)
Oxidisable compounds		
Phenols	Phenol	0.8
	Chlorinated phenols	0.6 to 0.9
	Hydroquinone	0.5
	Catecholamines	0.8
	Estrogens	0.8
	Morphine	1.0
Aromatic amines	Aniline	0.9
	Benzidines	0.5
	Hydrazines	1.0
Heterocyclic compounds	Indoles	1.0
	Tryptophan	1.1
Other compounds	Ascorbic acid	0.6
	Isocyanates	0.8
	Thiols	0.8
	Cysteine	0.15
	Phenothiazine	1.2
Reducible compounds		
Nitro aromatics		−0.3 to −0.6
Diazo compounds		−0.5
Anthraquinones		−0.5

in either direction, which could be confusing if complicated samples are being analysed (Figure 11.17).

Three designs of refractive index detector have been used with comparable but relatively low sensitivities (Figure 11.18). All have a wide linear concentration range and can be used in preparative chromatography when

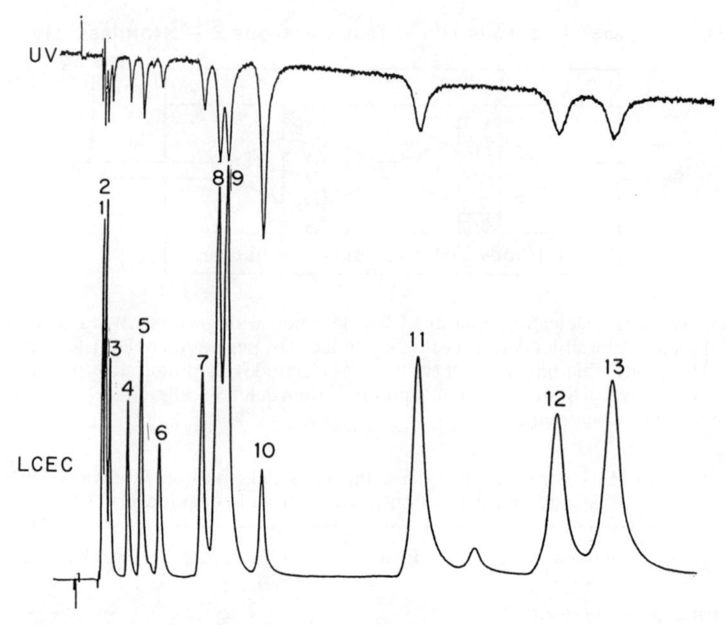

Figure 11.16 Separation and comparison of detection of catecholamines using a spectroscopic detector at 280 nm and thin-film electrochemical detector at 0.80 V. Peaks: 1, 3,4-dihydroxymandelic acid, 25 ng; 2, vanilmandelic acid, 25 ng; 3, 2,5-dihydroxyphenylacetic acid, 25 ng; 4, 3,4-dihydroxyphenylacetic acid, 25 ng; 5, 5-hydroxyindole-3-acetic acid, 25 ng; 6, homovanillic acid, 25 ng; 7, adrenaline, 125 ng; 8, 3,4-dihydroxyphenyl alanine (Dopa), 125 ng; 9, noradrenaline, 60 ng; 10, octopamine (norsynephrine), 750 ng; 11, metanephrine, 250 ng; 12, isoprenaline, 250 ng; 13, 3-hydroxytyramine, 250 ng. Injection volume 5 μl. (Reproduced by permission of Spark Holland.)

the high sample levels would swamp other detectors. Because refractive indices are temperature-dependent, these detectors are more sensitive to changes in the ambient temperature than other detectors. The Fresnel refractometer operates by detecting the loss of the intensity of a refracted beam of light when reflected near its critical angle (Figure 11.18a). The detector must therefore be aligned for each different eluent. White light from a tungsten lamp is usually employed and two cells, one eluent and one mobile phase, are operated in parallel and the outputs are compared.

The deflection refractometer uses a two-compartment cell with the reference mobile-phase and eluent solutions on either side of a window (Figure 11.18b). Initially it is adjusted so that the proportion of a light beam falling on either side of a prism is balanced. Any changes in the refractive index of the eluent will cause the light beam to bend. The proportion of

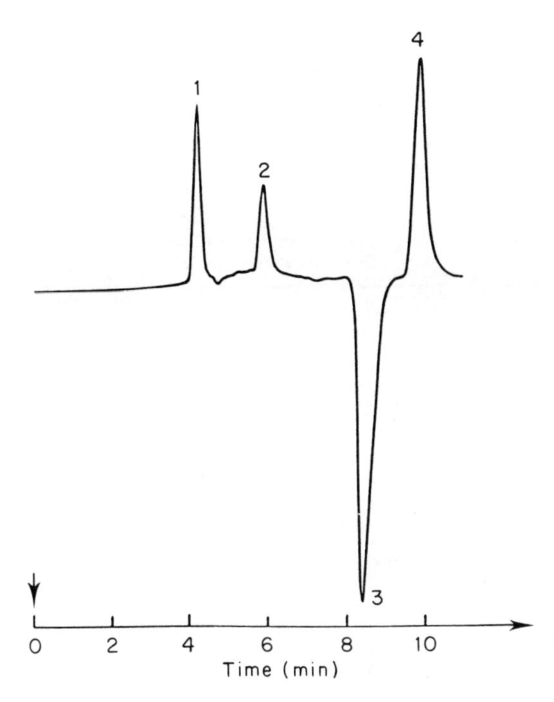

Figure 11.17 Separation of glyceryl esters using a refractive index detector, demonstrating the problem of both positive and negative peak formation. Column LiChrosorb Si 60, 250 mm × 4.6 mm. Eluent chloroform. Peaks: 1, glyceryl tripalmitate; 2, 1,3-glyceryl dipalmitate; 3, glyceryl monopalmitate; 4, cholesterin. (Reproduced with permission of E. Merck, Darmstadt.)

light falling on the two sides of the prism will alter and hence the response from the two photocells will change. The cell sizes are relatively large and this method cannot therefore be used for microbore chromatography.

The interferometer detector is probably the most sensitive method (Figure 11.18c). Changes in the refractive index of the eluent in one cell will cause the light beams to interfere partially on recombination and hence the intensity of the output from the photocell will change.

Although these refractive index detectors will give a nearly universal response and are potentially very widely applicable, their use has been limited by the poor sensitivity. Their application has also been restricted because they cannot be used with programmed elution, as any change in the eluent composition will alter its refractive index and hence change the baseline signal. They are principally employed only when ultraviolet–visible

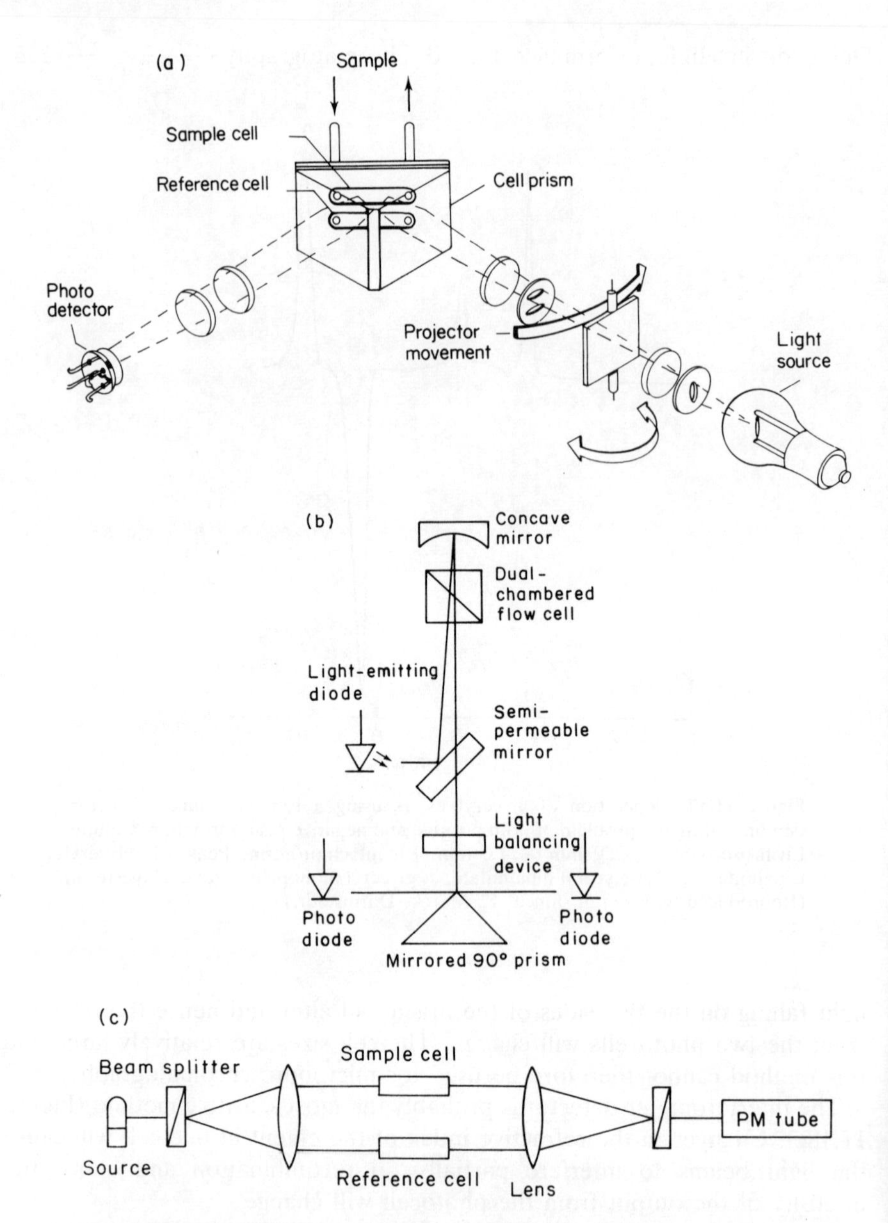

Figure 11.18 (a) Design of a Fresnel refractive index detector (Reproduced by permission of E.I. Dupont de Nemours & Co.) (b) Schematic diagram of a dual-cell refractive index detector. As the refractive index changes in the eluent side of the flow cell, the light beam is deviated onto the prism at various angles, and the two photodiodes receive varying amounts of light. Each photodiode sends a current to a differential amplifier according to the amount of light received. (Reproduced by permission of Gilson). (c) Design of an interferometer refractive index detector.

spectrophotometric detectors cannot be used. This is usually because the analyte of interest lacks a chromophore. Typical applications have been for aliphatic biomolecules that lack unsaturated groups, particularly sterols, fatty acids (see Figure 11.17) and carbohydrates (see Figure 12.8).

11.5 OTHER LIQUID CHROMATOGRAPHY DETECTORS

Other detectors are commercially available for HPLC but none has gained wide acceptance. One of the most interesting is the mass or evaporative detector, in which the eluent stream is sprayed into a heated air stream. The volatile eluent solvents evaporate, leaving any involatile analytes as particles or aerosols, which can be detected because they will scatter a beam of light. This method has found particular application for the detection of fatty acids, lipids and carbohydrates as an alternative to the refractive index detector.

In the photoconductivity detector, the eluent stream is irradiated with light. Selected analytes decompose to give ions, which can be detected using a conductivity cell. This detector has been used for organochlorine pesticides and triazine herbicides, but generally it is not specific enough for widespread application.

Because the eluent often interferes with detection in HPLC, a number of transport detectors have been developed. The eluent is fed onto a moving wire, belt, or ceramic disc and transported to a heated zone where the solvent is evaporated. The residual involatile analytes, free of eluent, are then detected by either passing the wire through the flame of a flame ionisation detector (FID) flame or by pyrolysis to give carbon dioxide, which is then detected. However, despite numerous attempts, none of the commercial systems has been completely successful and studies are still continuing. These detectors have been particularly popular for some applications, such as lipids and carbohydrates, as alternatives to refractive index detectors. Attempts have also been made to pass the eluent from an HPLC to thermionic ionisation or electron capture detectors, either via a transport system or directly if the eluent components do not give a response.

Numerous other concepts have been reported and a recent survey lists over 20 “minor” detector types, including viscosity, photoacoustic spectrometry, permittivity and streaming current, but only time will test whether any will prove more useful than the established detectors.

11.6 COUPLED LIQUID CHROMATOGRAPHY DETECTORS

In order to gain additional information about the analyte, the eluent from an HPLC system can be passed to a second analytical technique, such as to a radiochemical detector, to atomic absorption and inductively coupled plasma spectrophotometers for metal-ion analysis, to a polarimeter for the assay of chiral compounds, or to infrared, nuclear magnetic resonance, or mass spectrometers. Radiochemical detection is valuable in metabolism studies or for the speciation of radioactive compounds, and atomic absorption spectrometers are invaluable for metal speciation studies in environmental investigations.

Although potentially the most informative of all the coupled detectors, infrared and mass spectrometers have not gained as widespread an application as in GLC, because of experimental problems caused by the eluent.

a. *LC–infrared spectroscopic detection*

As in GLC, infrared detection has been transformed by the use of Fourier transform infrared (FTIR) spectrometers. Complete spectra can be measured at frequent intervals and the spectra of the analytes can be compared with standard solution spectra for identification purposes. However, unlike GLC, in HPLC the mobile phase often also possesses a strong infrared spectrum, which may mask the spectrum of the analyte. Organic solvents and normal-phase separations are most easy to handle as the spectra of the eluents can contain windows with only weak background absorbance. However, in reversed-phase separations the broad absorbance bands of water and methanol cause considerable problems and it appears likely that LC–FTIR will only find limited application.

Limited use has also been made of fixed-wavelength infrared spectrometric detection, such as the monitoring of the CH band at 3010 cm^{-1} from polyethylenes separated by size exclusion chromatography.

b. *LC–mass spectrometric detection*

Potentially this is the most powerful identification technique for use in HPLC. However, its adoption has been fraught with practical problems, because as the eluent evaporates it swamps the vacuum in the mass spectrometer. Ideally, the solvent should be removed so that only the analyte is measured. This can be achieved using a transport detector, in which the eluent is sprayed onto a circulating belt and the solvent is evaporated in stages so that only the analyte is carried into the mass spectrometer (Figure

11.19a). This method seemed to be satisfactory for some analytes but not for others, and can be used for chemical ionisation (CI) and electron impact (EI) mass spectroscopy. The former technique gives spectra with good molecular-ion information but little structural information from fragmentation ions. The EI method may only give weak molecular ions, which can be difficult to recognise, but it gives characteristic fragmentation patterns, which can be used for identification.

By using microbore columns with only a low flow rate of a few microlitres per minute, direct introduction of the eluent into the mass spectrometer can be carried out, but the sample capacity of the columns is so low that detection can be difficult.

A recent advance has been the introduction by Vestal of the thermospray source (Figure 11.19b). The eluent, usually containing ammonium acetate,

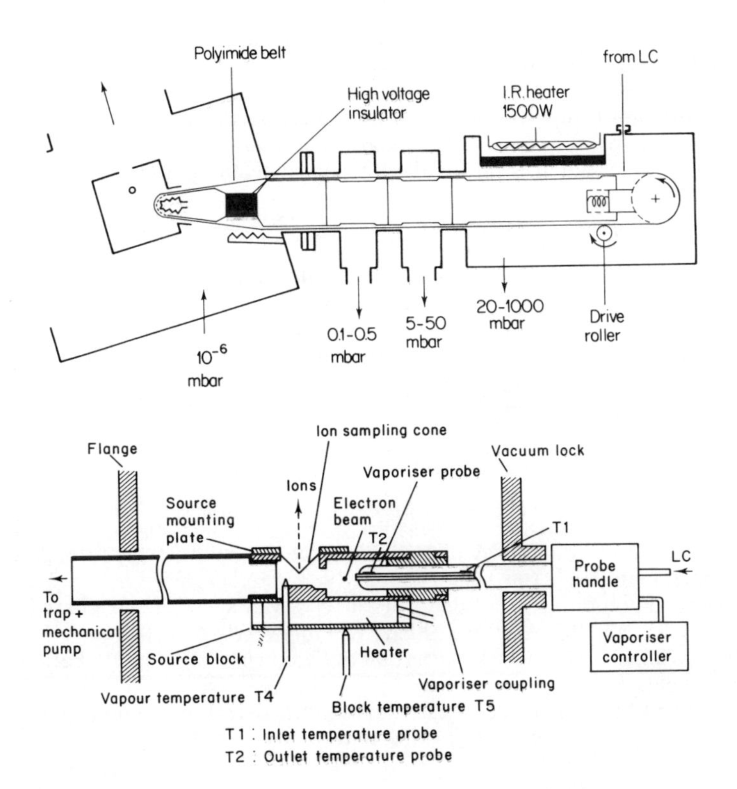

Figure 11.19 Interfaces for LC–mass spectrometry: (a) moving belt interface (Reproduced by permission of VG–Organic Ltd.) (b) thermospray interface (Reproduced by permission of Vestec Corp.)

passes through a heated jet into the evacuated source chamber. The eluent forms a supersonic mist of electrically charged particles. The eluent vaporises and simultaneously ionises the analyte. This method can only give CI spectra. A number of reports have suggested that this method of sample introduction has the widest potential but other mild forms of ionisation, including atmospheric pressure ionisation, are also being examined and this is still an area of current development.

In each case, the modes of application of the mass spectrometer are very similar to those used in GLC–MS and either single- and multiple-ion, or repeated scanning can be carried out.

BIBLIOGRAPHY

General HPLC detectors

Many general texts on HPLC also have extensive chapters on HPLC detectors (see Bibliography to Chapter 10), so only specific texts and references are included here.

J.L. DiCesare and L.S. Ettre, "New ways to increase the specificity of detection in liquid chromatography", *J. Chromatogr., Chromatogr. Rev.*, 1982, **251**, 1–16.

"LC detectors: the search is on for the ultimate detector", *Anal. Chem.*, 1982, **54**, 327A–333A.

R.P.W. Scott, *Liquid chromatography detectors*, 2nd ed., J. Chromatogr. Library, Vol. 33, Elsevier, Amsterdam, 1986.

T.M. Vickrey (Ed.), *Liquid chromatography detectors*, Chromatographic Science Series, Vol. 23, Marcel Dekker, New York, 1983.

P.C. White, "Recent developments in detection techniques for high performance liquid chromatography. Part I. Spectroscopic and electrochemistry detectors", *Analyst*, 1984, **109**, 677–697; "Part II. Other detectors", *ibid.*, 973–983.

E.S. Yeung (Ed.), *Detectors for liquid chromatography*, John Wiley, New York, 1986.

UV–visible/fluorescence detectors

S.A. George and A. Maute, "A photodiode array detection system: design concept and implementation", *Chromatographia*, 1982, **15**, 419–425.

D.G. Jones, "Photodiode array detectors in UV–VIS spectroscopy: Part I", *Anal. Chem.*, 1985, **57**, 1057A–1073A; "Part II", *ibid.*, 1207A–1214A.

A.T. Rhys Williams, *Fluorescence detection in liquid chromatography*, Perkin Elmer, Beaconsfield, 1980

E.S. Yeung and M.J. Sepaniak, "Laser fluorometric detection in liquid chromatography", *Anal. Chem.*, 1980, **52**, 1465A–1481A.

Electrochemical detectors

K. Bratin and P.T. Kissinger, "Reductive LCEC of organic compounds", *J. Liquid Chromatogr.*, 1981, **4** (Supp I.2), 321–357.

P.T. Kissinger, "Amperometric and coulometric detectors for high-performance liquid chromatography", *Anal. Chem.*, 1977, **49**, 447A–456A.
A.M. Krstulović, H. Colin and G.A. Guiochon, "Electrochemical detectors for liquid chromatography", *Adv. Chromatogr.*, 1984, **24**, 83–124.
J.B.F. Lloyd, "High-performance liquid chromatography of organic explosives components with electrochemical detection at a pendant mercury drop electrode", *J. Chromatogr.*, 1983, **257**, 227–236.
D.A. Roston, R.E. Shoup and P.T. Kissinger, "Liquid chromatography/electrochemistry: thin-layer multiple electrode detection", *Anal. Chem.*, 1982, **54**, 1417A–1434A.
R.J. Rucki, "Electrochemical detectors for flowing liquid systems", *Talanta*, 1980, **27**, 147–156.
T. Ryan (Ed.), *Electrochemical detectors. Fundamental aspects and analytical applications*, Plenum, New York, 1984.
K. Štulík and V. Pacáková, "Electrochemical detection in high-performance liquid chromatography", *CRC Crit. Rev. Anal. Chem.*, 1982–84, **14**, 297–352.

Other detectors

J.M. Charlesworth, "Evaporative analyzer as a mass detector for liquid chromatography", *Anal. Chem.*, 1978, **50**, 1414–1420.
J.B. Dixon, "A new flame ionization detector (FID) for liquid chromatography", *Chimia*, 1984, **38**, 82–86.
D.H. Mourey and L.E. Oppenheimer, "Principles of operation of an evaporative light-scattering detector for liquid chromatography", *Anal. Chem.*, 1984, **56**, 2427–2434.
D.J. Popovich, J.B. Dixon and B.J. Ehrlich, "The photo-conductivity detector—A new selective detector for HPLC", *J. Chromatogr. Sci.*, 1979, **17**, 643–650.

Coupled detectors

P.J. Arpino, "Ten years of liquid chromatography–mass spectrometry", *J. Chromatogr.*, 1985, **323**, 3–11.
P.J. Arpino and G. Guiochon, "LC/MS coupling", *Anal. Chem.*, 1979, **51**, 682A–701A.
C.R. Blakley and M.L. Vestal, "Thermospray interface for liquid chromatography/mass spectrometry", *Anal. Chem.*, 1983, **55**, 750–754.
T.R. Covey, E.D. Lee, A.P. Bruins and J.D. Henion, "Liquid chromatography/mass spectrometry", *Anal. Chem.*, 1986, **58**, 1451A–1461A.
H.C. Dorn, "^{1}H-NMR: a new detector for liquid chromatography", *Anal. Chem.*, 1984, **56**, 747A–758A.
L. Ebdon, S. Hill and R.W. Ward, "Directly coupled chromatography–atomic spectroscopy. Part 2. Directly coupled liquid chromatography–atomic spectroscopy. A review", *Analyst*, 1987, **112**, 1–16.
D.E. Games, "High performance liquid chromatography/mass spectrometry (HPLC/MS)", *Adv. Chromatogr.*, 1985, **21**, 1–40.
P.R. Griffiths and C.M. Conroy, "Solvent elimination techniques for HPLC/FT-IR", *Adv. Chromatogr.*, 1986, **25**, 105–138.
S. Hill and L. Ebdon, "Directly coupled HPLC/flame atomic absorption spectroscopy

for metal speciation", *Eur. Spectrosc. News*, 1985, **58**, 20–24.

W.M.A. Niessen, "A review of direct liquid introduction interfacing for LC/MS. Part 1. Instrumental aspects", *Chromatographia*, 1986, **21**, 277–287; "Part 2. Mass spectrometry and applications", *ibid.*, 1986, **21**, 342–354.

M.L. Vestal, "High performance liquid chromatography–mass spectrometry", *Science*, 1984, **226**, 275–281.

CHAPTER 12

HIGH-PERFORMANCE LIQUID CHROMATOGRAPHY SEPARATION METHODS: COLUMNS AND MOBILE PHASES

12.1 SELECTION OF A SEPARATION METHOD

The choice of the most suitable chromatographic conditions for a particular HPLC analysis is complex and depends on a number of factors. The great versatility of HPLC is that many different separation methods can be carried out using the same basic equipment simply by changing the stationary phase and/or mobile phase. The two principal considerations are the molecular sizes and polarities of the analytes of primary interest and the nature of any other components in the sample. Even though there is often a tendency to regard reversed-phase chromatography as the automatic method of choice for most samples, this mode includes a number of variations and the most suitable option must still be selected. The mobile phase may include buffers or additives and different reversed-phase column materials can be used to modify the separation.

In most cases fairly simple considerations based on the properties of the analyte can lead to an initial set of separation conditions, which can then be used as a starting point for the refining or optimisation of the separation (Figure 12.1). The first specific group of analytes that can be recognised are high-molecular-weight compounds, such as proteins or biopolymers (Section

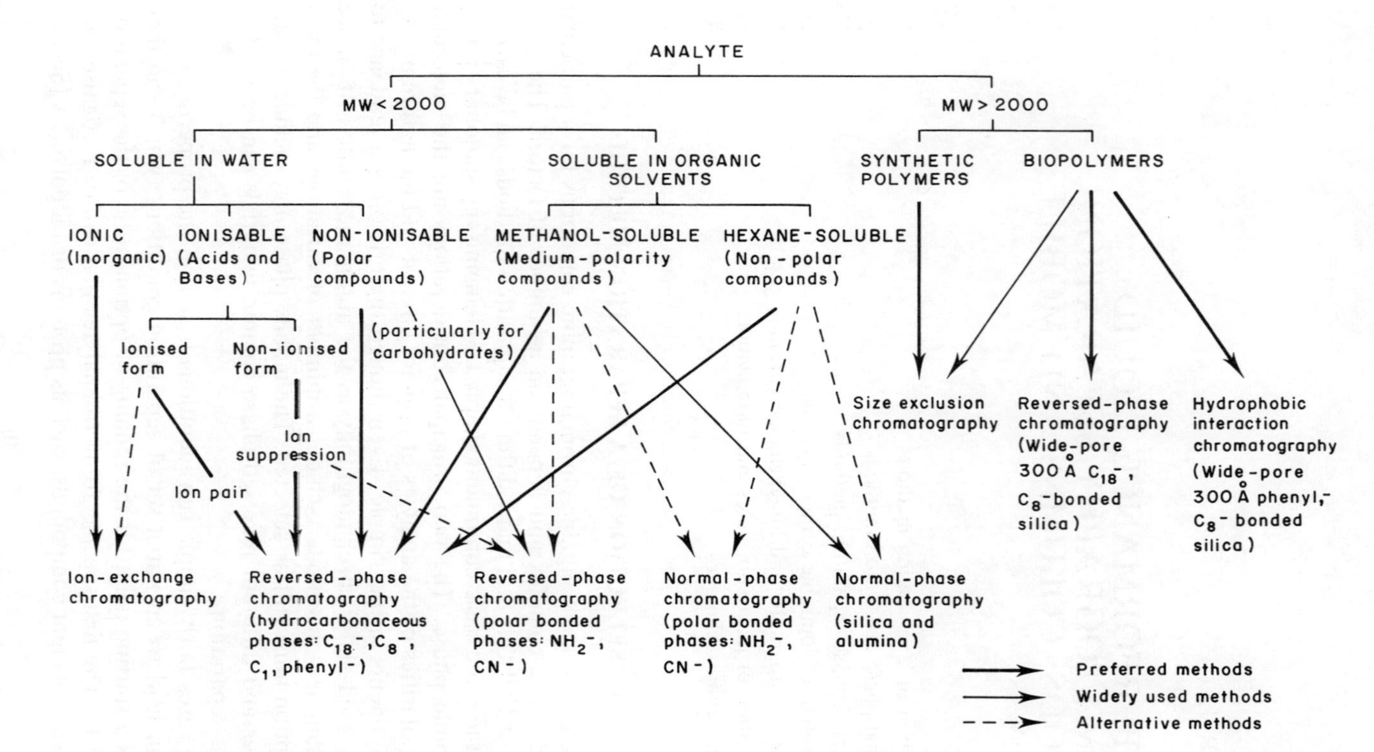

Figure 12.1 Selection of suitable liquid chromatographic methods for analytes depending on their structure and properties.

12.6). These can be separated on the basis of their molecular sizes using size exclusion chromatography. This method is most effective with molecular weights greater than 2000, although it has been used for smaller samples. The separation of many biological polymers, including proteins, can also be carried on wide-pore alkyl-bonded silica columns with aqueous–organic eluents using reversed-phase chromatography or hydrophobic interaction chromatography. A second group of clearly defined analytes are anions and cations, such as the alkali metals, halides or other inorganic salts, which can only be separated using ion-exchange chromatography (see Section 12.5). For most other analytes, which includes the majority of organic compounds, the separation method of choice depends very much on the other components in the sample or the sample matrix. The two principal alternatives are normal- or reversed-phase chromatography (Sections 12.2, 12.3 and 12.4).

Samples that contain a mixture of compounds with different functional groups can be analysed by normal-phase chromatography as, in this mode, separations depend on the differences in the polar interactions of the analytes with the polar stationary phase. However, these polar–polar interactions so dominate the separation that homologous or isomeric compounds with the same polar functional group will have very similar interactions and will only be poorly resolved. Similarly, very non-polar compounds, such as hydrocarbons, which lack a functional group, will have such a weak interaction with the column that they will barely be retained even with very mild eluents. In contrast, very polar amines and acids may interact so strongly that they cannot be eluted from the column within a reasonable time.

These limitations are a primary reason for the popularity of reversed-phase chromatography, in which the major factor governing retention is the hydrophobicity of the analyte. This is dependent on both the polarity and the molecular size of the analyte, and thus even homologous and isomeric compounds with similar polarity can be separated. Moderately polar compounds can thus be separated by either normal- or reversed-phase chromatography. The polarity range of reversed-phase chromatography is also wider than normal-phase chromatography and even very non-polar analytes can be readily separated. In addition, reversed-phase chromatography is suitable for more polar or ionisable compounds that might be highly retained on normal-phase columns. The interchange between the ionised and neutral forms of acids or amines can cause particular problems but the ionisation can be suppressed by controlling the pH of the mobile phase with a buffer. Alternatively, modifiers can be added to the mobile phase to form neutral ion pairs, which will be retained (see Section 12.4). Ionised compounds can also be separated by ion-exchange chromatography, although this is much less popular than reversed-phase chromatography and is now rarely used for organic compounds (see Section 12.5). An ionisable analyte can also be converted with a suitable reagent into a neutral derivative, which can be readily examined and will probably also be more readily detected (Section 14.1).

Although the nature of the analyte is the primary factor in the selection of a chromatographic method, a second factor is the matrix of the sample. For example, aqueous samples, such as plasma, food, or many environmental solutions, cannot be injected directly onto a normal-phase column, because even small traces of water will adversely affect the reproducibility of the separation. The analytes must therefore first be extracted into a suitable organic solvent, the solvent evaporated, and the sample redissolved before analysis. In contrast, as long as all the components of the sample are soluble in the mobile phase, aqueous solutions can be directly injected in reversed-phase separation systems, thus saving considerable effort and time.

A further consideration is that the time taken to complete an assay may become excessive and uneconomic, if the sample contains other components that have much longer retention times than the analyte of interest. Thus the presence in the sample of very non-polar components, such as hydrocarbons or lipids, could lengthen a reversed-phase separation but in a normal-phase separation they would only be weakly retained and would probably be eluted with the solvent front well before the analyte. Again reversed-phase separations are more versatile as the eluent can be programmed to speed up the elution of late components, whereas normal-phase separations are limited to isocratic elution because of the time taken to re-equilibrate the column at the end of the run. Complex samples may therefore require a pre-analysis solvent extraction to simplify the matrix before chromatography, or column switching may need to be used (Chapter 14).

As seen earlier, in practice most separations are carried out by reversed-phase chromatography, largely on C_{18}-bonded silica columns, and only a much smaller number of separations use normal-phase or other methods of separation (Table 10.1).

12.2 NORMAL-PHASE SEPARATION METHODS

Although less popular than reversed-phase methods, normal-phase separations are still widely used. It is often useful to be able to relate HPLC separations to thin-layer chromatographic methods, or to column and flash chromatography, which are invariably carried out on silica stationary phases.

Normal-phase chromatography is also often used for large scale preparative separations, partly because the stationary phase is cheaper than for reversed-phase chromatography. It is also usually easy to evaporate the organic eluent after the separation, whereas the aqueous eluents used in reversed-phase chromatography often have to be removed by freeze-drying to recover the analyte. Normal-phase chromatography therefore has an important role in the synthetic organic chemistry laboratory for the monitoring and separation of reaction products.

a. *Stationary phase materials*

Almost all normal-phase separations have been carried out using silica gel as the stationary phase. Originally irregular packing materials prepared by crushing larger particles were used. Subsequently spherical particles of porous silica were developed, which usually give higher efficiency columns, and both types are commercially available (Figure 12.2 and Table 12.1)

Silica gel beads are prepared by precipitation, under controlled conditions, from a sol of sodium silicate. These are then dried at 150°C, which removes the encased water, and then at higher temperatures to activate the surface, producing siloxanes (Si–O–Si) from hydrogen-bonded silanol groups. The silica is then sized in an air centrifuge to give a grade with a narrow range of particle sizes. The surfacc of the silica gel is very active and any water in the eluent will strongly interact with free silanol (SiOH) groups. This alters the activity of the surface and the retention properties, so that particular care must be taken to ensure the dryness or controlled humidity of the eluents. These remaining silanol groups on the surface of the silica surface are acidic and particularly strong interactions can occur with basic compounds. Silica gel can also behave as an ion-exchange stationary phase (see Section 12.5.b). These silica packing materials also form the basis of the bonded stationary phases discussed later.

Although early HPLC studies used 30–100 μm particles, almost all columns now use the more efficient 3, 5, or 10 μm particles. The last of these gives the least efficient columns but is considered the easiest to pack and the columns are more robust in use. Although 3 μm packings give the highest efficiency (up to 50 000 plates per metre) and can be used in short 3 cm columns (so-called fast LC columns), most 10 or 25 cm columns are packed with 5 μm packing material, giving efficiencies of 20 000 or 30 000 plates per metre (i.e. 4000–7000 plates for a 25 cm column), and this has become the usual size of packing material for routine analytical use. The smaller particles are harder to pack efficiently, particularly in the individual laboratory, and the columns appear to be less robust.

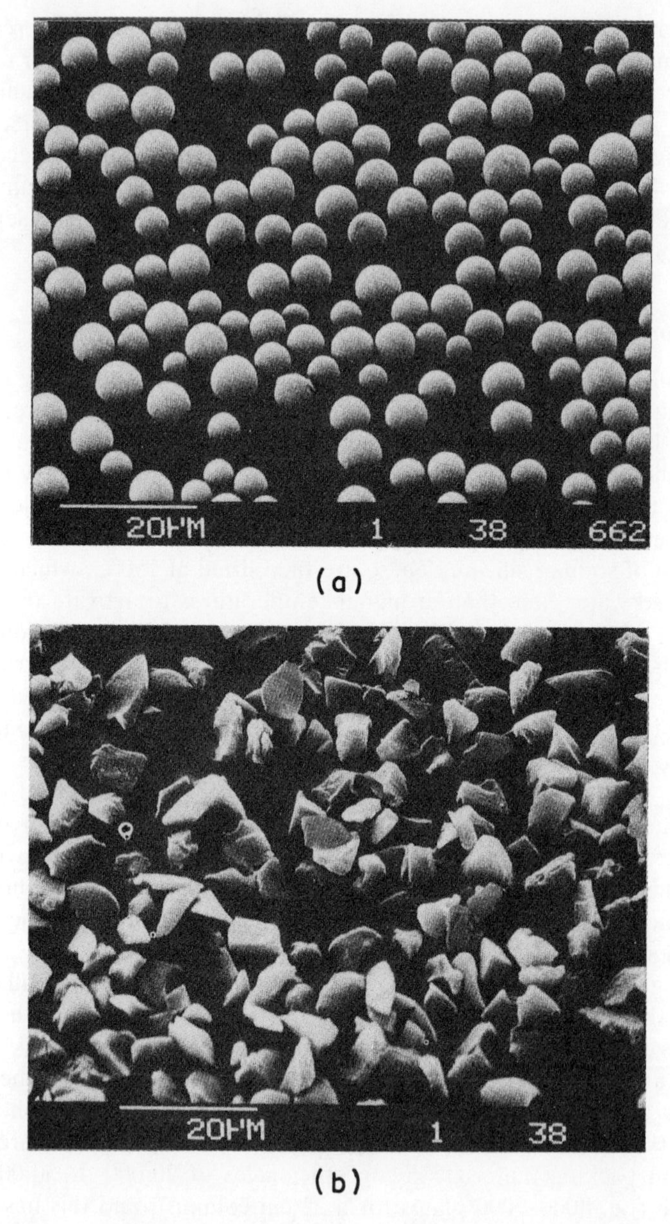

Figure 12.2 Scanning electron micrographs of silica gel materials used in HPLC: (a) spherical particles of 5 μm Nucleosil; (b) irregular silica particles of 5 μm Polygosil materials. (Reproduced by permission of Macherey-Nagel & Co. GmbH.)

Table 12.1 Silica gel stationary phase materials for HPLC.

Material	Source	Shape[a]	Particle sizes (μm)	Pore diameter (nm)	Surface area ($m^2\ g^{-1}$)
Hypersil	Shandon	S	3, 5, 10	10	200
LiChrosorb Si 60	Merck	I	5, 10	6	500
LiChrosorb Si 100	Merck	I	5, 10	10	300
LiChrospher Si 100	Merck	S	5, 10	10	250
Novapak	Waters	S	4	6	120
Nucleosil 50	Machery Nagel	S	5, 7, 10	5	500
Nucleosil 100	Machery Nagel	S	5, 7, 10	10	300
Partisil	Whatman	I	5, 10	4.5	400
Polygosil	Machery Nagel	I	3	6	500
μ Porasil	Waters	I	10	12.5	300
Resolve	Waters	S	5, 10	9	200
Superspher 60	Merck	S	4	6	650
Superspher 100	Merck	S	4	10	350
Spherisorb	Phase Sep	S	3, 5, 10	8	220
Ultrasphere	Altex	S	3, 5		190
Vydac HS	Separations	S	5, 10	8	500
Vydac TP	Separations	S	5, 10	30	90
Zorbax SIL	Du Pont	S	3, 5	7	350

[a] Particle shape: S, spherical; I, irregular.

There are significant differences between the silica gels from different manufacturers in their surface areas, surface pH and particle shapes. Important factors in the quality of the particles are the uniformity of the diameters of the internal pores (usually nominally 100 Å for analytical columns) and the size distribution of the particles (typical distributions are given in Figure 12.3). Columns for macromolecular separations usually use particles with wider (300 Å) pores to avoid size exclusion effects. A material with a closely defined particle size range will lead to a more uniform packing and lower back-pressures than materials with large particles or fines. Wide distributions in either the particle or pore size of a stationary phase will give columns with reduced efficiency and higher back-pressures.

The only other stationary phase material that has been used to any extent in normal-phase separations is neutral alumina (Table 12.2). Unlike the acidic silica surface, it is a fundamentally basic material and it therefore has some advantages for the separation of basic compounds. Cyano-, amino- and diol-bonded silica phases (see Section 12.3.b) can be used as stationary phases in both normal- and reversed-phase methods and in the former mode behave much as silica columns. They have the advantage that the surface equilibrates much more rapidly than silica itself. However, they have not

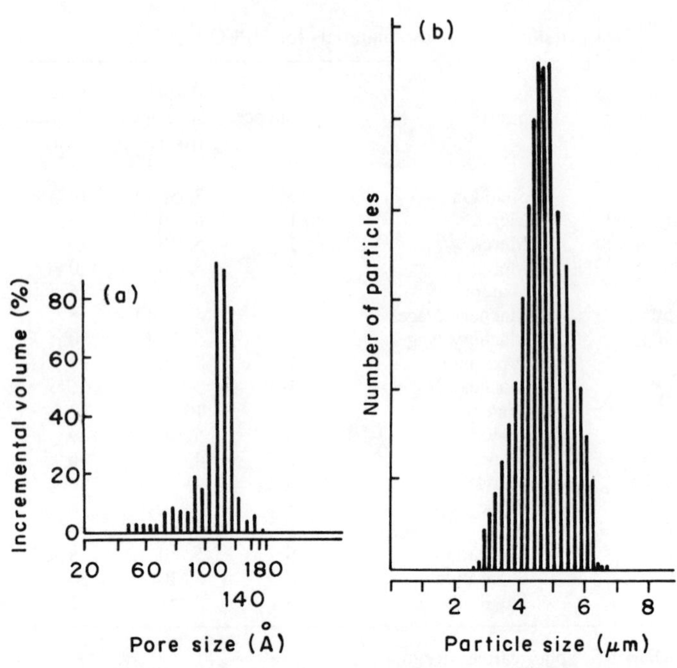

Figure 12.3 Typical ranges of (a) pore sizes (nominally 100Å) and (b) particle sizes (nominally 5 μm) of Hypersil, a spherical silica material for HPLC. (Reproduced by permission of Shandon Southern.)

been widely applied but could be useful alternatives to silica as they have different retention selectivities, which may offer an advantage in achieving the separation of complex mixtures.

b. Mobile phases for normal-phase separations

The elution power of the mobile phase in normal-phase chromatography depends very much on its polarity. As with TLC, most separations are carried out using combinations of organic solvents but many potential components, such as aromatic hydrocarbons, cannot be used because they would interfere with spectrophotometric detection (Section 11.2) or are too viscous to be readily pumped through the column. The individual solvents in common use can be classified in order of their increasing elution strength (Table 12.3). Eluents with intermediate strengths can be prepared by mixing solvents (Figure 12.4). Even small proportions of a polar hydrogen-bonding solvent such as methanol or ethanol can have a marked effect on the overall

Table 12.2 Alumina stationary-phase materials for HPLC.

Material	Source	Particle sizes (μm)	Pore diameter (nm)	Surface area ($m^2\ g^{-1}$)
Spherisorb A5Y	Phase Sep	5	13	93
Alumina Woelm	Woelm	5, 10	6	200
LiChrosorb Alox T	Merck	5, 10	–	70
Aluminium oxide (Alox 60)	Machery Nagel	5, 10	6	155

Table 12.3 Properties and elution strengths of eluents for normal-phase HPLC.

Solvent	Eluent strength[a], $\epsilon°$	Viscosity at 20°C (cP)	Boiling point (°C)	Dielectric constant
Pentane	0.01	0.23	35	1.8
Isooctane (2,2,4-trimethylpentane)	0.01	0.51	99	1.9
Cyclohexane	0.01	1.00	81	2.0
Diethyl ether	0.38	0.23	34	4.3
Chloroform	0.26	0.57	62	4.8
Dichloromethane	0.32	0.43	40	9.1
Tetrahydrofuran	0.44	0.55	66	7.4
1,4-Dioxane	0.43	1.54	101	2.2
Ethyl acetate	0.38	0.45	77	6.0
Acetonitrile	0.50	0.36	82	37.5
Propan-2-ol	0.63	2.30	82	18.3
Methanol	0.73	0.55	65	32.6

[a] Solvent strengths $\epsilon°$ based on separations on silica gel. Changes in solvent strength of 0.05 will alter capacity factors by a factor of 3–4.

polarity because of their interaction with the surface silanols on the silica column. Water has a proportionally even larger effect and, if present in even trace amounts, can totally deactivate the silica, dramatically altering the separation.

Mixed solvent eluents should if possible avoid combinations based on a small amount of a very polar component in a non-polar major component, as only small variations in the amounts of the minor component used in the preparation of the eluent would have a marked effect on the elution strength. Such combinations would therefore be difficult to reproduce accurately. Small concentrations of alkanols (1–5%) are sometimes included in mobile

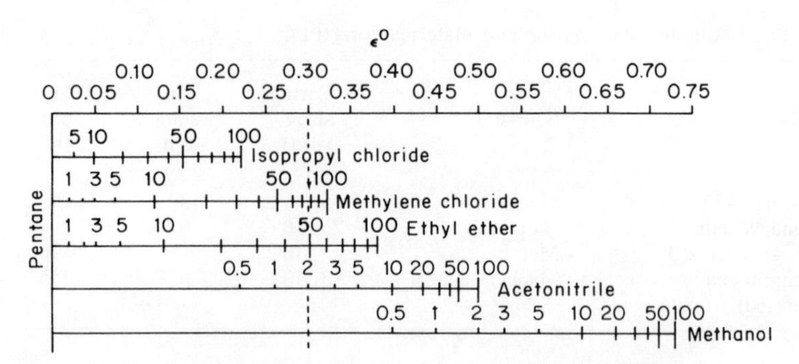

Figure 12.4 Combination of pentane with other organic solvents to give isoeluotropic eluents for normal-phase separations on silica gel. (Adapted from *Anal. Chem.*, 1974, **46**, 470 by permission of American Chemical Society.)

phases to mask the effect of trace amounts of atmospheric moisture and maintain a constant column activity.

The sensitivity of a silica column is such that care must be taken not to inject aqueous solutions of samples, as only a few microlitres of water can markedly alter the column activity. If a silica column does become contaminated with water, the activity can be restored by washing the column with a succession of eluents of reducing polarity in a reverse step gradient (methanol, acetone, dioxane, dichloromethane, then hexane). The sequence will first wash out the water and then the polar eluents. If just a low-polarity mobile phase was used, it would have such a low elution power for the water that a considerable volume would be needed to remove the water and restore the column activity. Most commercial solvents contain traces of water and will rapidly pick up more from the atmosphere, but as long as the level is constant, reproducible results can be obtained. However, after any disturbance to the system, it can take some time before consistent conditions can be re-established. Consequently, eluent programming is rarely used with normal-phase separation as the re-equilibration time usually exceeds any saving in time that might have been gained by speeding up the separation.

The relative separation of a mixture of compounds can be altered by examining eluents with different components but similar elution strengths (for example, dichloromethane–pentane 50:50, diethyl ether–pentane 2:98, or acetonitrile–pentane 0.3:99.7) (Figure 12.4). These isoeluotropic (same-strength) eluents should give similar overall retention times but will have different interactions, particularly with analytes containing different functional groups, which may lead to a better separation. The maximum differences have been shown to be obtainable by interchanging dichloromethane, methyl t-butyl ether and acetonitrile in the eluent with overall adjustment of the elution strength with hexane.

12.3 REVERSED-PHASE SEPARATION METHODS

Reversed-phase chromatography makes use of a non-polar stationary phase and a polar mobile phase to separate compounds according to their hydrophobicity. This can be regarded as the extent to which the analyte is excluded from the aqueous mobile phase into the stationary phase, because of its disruption of the hydrogen-bonded liquid structure. The stationary phase therefore largely plays a passive role in the separation and, if it is a non-polar hydrocarbon it should not significantly interact with the analyte, unless residual active sites are present on the underlying silica surface.

In the early development of chromatography, alternatives to normal-phase adsorption chromatography were investigated in which liquids, such as β,β-oxypropionitrile (OPN), were statically coated on silica particles to provide a non-polar liquid stationary phase. This technique enabled separations, particularly of polar compounds, to be carried out by liquid–liquid partition chromatography. However, as these stationary liquid phases were not bound to the supporting silica material, they could be washed out of the column by the mobile phase, causing poor column stability and reproducibility.

The step that enabled partition chromatography to be successful was the realisation that, even if a stationary phase was chemically bonded to the surface of the silica, it would still behave in much the same way as a liquid but would be stable and resistant to elution. This development has led to a range of bonded phases with different substituent groups, among which the hydrocarbonaceous alkyl-bonded materials (see below) now dominate HPLC separations. So far all these materials are based on silica gel particles, although other substrates have been reported in the experimental stage. This dependence on silica as the support has the disadvantage that the stationary phases are limited to use with mobile phases in the pH range 2.5 to 8. Outside this range the underlying silica matrix will be chemically degraded and column efficiency rapidly lost as voids and cracks form in the column bed.

a. Hydrocarbonaceous bonded-phase materials

A number of reactions have been used to bond alkyl or aryl groups onto the surface of silica particles. In the earliest materials, a C–O–Si ether linkage was formed between the silanol groups on the silica and the alkyl-

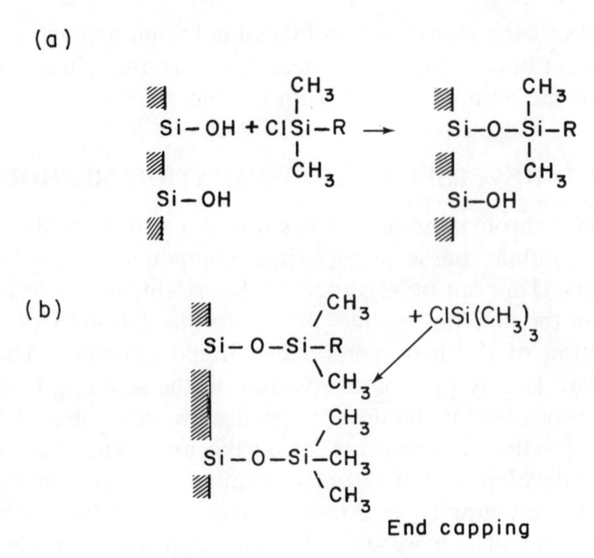

Figure 12.5 Reaction for (a) the formation of monomeric alkyl-bonded stationary phase. (b) end capping of residual silanol groups

bonded group, but these stationary phases were chemically unstable and this method is now rarely used.

Most modern column materials are based on the silica particles used for normal-phase chromatography (see Table 12.1) and are prepared by the reaction of the silanol groups on the silica surface with an alkylsilyl halide to form highly stable Si–O–Si linkages. Two main reactions are used based either on alkylmonohalosilanes or -monoethersilanes (Figure 12.5) or on alkyltrihalosilanes (Figure 12.6). The extent of reaction is measured by the proportion of carbon that can be bonded to the surface, which in turn determines the overall retention capacity of the column.

The monofunctional trialkylsilanes usually react to give a surface coverage of about 7–10% carbon (Figure 12.5a). The polyfunctional reagents react initially with the surface silanol groups (Figure 12.6), and then the remaining functional groups are partially hydrolysed to form additional silanol groups on the side chains. These can react further with the reagent and build up a thicker polymeric film on the surface of the silica to give a carbon loading of about 12–25%.

Neither reaction can derivatise more than about 50% of the silanol groups on the silica surface, because as each silanol group reacts, the adjacent groups are sterically protected from further reaction. However, these unsubstituted silanol groups are still able to interact with analytes. Many

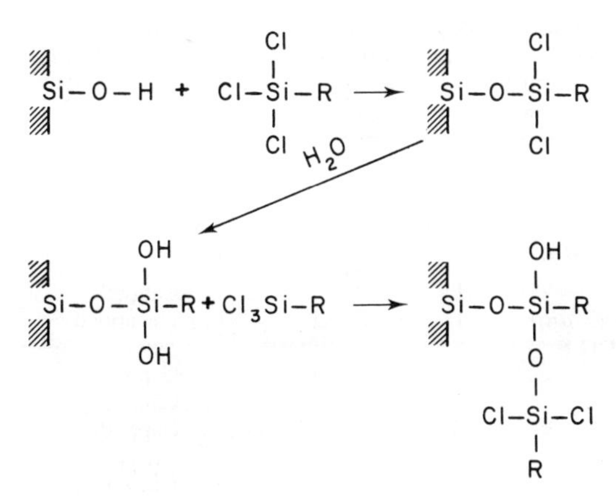

Figure 12.6 Reaction for the formation of polymeric alkyl-bonded stationary phases from alkyltrihalosilyl reagents for reversed-phase HPLC.

manufacturers attempt to cap these groups by reaction with trimethylsilyl chloride (Figure 12.5b). In addition, this reagent will also react with any residual secondary silanol groups formed by the hydrolysis of the trifunctional reagents. Usually this step adds about 1–2% to the carbon loading. Capped and uncapped stationary phases may therefore show very different properties towards polar or basic analytes.

Using these reactions, different alkyl and aryl side chains have been introduced onto the silica surface to provide stationary phases for HPLC. The most popular of these materials is the octadecylsilyl- (ODS) substituted silica (C_{18}) (Table 12.4), but stationary-phase materials with methyl, hexyl, octyl, doeicosanyl (C_{22}) and phenyl side chains are also commercially available (Table 12.5). The popularity of ODS-silica came initially from a comparison of the efficiency and selectivity of a series of phases with different chain lengths. Its suitability as a general-purpose phase has been confirmed by later studies.

Unfortunately, because different manufacturers use different chemical reactions for their column materials and not all use capping reactions, columns that are apparently equivalent may have different carbon loadings and numbers of free silanol groups (Tables 12.4 and 12.5). As a result apparently equivalent stationary phases, such as the octadecylsilyl (ODS) phases from different companies, often possess different retention properties both in their overall retention power and in the relative retentions of different analytes (Table 12.6).

Generally the retentive power of a column increases with the total carbon loading (C%). An ODS-silica column will usually give longer retentions

Table 12.4 Octadecylsilyl-bonded silica stationary phase materials for HPLC. The pore diameter and surface areas correspond to the original silica materials (Table 12.1)[a]. All the materials are available as 5 μm packing material (unless noted). Other sizes (3, 10 μm) may also be available.

Material	Shape[b]	Carbon loading (%)	Capping	Typical efficiency[c]
μ Bondapak C_{18}	I	10	Capped	45 000
LiChrosorb RP18 (7 μm)	I	17	Uncapped	40 000
LiChrospher 100 CH-8	S			
Hypersil ODS	S	10	Capped	100 000
Nova-Pak C_{18} (4 μm)	S	7	Capped	120 000
Nucleosil ODS	S	14	Capped	60 000
Partisil 10 ODS 1[d]	I	5	Uncapped	20 000
Partisil ODS 2	I	15	Uncapped	25 000
Partisil ODS 3	I	10.5	Capped	20 000
Resolve C_{18}	S	12	Uncapped	100 000
Spherisorb ODS 1	S	7	Partially	60 000–80 000
Spherisorb ODS 2	S	12	Capped	60 000–80 000
Ultrasphere ODS	S			65 000
Vydax 201 HS	S	13.5	Capped	–
Vydax 201 TP	S	8	Uncapped	–
Zorbax C-18	S	15		65 000

[a] μ Bondapak columns are based on μ Porapak silica materials.
[b] Particle shape: S, spherical; I, irregular.
[c] Based on 5 μm packing material unless noted.
[d] 10 μm particle size.

than a similarly packed octyl- or methyl-silica phase (Figure 12.7). Because in each case the alkyl chain is inert and does not interact with the analytes, the selectivity or relative elution order of analytes on stationary phases with different lengths of alkyl chains are relatively similar, except for analytes that can interact with the free silanol groups.

Because a sequence of chemical reactions is involved in the preparation of the bonded phases, some manufacturers have problems maintaining a consistent product and there may be batch-to-batch variations in the properties of the samc stationary phase. Over the last few years better quality control and more rigorous testing of the columns have largely eliminated the most serious variations. Because of these problems, laboratories needing highly reproducible results for critical assays over a prolonged period often purchase large quantities of a single homogeneous batch of packing material sufficient for the complete study.

As yet there is no widely used equivalent to the Rohrschneider or McReynolds constants to describe stationary phases for HPLC, although a

Table 12.5 Other alkyl- and aryl-bonded silica stationary phases for HPLC. The pore diameter and surface areas correspond to the original silica materials (Table 12.1).

Material	Carbon loading (%) (if reported)
Methyl- or trimethyl-bonded silica	
Hypersil SAS	
LiChrosorb RP-2	
Spherisorb C1	2
Zorbax TMS	
Hexyl-bonded silica	
Spherisorb 5 C6	6
Octyl-bonded silica	
Hypersil MOS	7
LiChrosorb RP-8	
LiChrosorb Select B	
LiChrospher CH-8	
Nucleosil 5 C8	9
Partisil C8(CCS)	9
Resolve C_8	6
Spherisorb C8	6
Ultrasphere Octyl	
Zorbax C8	
Aryl-bonded silica	
μ Bondapak phenyl	8
Nova-Pak phenyl	4
Nucleosil 7 C_6H_5	8
Phenyl Hypersil	
Spherisorb phenyl	3
Vydax 219 TP diphenyl	5
Zorbax phenyl (dimethylaryl)	

system based on retention indices derived from homologous alkyl aryl ketones has been proposed [1].

b. *Other bonded-phase materials*

As well as alkyl- and aryl-bonded phases, other side chains can be bonded onto silica to give a range of stationary phases with different polarities and interactions. These include commercially available stationary-phase materials substituted with amino, cyano, nitro, or diol functional groups (Table 12.7) and numerous experimental phases. These methods have also been used to prepare chiral stationary phases (see Section 14.2).

Table 12.6 Comparison of capacity factors of a dimethyl phthalate, di-n-butyl phthalate and pyrene test mixture on different ODS-silica packing materials. Eluent methanol–water 90:10. The chromatograms show the differences in retention and selectivity (relative retention) properties.

Commercial name	Carbon load (%)	Separation of test mixture
Partisil 10 ODS	5	
Hypersil ODS	9	
Spherisorb ODS	7	
Partisil 10 ODS 3	10	
μBondapak C_{18}	10	
Zorbax ODS	15	
Spherisorb S5ODS 2	10	
LiChrosorb RP18		
Partisil 10 ODS 2	15	
Nucleosil 5 C_{18}		

(Reproduced by permission of Hichrom Ltd, Reading, UK.)

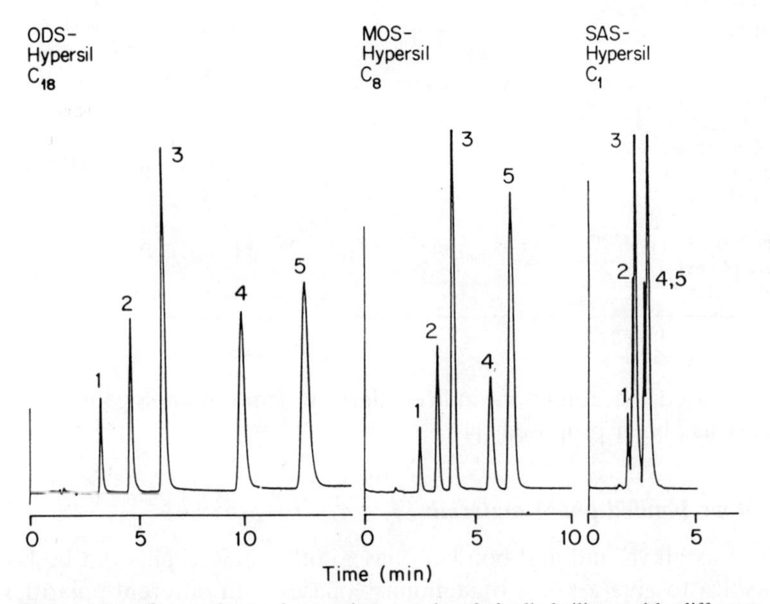

Figure 12.7 Comparison of retentions on bonded alkyl-silicas with different chain lengths, C_{18}, C_8 and C_1, showing changes in overall retention but little differences in retention order. Column, 100 mm × 5 mm; mobile phase, methanol–H_2O 70:30; flow rate, 0.8 ml min^{-1}. Solutes: 1, benzene; 2, toluene; 3, naphthalene; 4, biphenyl; 5, fluorene. (Reproduced with permission from Shandon Southern.)

Table 12.7 Substituted alkyl-bonded stationary phases for HPLC.

Material	Carbon loading (%)
Aminopropyl- and aminodimethylpropyl-bonded silica	
μ Bondapak NH_2	
Hypersil APS	
Hypersil APS-2	
LiChrosorb NH_2	
Nucleosil 5 NMe_2 (weakly basic)	
Nucleosil 5 NH_2	
Spherisorb NH_2	
Zorbax NH2	
Special amino-bonded silica	
Carbohydrate Analysis (Waters) (amino-bonded silica)	
Cyanopropyl-bonded silica[a]	
μ Bondapak CN	6
Hypersil CPS	
Nova-Pak CN	3
Nucleosil 5CN	
Spherisorb S5CN	3.5
Resolve CN	4
Ultrasphere Cyano	
Zorbax CN	
Mixed bonded phase	
Partisil 10 PAC (amino-cyano)	
Diol-substituted silica[b]	
LiChrosorb DIOL	
Nucleosil 7 OH	
Nitropropyl-bonded silica[c]	
Nucleosil 5 NO_2 (for unsaturated organics)	

[a] The group is $-(CH_2)_3-CN$.
[b] The group is $Si-(CH_2)_3-O-CH_2-CH(OH)-CH_2OH$.
[c] The group is $(CH_2)_3-C_6H_4-NO_2$.

Although many of these stationary phases have been available for some time, none has gained widespread application (see Table 10.1). They can often be used in either normal- or reversed-phase modes and can show useful selectivity properties. In the former mode they appear to offer the advantage of greater reproducibility and a more rapid equilibration than silica columns, with less susceptibility to changes caused by traces of water. The amino- and cyano-bonded columns have been particularly used for the analysis of carbohydrates (Figure 12.8). However, amino columns should not be used with samples containing aldehydes or ketones as these may irreversibly form Schiff's bases with the stationary phase.

Potentially the most useful material appears to be the cyano-bonded phase, which has been used for the separation of polar drug analytes by reversed-phase chromatography as an alternative to ODS-silica columns (Figure 12.9). However, for most of these polar phases, not enough practical experience has been gained by different laboratories to establish clearly their advantages and disadvantages compared to the alkyl-bonded or silica columns.

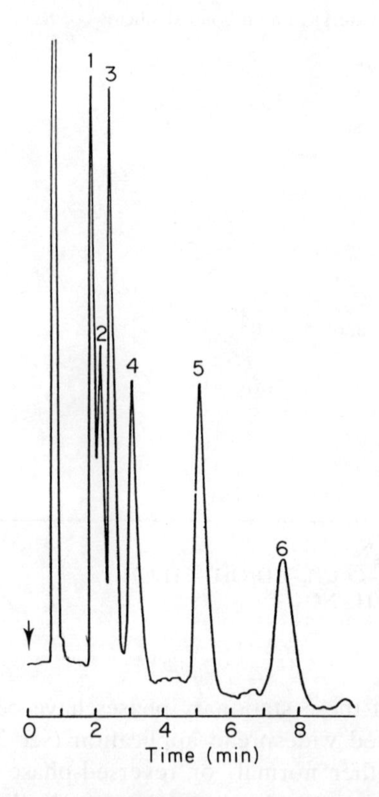

Figure 12.8 Separation of carbohydrates on an amino-bonded column with refractive index detection. Column LiChrosorb NH_2, 10μm. Eluent acetonitrile–water 80:20 at 2 ml min^{-1}. Peaks: 1, xylose; 2, arabinose; 3, fructose; 4, glucose; 5, sucrose (saccharose); 6, lactose. (Reproduced by permission of E. Merck, Darmstadt.)

c. *Polymer stationary phases*

A constant limitation in the use of bonded silica columns is that the eluent is restricted to the range pH 2.5 to 8 because of the instability of the silica gel matrix. This problem has led in recent years to considerable interest in the use of beads of crosslinked porous synthetic polymers as stationary phases. Originally organic polymers were used in chromatography as the basis of ion-exchange resins but most of these materials were too soft and were compressed or collapsed under the flow rates and pressures used in HPLC. Developments in polymer technology have now led to more rigid materials and smaller particle sizes. Most of the commercially available

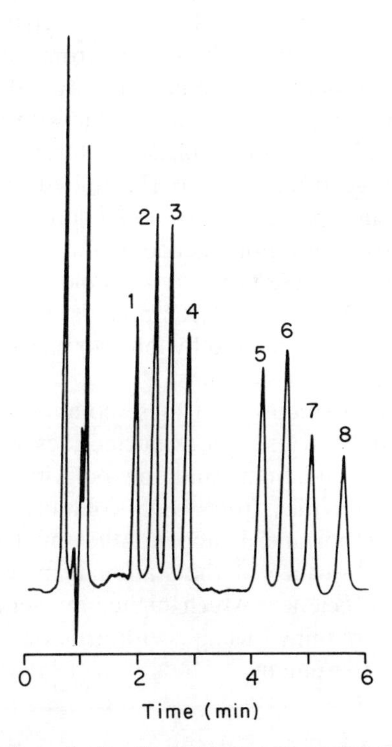

Figure 12.9 Separation of tricyclic antidepressants on a cyano-bonded silica column. Column Supelcosil LC-PCN, 5 μm, 150 mm × 4.6 mm. Eluent acetonitrile–methanol–0.01 M K_2HPO_4 buffer, pH 7.0, 60:15:25 at 2 ml min^{-1}. Ultraviolet spectrophotometric detection at 215 nm. Peaks: 1, trimipramine (internal standard); 2, doxepin; 3, amitriptyline; 4, imipramine; 5, desmethyldoxepin; 6, nortriptyline; 7, desipramine; 8, proptriptyline (internal standard). (Reproduced with permission of Supelco, Inc., Bellefonte, PA 16823.)

Table 12.8 Polymer stationary-phase materials used in HPLC.

Material	Manufacturer	Polymer	Particle sizes (μm)	Pore size (nm)	Surface area ($m^2 g^{-1}$)
PRP-1	Hamilton	PS–DVB	5, 10	7.5	415
PLRP-S 100	Polymer Labs	PS–DVB	5, 10	10	550
ACT-1	Interaction chemicals	PS–DVB C18			
ACT-2	Interaction chemicals	PS-DVB pyridyl			

materials are based on crosslinked polystyrene–divinylbenzene (PS–DVB) but polymeric materials with bonded ODS side chains are also available (Table 12.8). Experimental materials derived from polyacrylates have also been described and may become commercially available.

The PS–DVB column materials are non-polar with a very high retention capacity. They make ideal reversed-phase materials as they can be used over a wide eluent range from pH 1 to 13, limited only by the stability of the materials of the pumps, detectors and columns. The surface of the stationary phase is essentially homogeneous and is free from highly polar and acidic silanol groups. These properties enable the direct analysis of very polar samples (Figure 12.10) and of compounds, such as amines or strong acids, that must normally by analysed by ion-pair chromatography (see later) or suffer tailing on ODS-silica columns.

Many of the published studies with these columns have concentrated on the separation of polar analytes such as nucleotides, but these columns can also be used to analyse medium- and low-polarity analytes and provide different retention selectivities from ODS columns. The selection of the mobile phase can be important. It appears that methanolic eluents do not appear fully to wet the surface of the stationary phase and can give poor peak shapes and low efficiency. Much higher efficiencies and symmetrical peak shapes have been obtained using acetonitrile or tetrahydrofuran as the organic component of the eluent.

d. Dynamically coated columns

When alkyl-bonded silica stationary phases became commercially available, little further use was made of the original unbonded liquid phases. However, if a mobile phase incorporating a constant proportion of a potential liquid phase, such as iso-octane, is used with a silica column, the liquid will form

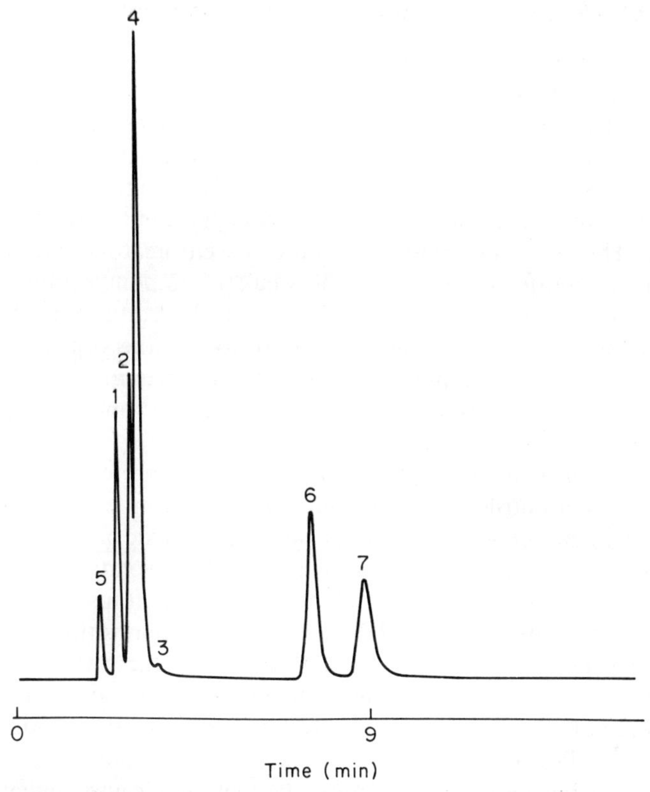

Figure 12.10 Separation of polar purine and pyrimidine bases on a polystyrene–divinylbenzene column at pH 8.5. Column PLRP-S, 150 mm × 4.6 mm. Eluent acetonitrile–ammonium formate, pH 8.5, 1:100 at 1 ml min^{-1}. Peaks: 1, cytosine; 2, uracil; 3, guanine; 4, hypoxanthine; 5, xanthine; 6, adenine; 7, thymine. (Reproduced by permission of Polymer Laboratories Ltd from *PL applications note*, No. 314.)

a layer on the surface of the silica [e.g. 2]. A number of research groups have studied this approach and claim that these dynamically coated columns give more efficent separations than bonded silica reversed-phase chromatography. Because the stationary phase can be readily changed, it also provides great flexibility.

However, these methods have not gained wide acceptance in routine analysis, probably because of the need to maintain both a constant concentration of the liquid phase in the eluent and a constant column temperature to ensure a reproducible and stable coating of the stationary phase.

e. Solvent selection for reversed phase chromatography

One of the reasons for the popularity of reversed-phase HPLC is the relative simplicity of eluents that will separate a wide range of analytes with very different polarities. The great majority of separations can also be carried out using a single octadecylsilyl-bonded stationary phase. The retentive power of the shorter alkyl-bonded phases is correspondingly much weaker, whereas the fully organic polymer columns have much higher retention capacities. The usual eluents for reversed-phase chromatography are mixtures of an aqueous buffer or water, as the weaker component, and an organic solvent, either methanol, acetonitrile or tetrahydrofuran (THF), as the stronger component. Increasing the proportion of the organic component increases the elution strength and reduces the retention time. Buffered solutions are usually only needed if the analyte can ionise and pH control is required (see Section 12.4).

Of the three organic solvents, methanol is often preferred because of its lower cost. Acetonitrile is also popular because it is less viscous and will usually give higher efficiencies, but its vapours are toxic and good ventilation is needed. THF is the least popular as it has a high volatility that can cause operating problems. There is a tendency for vapour traps to form in pump heads and evaporation of the THF can alter the composition of the eluent. All three solvents have virtually no absorbance in the ultraviolet spectrum (see Table 11.2) and are readily available in high purity and reasonable cost as commercial HPLC-grade solvents. Related compounds including ethanol, propanol and dioxane are usually only used as a minor component in an eluent, partly because they are more viscous and cause higher pumping back-pressures but mainly because they offer little advantages in separation selectivity.

The three organic solvents increase in elution strength in the order methanol, acetonitrile, and THF. The overall strength of an eluent mixture can be estimated from the proportions of each component by using the solvent strength factors (Table 12.9). Thus methanol–water 70:30, acetonitrile–water 50:50 and tetrahydrofuran–water 35:65 are roughly isocluotropic and have similar elution strengths. However, these three different eluent combinations will have markedly different selectivities for different compounds, because they will interact in different ways with proton-accepting and proton-donating groups and by dipole–dipole interactions. The separation of mixtures with different organic components may therefore differ markedly (Figures 12.11 and 2.4). By exploiting these differences, the operator can often achieve the resolution of a complex mixture by selecting the most appropriate binary (two-component) eluent or by using intermediate ternary (three-component) eluents of similar strength. Quaternary (four-

Table 12.9 Properties and eluent strengths of solvents used in reversed-phase separations.

Solvent	Solvent strength S_i	Viscosity at 20°C (cP)	Boiling point (°C)	Dielectric constant
Water	0.1	1.00	100.0	80.1
Methanol	2.6	0.55	64.7	32.6
Acetonitrile	3.1	0.36	81.6	37.5
THF	4.4	0.55	66.0	7.4
1,4-Dioxane		1.54	101.1	2.2
Propan-2-ol		1.35	82.3	18.3

Total eluent strength = $\Sigma\, S_i \times \%i$, where $\%i$ = proportion of each component in the eluent.

component) eluents have also been proposed but these are rarely needed in practice and usually offer little advantage over the nearest equivalent ternary eluents.

Normally mobile phases are selected so that the capacity factors of the analytes are in the range k' = 2 to 20. Typical combinations range from methanol–water 40:60 for the retention of the relatively polar barbiturates to 70:30 for the separation of simple aromatic ketones and esters up to 100% methanol for aromatic hydrocarbons. In each case significant changes in retention times can be obtained by altering the mobile phase. Typically, for the range of 20 to 80% of organic component in the mobile phase, a 10% reduction in the organic component will double the capacity factors of an analyte. For many compounds, there is roughly a linear relationship between the proportion of water in the eluent and the logarithm of the capacity factor. At low proportions of organic modifier or aqueous phase, small changes in the eluent can have a more marked effect.

As none of the common eluent components has a significant ultraviolet absorbance, the composition of the eluent can be programmed during the separation in order to increase the elution rate of the more highly retained compounds, without a significant baseline shift (see Figure 10.2). This method is often used to determine suitable initial separation conditions. However, with a little experience, a possible eluent can be readily determined by trial and error and a knowledge of the properties of the analyte. One advantage of bonded stationary phases is that the column rapidly equilibrates on changing the eluent or on returning to the initial conditions after a gradient separation. Stable conditions can usually be achieved after 10–20 column volumes (10–20 column void volume) of a new eluent have passed through the column.

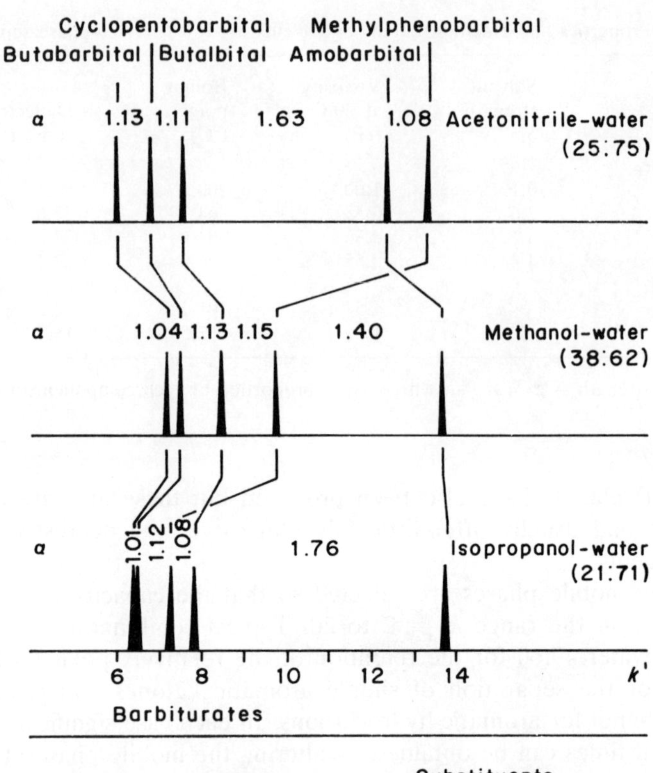

Structure	Name	Substituents R, R′	R″
	butabarbital	ethyl	s-butyl
	cyclopentabarbital	allyl	cyclopentenyl
	butalbital	allyl	isobutyl
	amobarbital	ethyl	isoamyl
	methylphenobarbital	ethyl	pnenyl

Figure 12.11 Changes in retentions and relative retentions (α) of barbiturates on RP-8 column with different organic components in the mobile phase. (Copyright © Hewlett–Packard Co. (1979). Reproduced with permission.)

Very non-polar samples, such as triglycerides or aliphatic hydrocarbons, may be very highly retained on an ODS-silica column even with 100% methanol or acetonitrile as the eluent. Instead, non-aqueous eluents, such as tetrahydrofuran or dichloromethane, may be needed to give elution in a reasonable time.

12.4 SEPARATION OF IONISABLE COMPOUNDS

Compounds which can ionise in solution often give poor peak shapes on reversed-phase chromatography and their retention times are frequently irreproducible or vary with sample size. Because the neutral and ionised forms of a compound have very different polarities, any interchange between the two forms on the column leads to considerable band spreading. Usually these analytes are so polar that they cannot be eluted from normal-phase systems with conventional eluents but it may be possible to use silica as an ion-exchange medium (Section 12.5.b). Even on reversed-phase columns, amines and other basic compounds cause particular problems, because they interact with the acidic silanol groups on the the surface of the silica support material.

In order to eliminate the problems of ionisation, the separation conditions can be modified either by ion suppression, so that only the neutral form of the analyte is present, by the inclusion of masking agents in the eluent to neutralise the acidic groups on the column, or by the addition of an ion-pair reagent to the eluent to convert the ionised analyte into a neutral complex.

In extreme cases the retention of the amine is a consequence of a combination of both reversed-phase and ion-exchange interactions. As the proportion of organic modifier is increased, the capacity factors of the basic compounds may increase, whereas the retentions of compounds being retained only by the reversed-phase mechanism are reduced. The graph of k' against proportion of modifier can show a distinct minimum (Figure 12.12). These anomalous effects can be reversed by masking the silanol groups with butylamine (see below).

a. *Ion suppression*

The simplest approach to the problem of analyte ionisation is to control the pH of the mobile phase so that the ionisation of the analyte is suppressed and it can be analysed as its neutral form. However, as already noted, silica-based bonded phases are restricted to the range pH 2.5 to 8 and outside this range rapid column degradation can occur. The choice of the buffer is also limited by the widespread use of spectrophotometric detection. Phosphate, ammonium, acetate and citrate buffers are popular but phthalate buffers cannot be used. The buffer salt concentrations must be kept low (0.1–0.5 M) to prevent precipitation when the aqueous solution is mixed with the organic component of the eluent. Particular care must be taken to prevent precipitation on programming the eluent as the crystals could block

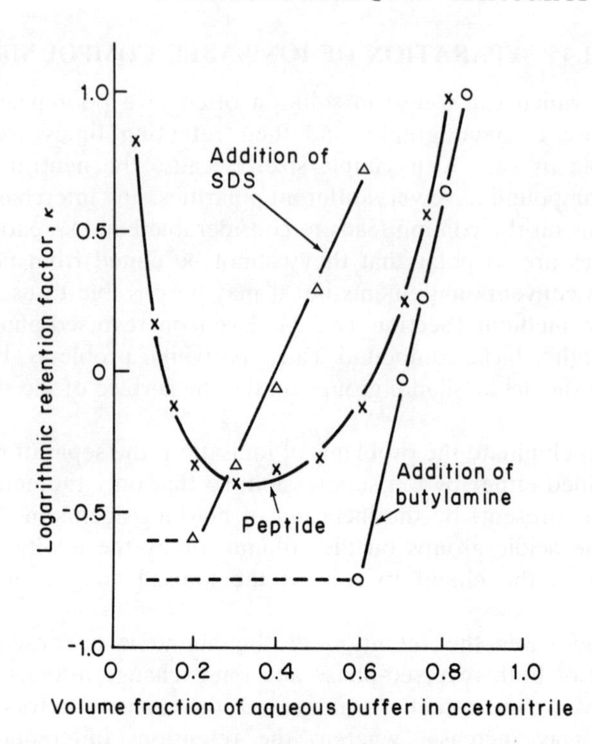

Figure 12.12 Changes in capacity factors of the peptide NH_2-Val-Met-Ala-Gly-Val-Ile-Gly-OEt with the fraction of acetonitrile in the eluent. Effect of the presence of 10 mM n-butylamine or 20 mM sodium dodecylsulphonate as masking agents in the mobile phase. Column Supelcosil LC-8, 150 mm × 4.6 mm. Eluent acetonitrile–50 mM sodium phosphate, pH 2.25. Ultraviolet detection at 220 nm. (Reproduced by permission of Elsevier Science Publishers from *J. Chromatogr.*, 1981, **203**, 65–84.)

the narrow-bore connecting tubing, frits, or injection valve loop. The HPLC system should always be flushed with water or aqueous–organic eluents to remove residual buffer solutions before washing the column with a high proportion of organic solvent.

Most carboxylic acids, phenols, or weakly acidic compounds, such as barbiturates, can be neutralised by using buffers in the range pH 2 to 5 or by the addition of 0.1% of an aliphatic acid, such as propionic acid, to the mobile phase (Figure 12.13). It is not usually possible to suppress the ionisation of the more acidic sulphonic acids within the permitted pH range and either polymer columns or alternative techniques such as ion-pair chromatography must be used.

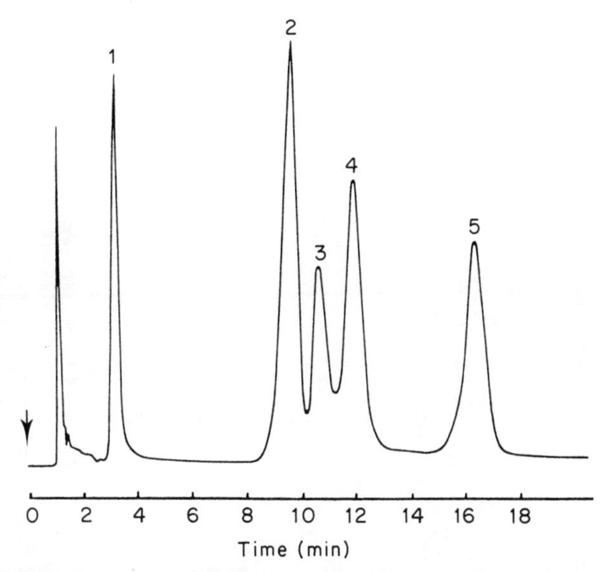

Figure 12.13 Chromatography of free aromatic carboxylic acids by ion suppression. Column LiChrosorb RP-8 (10 μm) 250 mm × 4.6 mm. Eluent methanol–water–propionic acid 20:80:0.1. Peaks: 1, oxalic acid; 2, phthalic acid; 3, terephthalic acid; 4, isophthalic acid; 5, benzoic acid. (Reproduced by permission of E. Merck, Darmstadt.)

If the pH of the buffered eluent is greater than 7, most aromatic amines are neutralised and show no ionic interactions with the silica surface. More basic aliphatic amines would often require a mobile phase with a pH greater than 8 for neutralisation. Thus ion suppression cannot be used for the chromatography of these amines with silica-based columns but separations can be carried out on polymer columns.

b. Masking agents

The residual silanol groups on the surface of an alkyl-bonded silica column can be masked from basic analytes by the addition of 1% of a simple aliphatic amine, such as hexylamine, to the mobile phase [3]. This modifier can greatly improve the separation of basic compounds, which would otherwise interact with the acidic silanols and give severely tailed peaks (see Figure 12.12). As long as sufficient amine is present to neutralise the surface, the proportion is not critical.

A number of manufacturers are now also offering special "base-deactivated" reversed-phase column materials for the analysis of basic

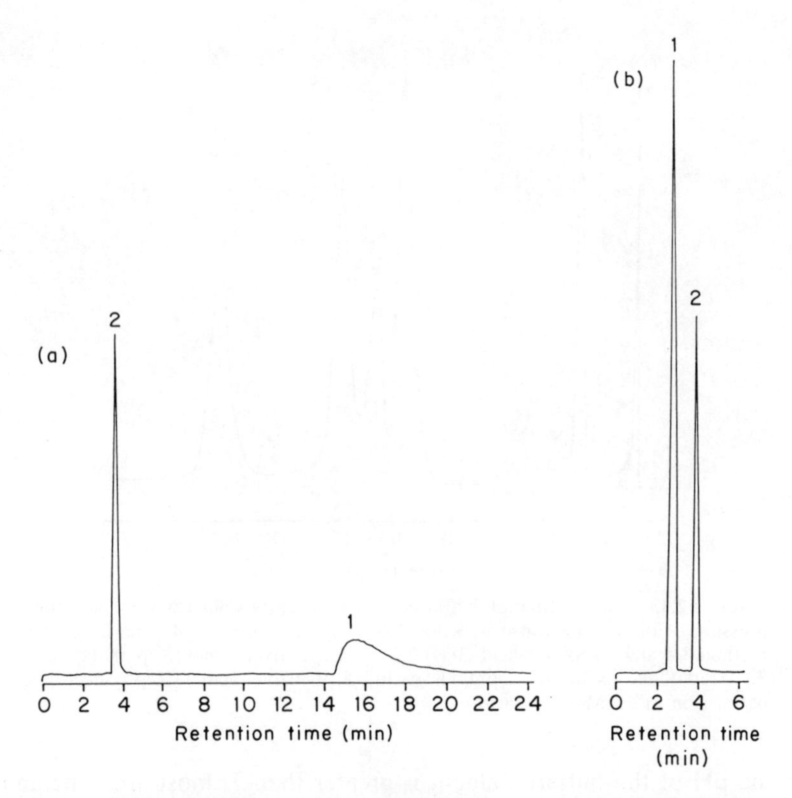

Figure 12.14 Comparison of conventional and base-deactivated columns for the separation of chlorhexidine (1) and *p*-chloroaniline (2). (a) Conventional RP-8 column, 7 μm; (b) RP Select B (C_8 bonded) column, 7 μm. Eluent methanol–50 mM phosphate buffer, pH 3, 70:30 at 1 ml min^{-1}. UV spectroscopic detection at 254 nm. (Reproduced by permission of E. Merck Darmstadt.)

analytes. The preparation of these materials is still proprietary but they are not prepared by fully capping the silanols. They have probably been chemically treated to reduce the acidity. These columns generally show much lower interactions with bases than conventional reversed-phase packings and the retention results are more reproducible (Figure 12.14).

c. *Ion-pair chromatography*

Ion-pair formation is probably the most widely applied secondary modification method in reversed-phase HPLC, although the technique was originally developed for extraction chemistry. It was found that if an ionised water-soluble compound was mixed with a solution containing a counter-ion of opposite charge, the two ions would form a neutral ion-pair complex, which could then be extracted into an organic solvent. These complexes are weak interactions involving outer shells of electrons and an equilibrium is present between the individual ions and the ion-paired complex. The adaptation of this concept to HPLC is usually attributed to Schill and his coworkers, many of the early studies being carried out using normal-phase chromatography. However, almost all current applications are in reversed-phase chromatography. The addition of ion-pair reagents to the mobile phase can have a dramatic effect on the resolution and retention of the ionised components of a mixture. It has been used principally to improve the retentions and peak shapes of strong acids or bases but may also be used in the resolution of enantiomers (Section 14.2).

i. Mechanism of ion-pair separations

In order to carry out an ion-pair separation, the pH of the buffered mobile phase is adjusted so that the analyte will be ionised. A counter-ion of opposite charge is added in a fixed concentration to the mobile phase. As an acidic analyte (RH) is injected, it is ionised to give the anion (R^-), which forms a neutral ion pair (RC) with the positively charged counter-ion (C^+) (Equation 12.1).

$$R^-_{aq} + C^+_{aq} \rightleftarrows RC_{aq} \rightleftarrows RC_{stat} \tag{12.1}$$

(R^- = analyte, C^+ = counter-ion, RC = neutral ion pair in aqueous and stationary phases). The neutral ion pair is then retained on the column in the same way as any neutral analyte and its retention is governed by the elution strength of the mobile phase. The observed capacity factor of the analyte depends on the proportions of the analyte that are present as the ion pair and as the free anion. Under the conditions of the separation the free anion (R^-) would effectively be unretained in the absence of the ion-pair reagent.

Alternatively, a basic analyte is converted into a cation (i.e. RH^+) with a low pH buffer in the mobile phase, and a corresponding counter-ion (C^-) with a negative charge is included in the mobile phase to give the neutral ion pair (RHC).

The exact method of operation of ion-pair chromatography is still in some doubt and two mechanisms have been proposed. The first suggests that the ion pair is formed primarily at the interface between the mobile and stationary phases. The neutral ion-pair complex is only present in a significant amount in the less polar stationary phase (RC_{stat}) and is effectively absent from the mobile phase. The second mechanism, which has been particularly proposed when counter-ions containing long aliphatic side chains are being used, is that the alkyl chain of the counter-ion is incorporated into the bonded phase to give an active surface of ionised groups. These groups then act as if they were a dynamic ion-exchange phase and the ion-pair complexes are formed on the surface of the stationary phase. As this is effectively part of the stationary phase, it is experimentally very difficult to distinguish between these mechanisms. Both lead to the same conclusions and the same observable effects of the counter-ion concentration. In operation the distinction is largely academic.

The proportion of the analyte (R) which is present as the ion pair (RC) depends on the equilibrium constant (K_{IP}) for the formation of the ion pair and the concentrations of the analyte and counter-ion (Equation 2.2).

$$K_{IP} = \frac{[RC_{stat}]}{[R^-_{aq}]\,[C^+_{aq}]} \tag{12.2}$$

In deriving this equation, it is assumed that the ionised analytes and counter-ions are present only in the aqueous mobile phase and that the neutral ion-pair complex is present only in the non-polar stationary phase (RC_{stat}). The value of K_{IP} depends on the composition of the mobile phase.

If the concentration of the counter-ion (C^+_{aq}) is high compared to that of the analyte (R^-_{aq}), it will remain effectively constant as ion-pair formation occurs. Equation 12.2 can be rearranged to give Equation 12.3.

$$\frac{[RC_{stat}]}{[R^-_{aq}]} = K_{IP}\,[C^+_{aq}] \tag{12.3}$$

The right hand side of the equation will be a constant for a fixed concentration of a particular counter-ion and eluent. The left-hand side of the equation is then the ratio of the concentrations of the analyte (as R^- and RC) in the stationary and mobile phases. This ratio is the capacity factor k' of the analyte (see Equation 2.4).

Thus the retention time of the ion-pair complex depends only on the equilibrium constant (K_{IP}) and the concentration of the counter-ions $[C^+_{aq}]$ and is independent of the concentration of the analyte. The retention of the ion-pair complex can therefore be adjusted to give the best separation of the components of a mixture, by altering the concentration of counter-ions or by using a different counter-ion, which will have a different equilibrium constant.

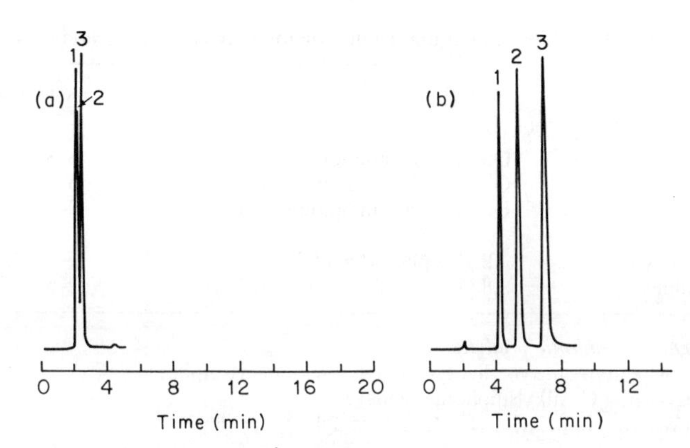

Figure 12.15 Ion-pair chromatographic separation of morphine and homologues. Column LiChrosorb RP Select B, 250 mm × 4.6 m. Eluent: (a) methanol–0.05 M phosphate buffer, pH 3.5, 60:40 without ion-pair reagent; (b) as (a) plus 0.05 M dodecanesulphonic acid sodium salt as ion-pair counter-ion. Peaks: 1, morphine; 2, methylmorphine; 3, ethylmorphine. (Reproduced by permission of Merck, Darmstadt.)

ii. Typical separation systems

Ion-pair chromatography can be used for the separation of any ionisable analyte and is particularly valuable for compounds which are too strongly acidic or basic for separation using ion-suppression methods. Typical examples are sulphonic acid dyestuffs, pharmaceuticals (such as morphine derivatives; Figure 12.15) and surfactants. By varying the conditions and the counter-ions the method can be very flexible and the retention of the analyte can be carefully adjusted.

The counter-ions are usually aliphatic acids or bases to avoid any interference with spectrophotometric detection. Alkylsulphonic acids, with chain lengths from methyl to decyl, are often used as counter-ions for the separation of bases. For the separation of acidic analytes, such as carboxylic acids, phenols and sulphonic acids, the counter-ions are usually tetraalkylammonium salts, typically tetrabutylammonium bromide, or quaternary ammonium salts containing a long side chain, in particular trimethylhexadecylammonium salts (cetrimide) (Table 12.10). As many of these counter-ions are also surfactants, in some early papers ion-pair chromatography was also referred to as "soap" chromatography. Other counter-ions have also been used, including inorganic ions such as perchlorate, but these have been less popular. Generally, counter-ions with longer alkyl side chains give higher retentions. However, smaller counter-ions give greater discrimination between different analytes.

Table 12.10 Typical analytes and counter-ions for ion-pair chromatography.

Analyte	Counter-ion	pH of eluent
Sulphonic acids	Quaternary ammonium salts	3–8
Carboxylic acids	Quaternary ammonium salts	5–7
Phenols	Quaternary ammonium salts	7–9
Aliphatic amines	Alkylsulphonic acids	2–7
Aromatic amines	Alkylsulphonic acids	2–5

Typical reagents for anionic analytes
Sodium salts of pentyl-, hexyl-, heptyl-, octyl-, or dodecylsulphonic acids
S Series (Regis) (C_5–C_8 alkylsulphonic acids)
PIC B5–B8 (Waters Associates) (C_5–C_8 alkylsulphonic acids)

Typical quaternary amines for cationic analytes
Cetrimide (cetyltrimethylammonium bromide/hexadecyltrimethylammonium bromide)
Triethylammonium chloride
Tetrabutylammonium hydroxide
Tetraethylammonium hydroxide
Octadecyldimethylammonium sulphate
PIC A (Waters Associates) (tetrabutylammonium phosphate)
PIC D-4 (Waters Associates) (dibutylamine phosphate)
Q Series (Regis) ($C_{5,6,7,8,12}$-N^+Et_3)

The concentrations of the counter-ions are usually quite low, 0.05 to 0.005 M. Higher levels are often limited by the solubility of the counter-ions or by micelle formation in the mobile phase (this produces an alternative and more specialised mode of separation for ionised compounds). The eluent composition can be programmed as long as the counter-ions are present in both eluent components in the same concentrations so that the total concentration remains constant. The proportion and type of organic modifier in the mobile phase will have a similar effect on the retention of the ion pair as on the other neutral compounds and can also be used to refine the separation.

The pH of the eluent must be adjusted so that the analyte is fully ionised and will thus interact completely with the counter-ion. However, the usual restrictions on the pH stability of silica-based columns must still be observed. If possible, the mobile phase should be degassed before the counter-ions are added, because many of the ion-pair reagents are surfactants and the solution will foam.

Any bonded silica or polymer column can be used as the stationary phase for ion-pair chromatography. In practice, shorter alkyl-bonded phases, such as C_2- or C_6-bonded silicas, appear to give better results than ODS-silica columns. Normally it takes longer for an ion-pair separation system to equilibrate and give reproducible retentions than with a simple reversed-phase eluent. At the end of a separation run, it is usually not possible to wash ion-pair reagents completely out of the column. Consequently, once a column has been used for ion-pair chromatography, it should be reversed for that system, otherwise the residual reagent may cause anomalous results. Because changes in the column temperature would alter both the equilibrium constant of ion-pair formation (K_{IP}) as well as the usual effects on the retention times, ion-pair separations are particularly sensitive to temperature changes, and the column and eluent reservoirs should be thermostated for reproducible results.

12.5 ION CHROMATOGRAPHY AND ION-EXCHANGE CHROMATOGRAPHY

Inorganic cations and anions cannot be separated by reversed-phase chromatography using ion suppression or ion-pair chromatography but they can be separated using ion-exchange chromatography. Although this method was also popular for ionised organic compounds during the early days of HPLC, because the ion-exchange stationary phases then available were based on relatively soft polymers, the separations were inefficient and most of these analytes are now examined by ion-pair chromatography. Subsequently, advances in polymerisation methods have led to more rigid ion-exchange column materials but these have not yet proved popular in general use.

Most ion-exchange chromatography is a specialised area, primarily of interest to the transition-metal inorganic chemist, and will not be discussed here. However, in recent years the analytical separation of many common anions and cations has been revolutionised by the introduction of "ion chromatography" based on the technology of HPLC equipment and offering similar resolution and speed of analysis. There has also been a limited but useful application of silica gel as an ion-exchange column for the separation of basic drug compounds.

a. *Separation of cations and anions*

Ion-selective electrodes are now widely used for the determination of specific ions in solution and have found widespread application in clinical and environmental analysis. However, they can only detect one ion at a time and cannot provide a profile or screening method for the determination of the composition of a sample. Chromatographic methods using ion-exchange columns can readily separate the ions but problems are encountered with the detection of the ions at low concentrations. A few ions such as bromide, nitrate and nitrite can be detected spectrophotometrically at short (200–210 nm) wavelengths but most have no absorbance. Conductometric detection can be used but a weak ion-exchange resin column and an eluent with a low ionic strength must be used to avoid swamping the analyte signal with the background. The separation power and capacity of the columns are rather limited. Alternatively, a stronger ion-exchange resin and eluent can be used as long as the background ionisation can be suppressed in a post-column reactor before conductometric detection or if indirect spectrophotometric detection following post-column derivatisation can be employed.

i. Weak ion-exchange columns

Both inorganic and organic anions and cations can be separated by ion-exchange chromatography using weak resins. These are prepared from rigidly crosslinked polymers substituted with a low surface coverage of quaternary amines or carboxylic acids. The analytes are eluted with relatively weak eluents, such as 0.005 M phosphate buffer, and can be detected by conductometric or spectrophotometric detection. This method has been applied both to anions, such as halide and nitrate, and to cations, including sodium and potassium ions and organic acids (see Figure 11.13).

ii. Ion chromatography with suppressed detection

Although the term "ion chromatography" could apply to all three ion-exchange techniques, it has become a synonym for the use of a suppressor system, which removes the eluent buffer ions from the mobile phase and thereby enables the analyte ions to be readily detected by conductometry. The concept was developed by Small, who realised that the problem of the high background conductivity of the eluent buffer could be overcome by exchanging the buffer salts in a post-column reactor column to give weakly ionised species.

For example, aqueous sodium hydroxide can be used to elute anions, such as chloride and nitrate, from a strong anion ion-exchange column. The resulting eluent, containing sodium, chloride, nitrate and hydroxide ions, was passed through a cation ion-exchange column in the acid (H^+) form. The sodium ions were replaced by protons, converting the hydroxide ions to virtually un-ionised water but leaving the chloride and nitrate ions as the strongly ionised mineral acids.

Anion separation

$$NaOH + H^+ \rightarrow HOH \text{ (eluent)}$$
$$NaCl + H^+ \rightarrow H^+ + Cl^-$$
$$NaNo_3 + H^+ \rightarrow H^+ + NO_3-$$

If solutions of sodium carbonate or bicarbonate were used as the eluent, they would be converted into the weakly ionised carbonic acid and also effectively suppressed.

$$Na_2CO_3 + H^+ \rightarrow H_2CO_3$$

The corresponding suppression could also be carried out for cation analysis using hydrochloric acid as the eluent. In the suppression step using an anion-exchange column, the chloride ions would be exchanged for hydroxide ions, giving water, but the other analytes would form the corresponding hydroxides and could be detected.

Cation separation

$$HCl + HO^- \rightarrow HOH$$
$$NaCl + HO^- \rightarrow NaOH$$
$$KCl + HO^- \rightarrow KOH$$

These methods have been packaged and marketed as a dedicated automated system for ion chromatography using specially designed high-efficiency ion-exchange column materials. These materials are based on rigid non-porous polymer beads to which small polymer spheres carrying the ion-exchange groupings have been attached. Originally the anion resins were based on sulphonic acid groups and the cation resins on quaternary ammonium groups, but the range of specialised column materials has subsequently been expanded. By limiting the active sites to the surface of the beads, the efficiency of the column is increased and most of the mass transfer problems of conventional ion-exchange materials are avoided.

The first ion chromatographs used an ion-exchange analytical column and a second ion-exchange suppressor column, which had to be regenerated at intervals. This second column was subsequently replaced by a fibre suppressor made from a selective ion-exchange membrane. This allows the buffer ions to be exchanged between the mobile phase and an external regeneration solution, which can be replaced without stopping the analysis (Figure 12.16). This also diminishes the dead volume that was present in the older second ion-exchange column. In current instruments it has been replaced by a small volume thin-film micromembrane suppressor using the same concept. Using this method, multiple-ion analysis has become routine at the parts per

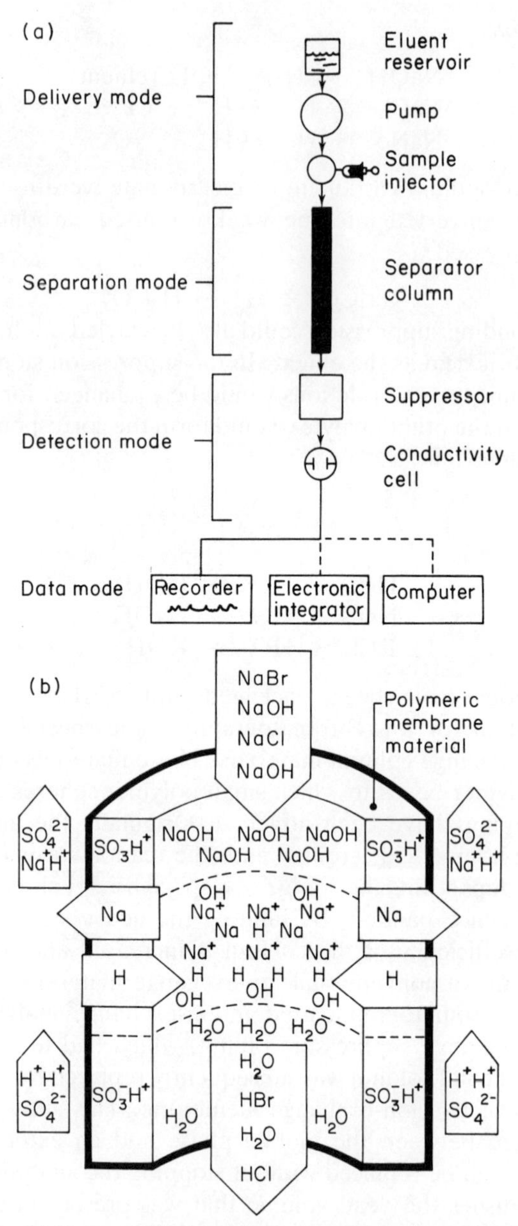

Figure 12.16 (a) Configuration of ion chromatograph with ion-exchange separation column and conductometric detector. (b) Ion chromatography suppressor fibre exchanging hydrogen ions for sodium ions during an anion separation. The halide ions remain as the mineral acids. (Reproduced by permission of Dionex Corporation.)

million level (Figure 12.17) and below, and has extended the application of ion-exchange chromatography, particularly for anion analysis, as a routine service into many new areas.

Similar high-efficiency separation columns have also become available for transition metal analysis but in this case detection is carried out by the post-column addition of pyridylazoresorcinol (PAR) to form coloured complexes which are detected spectrophotometrically.

iii. Indirect photometric detection

Once the viability of ion chromatography had been demonstrated, there was considerable interest in the use of conventional HPLC systems and alternative detection methods. It was shown that if a fixed concentration of an ionised ultraviolet-absorbing species, typically phthalate ions (absorbance maximum 280 nm), is included in the mobile phase it will give a steady background signal. As analyte ions, such as chloride and nitrate, are eluted from the

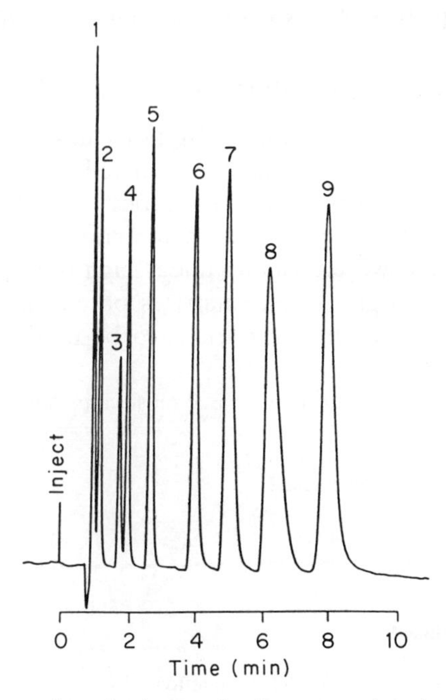

Figure 12.17 Separation of mixture of anions on an ion chromatograph using an anion-exchange column. Peaks: 1, F^-; 2, formate; 3, BrO_3^-; 4, Cl^-; 5, NO_2^-; 6 HPO_4^{2-}; 7, Br^-; 8, NO_3^-; 9, SO_4^{2-}. (Reproduced by permission of Dionesc Corporation.)

column, they replace phthalate ions in the eluent so that the overall neutrality is maintained at every point. The UV absorbance of the mobile phase will therefore decrease to give negative peaks, which can be recorded spectrophotometrically. One elegant aspect of this approach is that all ionic analytes will give the same magnitude of response as in each case the signal is due to the loss of phthalate ions. The sensitivity is comparable to that obtained by ion chromatography but the method has not had the same impact in the workplace.

b. Silica as an ion-exchange medium

As noted earlier, apart from ion chromatography, relatively little analytical use has been made of separations based on ion-exchange chromatography. However, it has been found that silica gel columns can be used as weak anion exchangers for the separation of basic drugs (Table 12.11). Either aqueous ammonium nitrate–methanol or perchloric acid–methanol mixtures are used as the mobile phase (Figure 12.18) [4, 5]. This technique is of particular interest in forensic analysis and drug monitoring, as it permits the direct analysis of aqueous samples with minimal work-up.

12.6 SEPARATION OF MACROMOLECULES AND BIOPOLYMERS

Although separations of biological macromolecules or synthetic polymers based on molecular size have long been carried out using column liquid chromatography, size exclusion chromatography was introduced only slowly into HPLC because the original column materials were principally made from soft polymers and were unstable under high pressures and flow rates.

The development of rigid porous matrices based on silica have enabled size exclusion chromatography separations to be carried out on non-polar

Table 12.11 Ion-exchange column materials based on silica packing materials.

Cation exchanger	
(Usually based on benzenesulphonic acids)	
LiChrosorb KAT	
Partisil 10 SCX	Benzenesulphonic acids
Nucleosil 5SA	
Anion exchanger	
(Based on quaternary amines)	
LiChrosorb AN	
Nucleosil SB	RN^+-trimethyl
Partisil 10 SAX	Strong anion exchanger
Spherisorb 5 SAX	*N*-propyl-*N*,*N*,*N*-trimethylammonium ligand
Zorbax SAX	

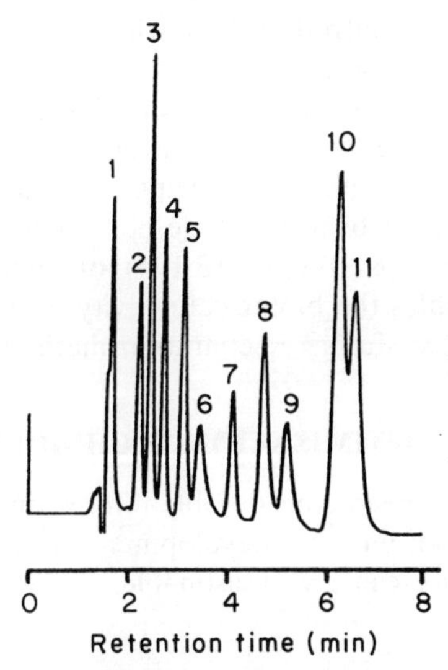

Figure 12.18 Separation of narcotic analgesic drugs and related compounds on a silica column. Column Spherisorb S5W, 250 mm × 5 mm. Eluent methanol–aqueous ammonium nitrate + ammonia buffer, pH 10.1, 90:10 at 2 ml min^{-1}. Detection at 254 nm. Peaks: 1, dextropropoxyphene; 2, dipipanone; 3, 6-monoacetylmorphine; 4, methadone; 5, morphine; 6, morphine-3-glucuronide; 7, norpethidine; 8, dihydrocodeine; 9, dihydromorphine; 10, norcodeine; 11, normorphine. (*J. Chromatogr.*, 1984, **301**, 165–172 (Crown copyright).)

synthetic polymer samples, such as plastics and resins. More recently, newer column materials using coated silica materials or more rigid crosslinked polymers have been introduced. It is now possible to use this mode of separation over a wide range of molecular sizes down to molecular weights of 200–2000. Each size exclusion column operates over a limited range of molecular sizes, depending on its pore size range, so that for a full range of analytes different columns may need to be used.

In the last few years the separation of biological macromolecules, proteins and peptides has become an important area, and HPLC is now routinely applied in biotechnology as both a preparative and analytical technique. Initially much of this work was carried out using size exclusion chromatography, but it was found that separations could also be carried out by conventional reversed-phase chromatographic methods using columns packed with alkyl-bonded wide-pore 300 Å silica (Figure 12.19a), and these materials are now available from many suppliers.

However, the organic solvents used in reversed-phase chromatography often cause the tertiary structure of proteins to be disrupted, with loss of their biological activity. To avoid this problem, hydrophobic interaction chromatography has been introduced in which an efficient separation of proteins can be carried out on a reversed-phase column by using a gradient of decreasing $(NH_4)_2SO_4$ salt concentration, giving very mild separation conditions. The protein is effectively salted out of the mobile phase above a certain salt concentration rather than being continuously eluted. The best results have been obtained by using wide-pore polymer column materials bonded with phenyl groups (Figure 12.19b) or short-chain alkyl-bonded silica. This method enables the biological activity of proteins to be retained and is proving very successful as a separation method for biopolymers.

12.7 OPTIMISATION TECHNIQUES

In recent years there has been considerable interest in the use of computer-based techniques to aid method development in HPLC. Although the concepts discussed earlier can give reasonable initial separation conditions,

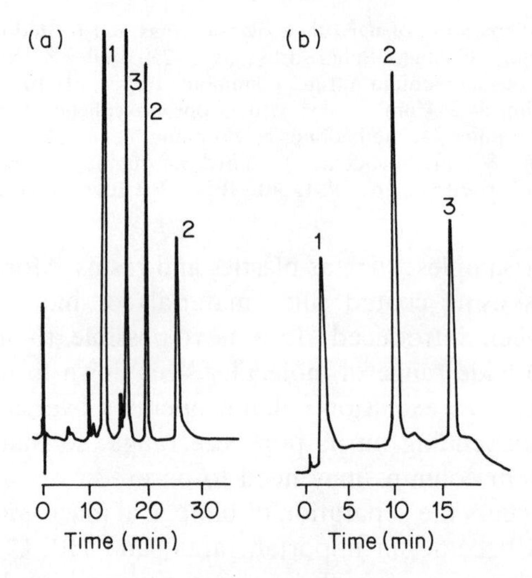

Figure 12.19 Separation of proteins by reversed-phase and hydrophobic interaction chromatography. Proteins: 1, cytochrome c; 2, myoglobin; 3, lysozyme. Spectroscopic detection at 280 nm. (a) Separation by reversed-phase chromatography on wide-pore column. Column Bio-Rad Reversed Phase Hi-Pore (RP-304), 250 mm × 4.6 mm. Eluent: A, 0.1% trifluoroacetic acid; B, 0.1% trifluoroacetic acid–acetonitrile, 95:5. Programmed elution linear gradient from 25% to 50% component B over 30 min. Flow rate 1.5 ml min^{-1}. (b) Hydrophobic interaction separation. Column Bio-Gel TSK Phenyl-5PW, 75 mm × 7.5 mm. Eluent: A, 1.7 M ammonium sulphate–0.1 M sodium phosphate, pH 7.0; B, 0.1 M sodium phosphate, pH 7.0. Linear gradient from 0 to 100% B over 15 min. Flow rate 1.0 ml min^{-1}. (Reproduced by permission of BioRad Laboratories.)

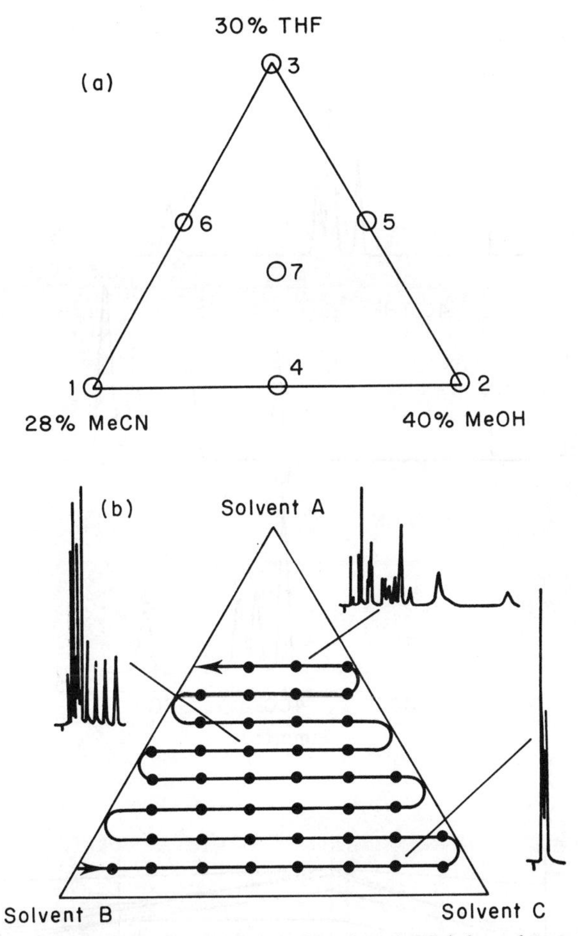

Figure 12.20 (a) Simple mapping pattern (Snyder triangle) for solvent optimisation based on three isoeluotropic eluents: 1, 2, 3. Points 4, 5 and 6 are 1:1 combinations and 7 is a 1:1:1 combination. (b) More detailed mapping pattern of multiple combinations between three solvents each modified with water, showing typical separations at three points in the map. (Reproduced by permission of Perkin-Elmer Corp.)

mixtures of analytes may not be completely resolved and a series of adjustments may be needed to optimise the separation and achieve the maximum resolution within a reasonable separation time. The aim of automated optimisation techniques is to reduce the time and effort that would be involved in a purely trial-and-error approach. These methods, which have bene extensively reviewed in recent years (see Bibliography),

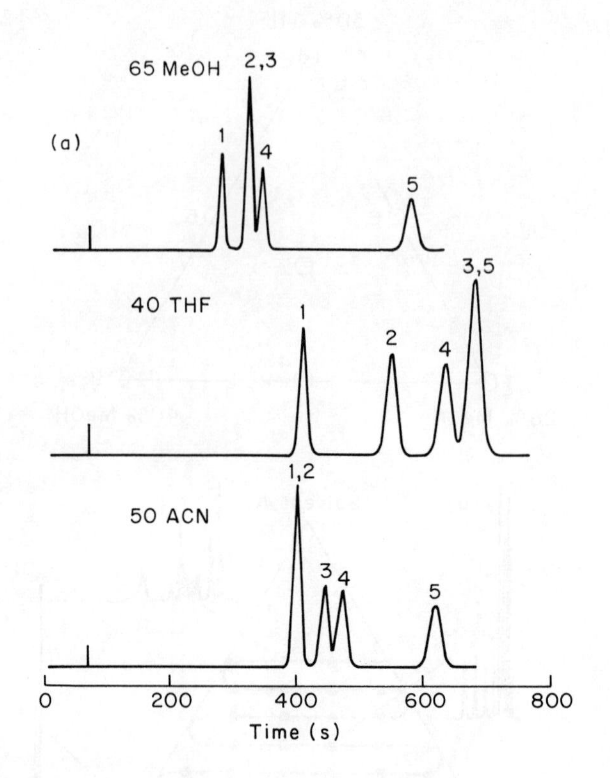
65 MeOH
2,3
(a)
1
4
5
40 THF
3,5
1
2
4
1,2
50 ACN
3
4
5
0
200
400
600
800
Time (s)

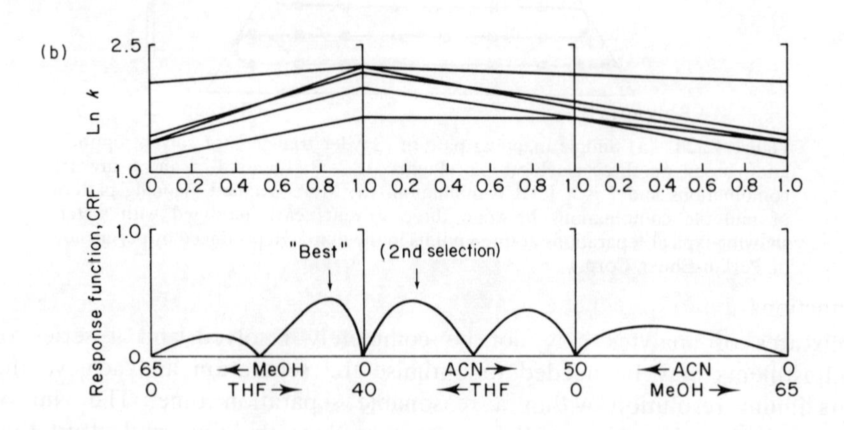
(b)
2.5
Ln k
1.0
0 0.2 0.4 0.6 0.8 1.0 0.2 0.4 0.6 0.8 1.0 0.2 0.4 0.6 0.8 1.0
1.0
Response function, CRF
"Best"
(2nd selection)
0
65
0
← MeOH
THF →
0
40
ACN →
← THF
50
0
← ACN
MeOH →
0
65

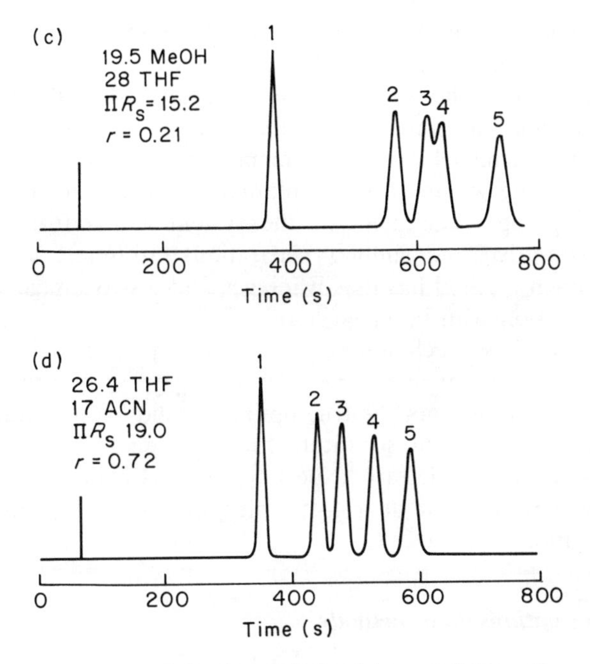

Figure 12.21 Iterative optimisation method. (a) (opposite) Separations achieved with three isoeluotropic eluent conditions. (b) (opposite) Initial linear interpolation of separations between isoeluotropic eluent and corresponding chromatographic response function (CRF) showing suggested "best" conditions at maximum response factor. (c) Separation at initial best conditions from (b) but results poorer than predicted. The retentions are then used to modify the interpolation diagram and thus the CRF results, which then suggests new best conditions (2nd selection). (d) Chromatogram using second proposed conditions, which are quite satisfactory as final conditions. Suitable separation achieved after five chromatograms plus initial programmed chromatogram to select isoeluotropic conditions. (Reproduced with permission from *Chromatographia*, 1982, **16**, 48.)

are usually based on the maximisation of a chromotographic response function, which is a numerical representation of the "goodness" of the separation. This can include elements for the number of resolved components, the resolution of the components and the overall separation time. Most functions aim to spread out the peaks evenly through the available separation time and in particular are heavily penalised if peaks overlap and are unresolved.

Optimisation systems have been applied to both normal- and reversed-phase separations and can be adapted to the selection of modifiers, such as buffer pH or ion-pair reagent concentrations. By looking objectively at the separation, they are often able to reduce the number of trial separations required. Some of the techniques only advise the operator on suitable conditions but others are linked to the chromatograph and can run unattended, changing the conditions automatically within defined limits, and repeating the analysis until specified criteria are reached or no further improvement is being achieved. Two main approaches have been developed in which either the whole of a possible field of conditions is systematically searched (mapping techniques) or the system learns from the previous separation (iterative and simplex separations) to select the next set of conditions. Each method has its advantages and disadvantages and only the briefest descriptions can be given here.

Although all these techniques are of current industrial and academic interest, in their present forms they are all imperfect and may give results which, while useful, are not the true optimum, and many chromatographers still feel that they are no substitute for experience. In the future more intelligent procedures allied to "expert systems" (Chapter 15) will probably give improved results but at a cost which may not be justified for most simple separations.

a. *Mapping optimisation methods*

The approximate conditions for a separation are estimated or are determined from a gradient separation. Using this starting point, a limited number of alternative solvent mixtures are selected based on isoeluotropic combinations of water with methanol, acetonitrile, or THF for reversed-phase systems, or on such combinations of hexane with chloroform, dichloromethane, or diethyl ether for normal-phase systems. These eluent combinations have been identified as having significant selectivity differences in each mode. A series of separations are then carried out using these eluents individually and in combination, with typically 1:1 and 1:1:1 mixtures (Figure 12.20a), but further intermediate combinations can also be used (Figure 12.20b). Each of the chromatograms is then analysed and chromatographic response functions calculated to give a map of the overall response of the separation. This can then be used to select possible optimum conditions, which can then be tested.

The disadvantage of this method is that all the separations must be carried out before any conclusions can be drawn, but it is easy to automate the collection of data and the changing of the separation conditions. As a method, it suffers if the order of the peaks is different in different eluents, as it is not then possible to interpolate between the measured points. The use of a large number of trial conditions is wasteful but if too few are used the optimum values may be missed.

b. *Iterative optimisation methods*

In this procedure, rather than plotting the whole response map, the separations are carried out using the three initial isoeluotropic binary solvent combinations. The results are examined and the separations for intermediate compositions are estimated by interpolation, usually by assuming a linear change in retention with composition. These results are then used to estimate the expected chromatographic response function values for these intermediate ternary eluents. The eluent combination with the best predicted separation is then examined experimentally and the results are used to refine the model (Figure 12.21). Usually in three or four steps a suitable combination can be identified. Thus this method concentrates only on those areas that appear to be successful, but it may fail to locate the global optimum conditions. However, as the search procedure is directed by the previous separations and only possible combinations are examined, it is more time-efficient than the mapping routine and is relatively easy to carry out.

c. *Simplex optimisation methods*

In its simplest form this method starts by taking three possible initial eluent conditions based on two eluent solvents and an overall polarity solvent (i.e.

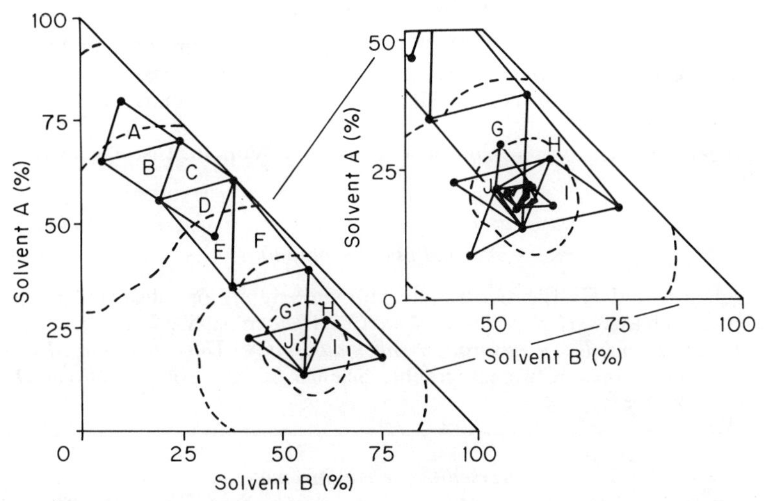

Figure 12.22 Simplex optimisation method. Variable compositions of solvent A and solvent B. Initial conditions A reflected in successive separation, until J is reached. Part expanded to show circling of conditions around optimum conditions. (Reproduced by permission of Bruker-Franzen Analytik GmbH, Bremen, West Germany.)

methanol, and acetonitrile with water) and examines the separations. The chromatographic response criteria are used to rank the eluents. The "worst" eluent combination is eliminated and a new separation condition is chosen, which is a "reflection" away from the originally worst conditions. The results of this new eluent combination are compared with the remaining original separations and again the worst combination is eliminated and new conditions selected. The process continues until no significant further improvement is obtained with a new eluent (Figure 12.22).

Again this method uses an "intelligent" or directed rather than a random approach but it suffers because it may follow an increasing improvement in the response function to a local maximum, which is not the global maximum for the separation. Usually a modified simplex procedure is used in which the size of the change is altered depending on whether the overall separation is improving or worsening. The optimisation can be extended to optimise simultaneously additional parameters, and factors such as pH and temperature can be included but additional steps will be required. These systems can be automated and unattended optimisation has been reported.

BIBLIOGRAPHY

General column materials

R.E. Majors, "Recent advances in high performance liquid chromatography packings and columns", *J. Chromatogr. Sci.*, 1977, **15**, 334–351.

R.E. Majors, "Recent advances in HPLC packings and columns", *J. Chromatogr. Sci.*, 1980, **18**, 488–511.

K.K. Unger, *Porous silica*, J. Chromatogr. Library, Vol. 16, Elsevier, Amsterdam, 1979.

Normal-phase separations

H. Engelhardt and H. Elgass, "Liquid chromatography on silica and alumina as stationary phases", *High-perform. Liq. Chromatogr.*, 1980, **2**, 57–111.

L.R. Snyder, *Principles of adsorption chromatography. The separation of nonionic organic compounds*, Chromatographic Science Series, Vol. 3, Marcel Dekker, New York, 1968.

Reversed-phase column materials

P.E. Antle and L.R. Snyder, "Selecting columns for reversed phase HPLC. Part I: Column selectivity", *LC Mag*, 1984, **2**, 840–846; "Part II: Column and packing configuration", *ibid.*, 1985, **3**, 98–109.

R.K. Gilpin, "The bonded phase: structure and dynamics", *J. Chromatogr. Sci.*, 1984, **22**, 371–377.

E. Grushka (Ed.), *Bonded stationary phases in chromatography*, Ann Arbor Press, Ann Arbor, MI, 1974.
I. Halász, "Columns for reversed phase liquid chromatography", *Anal. Chem.*, 1980, **52**, 1393A-1403A.
W.R. Melander and C. Horváth. "Reversed-phase chromatography", *High-Perform. Liq. Chromatogr.*, 1980, **2**, 113–304.
L.R. Snyder, "Liquid-solid chromatography. New insights into retention on bonded-phase packings", *LC Mag.*, 1983, **1**, 478–486.

Eluent properties

HPLC solvent reference manual, J.T. Baker Chemical Co., Phillipsburg, NJ, 1985.
P. Jandera and J. Churáček, *Gradient elution in column liquid chromatography: theory and practice*, J. Chromatogr. Library, Vol. 31, Elsevier, Amsterdam, 1985.
A.M. Krstulovic and P.R. Brown, *Reversed-phase high performance liquid chromatography: theory, practice and biomedical applications*, Wiley Interscience, New York, 1982.
L.R. Snyder, "Classification of the solvent properties of common liquids", *J. Chromatogr. Sci.*, 1978, **16**, 223–234.

Polymer stationary phases

J.R. Benson and D.J. Woo, "Polymeric columns for liquid chromatography", *J. Chromatogr. Sci.*, 1984, **22**, 386–399.
D.P. Lee, "Reversed-phase HPLC from pH 1 to 13", *J. Chromatogr. Sci.*, 1982, **20**, 203–208.

Ion-pair chromatography

B.A. Bidlingmeyer, "Separation of ionic compounds by reversed-phase liquid chromatography: an update of ion-pairing techniques", *J. Chromatogr. Sci.*, 1980, **18**, 525–539.
R. Gloor and E. L. Johnson, "Practical aspects of reverse phase ion pair chromatography", *J. Chromatogr. Sci.*, 1977, **15**, 413–423.
M.T.W. Hearn, "Ion-pair chromatography on normal- and reversed-phase systems", *Adv. Chromatogr.*, 1980, **18**, 59–100.
M.T.W. Hearn (Ed.), *Ion-pair chromatography: theory and biological and pharmaceutical applications*, Chromatographic Science Series, Vol. 31, Marcel Dekker, New York, 1985.
J.H. Knox and R.A. Hartwick, "Mechanism of ion-pair liquid chromatography of amines, neutrals, zwitterions and acids using anionic hetaerons", *J. Chromatogr.*, 1981, **204**, 3–21.
E. Tomlinson, "Ion-pair extraction and high-performance liquid chromatography in pharmaceutical and biomedical analysis", *J. Pharm. Biomed. Anal.*, 1983, **1**, 11–27.
E. Tomlinson, T.M. Jefferies. and C.M. Riley, "Ion-pair high performance liquid chromatography", *J. Chromatogr., Chromatogr. Rev.*, 1978, **159**, 315–358.

Ion chromatography

D.T. Gjerde and J.S. Fritz, *Ion Chromatography*, 2nd ed., Alfred Hüthig, Heidelberg, 1987.
H. Small, "Modern inorganic chromatography", *Anal. Chem.*, 1983, **55**, 235A-242A.
F.C. Smith and R.C. Chang, *The practice of ion chromatography*, John Wiley, New York, 1983.
J.G. Tarter (Ed.), *Ion chromatography*, Chromatographic Science Series, Vol. 37, Marcel Dekker, New York, 1986.

Optimisation methods

J.C. Berridge, *Techniques for the automated optimization of HPLC separations*, John Wiley, Chichester, 1985.
H.J.G. Debets, "Optimization methods for HPLC", *J. Liquid Chromatogr.*, 1985, **8**, 2725–2780.
L. de Galan and H.A.H. Billiet, "Mobile phase optimization in RPLC by an iterative regression design", *Adv. Chromatogr.*, 1986, **25**, 63–104.
J.L. Glajch and J.J. Kirkland, "Optimization of selectivity in liquid chromatography", *Anal. Chem.*, 1983, **55**, 319A-336A
R.E. Kaiser and O. Oelrich, *Optimisation in HPLC*, Alfred Hüthig, Heidelberg, 1981.
M. Otto and W. Wegscheider, "Multifactor model for the optimization of selectivity in reversed-phase chromatography", *J. Chromatogr.*, 1983, **258**, 11–22.
J.H. Purnell, "Window analysis: an approach to total optimization in chromatography" in F. Bruner (Ed.), *The science of chromatography*, J. Chromatogr. Library, Vol. 32, Elsevier, Amsterdam, 1985, pp. 363–379.
P.J. Schoenmakers, *Optimization of chromatographic selectivity. A guide to method development*, J. Chromatogr. Library, Vol. 35, Elsevier, Amsterdam, 1986.

REFERENCES

1. R.M. Smith, "Characterization of reversed-phase liquid chromatography columns with retention indexes of standards based on an alkyl aryl ketone scale", *Anal. Chem.*, 1984, **56**, 256–262.
2. J.F.K. Huber, M. Pawlowska and P. Markl, "Solvent-generated liquid–liquid chromatography with aqueous ternary systems", *Chromatographia*, 1984, **19**, 19–28.
3. R. Gill, S.P. Alexander and A.C. Moffat, "Comparison of amine modifiers used to reduce peak tailing of 2-phenylethylamine drugs in reversed-phase high performance liquid chromatography", *J. Chromatogr.*, 1982, **247**, 39–45.
4. B. Law, R. Gill and A.C. Moffat, "High-performance liquid chromatography retention data for 84 basic drugs of forensic interest on a silica column using an aqueous methanol eluent", *J. Chromatogr.*, 1984, **301**, 165–172.
5. R.J. Flanagan and I. Jane, "High-performance liquid chromatographic analysis of basic drugs on silica columns using non-aqueous ionic eluents. I. Factors influencing retention, peak shape and detector response", *J. Chromatogr.*, 1985, **323**, 173–189.

CHAPTER 13

QUANTITATIVE AND QUALITATIVE APPLICATIONS OF HIGH-PERFORMANCE LIQUID CHROMATOGRAPHY

13.1 APPLICATIONS OF HIGH-PERFORMANCE LIQUID CHROMATOGRAPHY

Virtually all compounds which will dissolve in an aqueous or organic solvent can be separated and quantified by HPLC, with a sensitivity dependent on their ease of detection. As these criteria include almost all non-polymeric and non-gaseous compounds of interest in present-day chemistry, a detailed list of applications would be virtually limitless. A bibliography of sources of application methods covering the main chemical types is therefore given in Appendix 1. Generally, HPLC has been most widely applied in areas where GLC is unsuitable, in particular the separation of relatively polar or involatile compounds. The most important groups of analytes are pharmaceutical compounds, including penicillins, tetracyclines, barbiturates, opium alkaloids and many other major drug groups. HPLC has also found application in the study of agrochemicals, although for the more volatile organohalogen and organophosphorus insecticides and herbicides, the high sensitivity and selectivity of the electron capture, thermionic and flame photometric detectors has meant that GLC is often still preferred. In addition, the need for the positive identification of pesticides and pollutants in complex matrices, will require the combination of high-resolution separation and a sophisticated detector provided by open-tubular GC–MS.

HPLC has proved especially useful for the separation of many biological samples and naturally occurring compounds, including carbohydrates, lipids and vitamins, and well as pollutants, food additives, food colours and preservatives. In recent years, there has been particular interest in the application of HPLC in the reversed-phase mode, with wide-pore column materials and hydrophobic chromatography for the separation and isolation of biopolymers, proteins and peptides, with the retention of biological activity. HPLC can also be applied to the analysis of inorganic complexes or chelates, organometallic compounds and simple anions and cations.

By scaling up the size of columns and pumps, HPLC can be carried out preparatively on the milligram to kilogram scale. Eluents used in normal phase chromatography can usually be readily removed but the aqueous–organic eluents used in reversed-phase chromatography normally have to be separated from the analyte by freeze-drying.

13.2 SAMPLE PREPARATION

One of the advantages of HPLC is that very little sample preparation is usually required before assay. Neat liquid samples are usually too concentrated to be injected directly without overloading the columns, so that both liquids and solids are prepared as solutions in a suitable solvent. The samples for analysis must be free from any insoluble material, which would be trapped on the top of the column and eventually degrade the separation. If suspended particles are present or suspected, the sample should be filtered using a millipore filter before injection. Care must also be taken that components of the sample will not precipitate on mixing with the eluent. This can often occur with aqueous samples containing salts or proteins, which may be insoluble in organic–aqueous eluents. However, in some cases diluted biological fluids, such as urine or serum, have been been injected directly on reversed-phase columns without problems.

The solvent used to dissolve the sample is important as it may produce peaks near the solvent front which interfere with early peaks. The ideal solvent has the same composition as the mobile phase and this will cause the minimum of disturbance. Problems can be found if the solvent is a stronger elute than the mobile phase, because the solvent will try to elute the sample components through the column more rapidly than the eluent (Figure 13.1). As a result, until the sample solvent and the eluent become completely mixed, different parts of the sample will travel at different speeds, causing peak broadening and reduced efficiency. The greater the difference between the strengths of the solvent and the eluent the worse will be the effect. If the solvent is effectively immiscible with the eluent

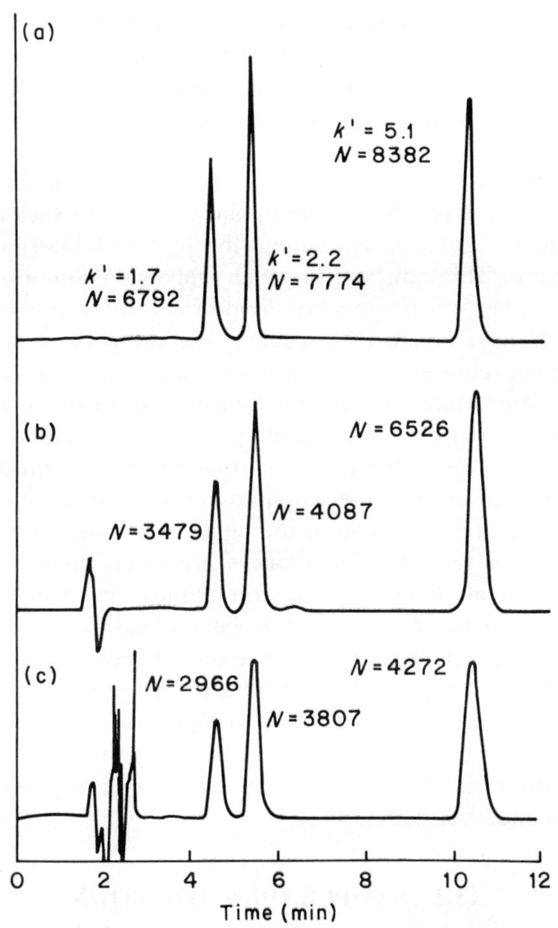

Figure 13.1 Effect of the sample solvent on the chromatographic efficiency. Separation conditions: ODS-Zorbax column, 250 mm × 4.6 mm, temperature 35°C. Mobile phase methanol–water 85:15. Flow rate 1 ml min^{-1}. Sample components in order of elution are: toluene, naphthalene and anthracene. Sample solvents: (a) mobile phase; (b) methanol; (c) isopropanol. (Reproduced by permission of Elsevier Science Publishers from Parris, *Instrumental liquid chromatography*, p. 300.)

(i.e. a sample in hexane being injected into methanol–water), the peak shapes may be severely distorted and the column efficiency may be reduced for some time. In both cases, the larger the volume of the sample solvent that is injected the worse will be the effects. Reducing the sample size can sometimes eliminate the problem. In contrast, a solvent that is a weaker eluent than the mobile phase can aid the separation. It will tend to focus the sample on the top of the column, increasing the efficiency, particularly for large volumes of dilute sample. However, a weaker eluent may be a poor solvent for the sample and care is needed to ensure that it is completely soluble.

The concentration of analytes with a weak eluent can be used for the analysis of very dilute solutions of non-polar compounds such as river water. The dilute aqueous solution is pumped through an ODS-silica or polymer PS–DVB column. This can be either the analytical column or a short pre-column held in place of the injection loop of a valve injector. Either a step change to a stronger eluent (by switching the pre-column into the eluent) or a programmed elution is then used to analyse the sample. In a typical example, all the polynuclear aromatic hydrocarbons from 1–10 ml of water could be concentrated onto the column as a sharp injection.

Even sample solvents lacking a chromophore can produce so-called refractive index peaks from a spectrophotometric detector (see Figure 11.10), because they will alter the angle of the light path through the detector flow cell. These solvent baseline disturbances are often taken to indicate the column void volume. However, the compounds may have been retained by the column and the retentions may be misleading. Care must also be taken as the negative peak of the baseline disturbance can be misinterpreted by an integrator as a minimum and be used to establish a new but false baseline. This may cause the integrator to report subsequent peaks as being unresolved and to assign them grossly misleading and overestimated peak areas. Integration should therefore be disabled or suppressed until the solvent disturbance has been eluted.

13.3 SAMPLE IDENTIFICATION

Relatively little use has been made of HPLC as a method for the identification of unknown compounds. As with GLC, there is the possibility that more than one compound will have a particular retention and thus identifications based solely on the comparison of a single retention time comparison can only be tentative. However, the chromatographic properties of a sample can often be used as a guide to the polarity and possible structural type. Although capacity factors have been published for many compounds, these are usually little help in identifications other than to suggest possible separation conditions. In each case it is necessary to compare the compound to be identified directly with an authentic standard on the same equipment and under the same conditions. Ideally comparison by two different separation modes, reversed- and normal-phase, are desirable for identification. The response from the detector should also match, and if a diode array spectrometer is available the full ultraviolet spectra can also be compared.

The reproducibility of separation times on HPLC can be very good, particularly on a single instrument under close temperature control. However, retentions measured on different instruments or even on the same instrument on different days may vary considerably. This is partly because the interactions are very dependent on the exact composition of the mobile phase and temperature, and these are often difficult to reproduce accurately. The method for the preparation of an eluent needs to be particularly carefully specified. For example, the elution strength of a 50:50 methanol–water mixture will vary depending on whether it is made up by volume, by weight, or if 50 ml of one component is made up to a total of 100 ml with the other component (Table 13.1).

In addition, differences in retention are often found if different chromatographs are used, as these may have slightly different volumes of connecting tubing, pump flow rates, or detector and column void volumes. The column is a particular source of variation between instruments. There are large differences between different brands of nominally equivalent column packing materials (e.g. ODS-Silica Table 12.6) and frequently there will be smaller batch-to-batch variations in retention power and selectivity within one brand of material.

A further problem in comparing capacity factors reported in the literature with measured values is that the calculation of capacity factors in different laboratories is a source of inconsistency. Capacity factors (k') are defined as the ratio of the adjusted retention time ($t_R - t_0$) to the column void volume (t_0) (equation 2.2). However, although the column void volume is defined as the time for an unretained analyte to pass through the column, different methods can be used for its measurement and these frequently give different values on a single column. Typical void volume markers for reversed-phase separations are uracil, water, aqueous sodium nitrate or

Table 13.1 Effect of small variations in the preparation of the mobile phase on retentions. Nominal composition methanol–water 50:50.

Method of preparation	k
Volume–volume	2.05
Mass–mass	1.32
500 ml methanol made up to 1000 ml with water	1.90
500 ml water made up to 1000 ml with methanol	2.24

(H. Engelhardt, personal communication.)

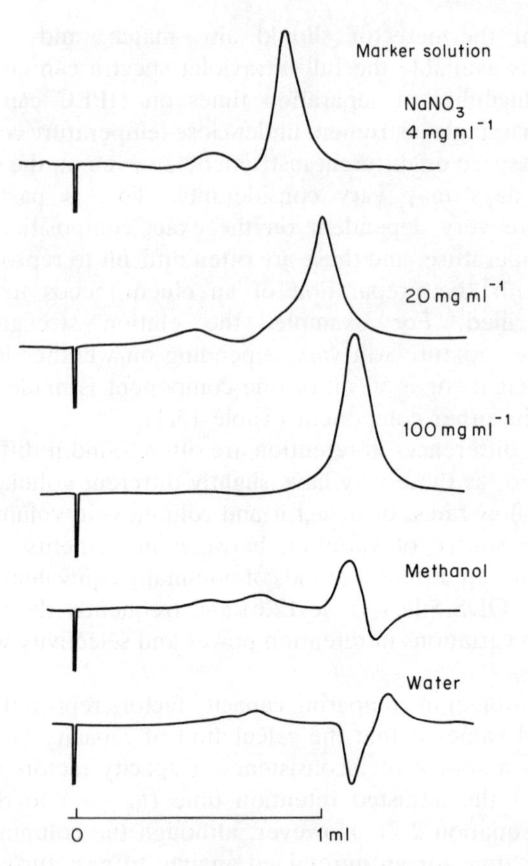

Figure 13.2 The separations of typical marker solutions for the column void volume. Column ODS-Hypersil, 5 μm, 100 mm × 5 mm. Mobile phase methanol–water 60:40. Flow rate 1 ml min^{-1}.

sodium nitrite solutions and deuterated analogues of the eluent. The selection is limited because it is usually desirable to detect the markers using a spectrophotometric detector. The different void volume markers will usually give approximately the same retention time but with small absolute differences of up to ±10–15% (Figure 13.2). However, from equation 2.2, these differences will result in a significant 10–15% variation in the calculated capacity factors for compounds with the same retention times. For some of the column void volume marker solutions, it is difficult to decide whether the retention time should be measured to the start of the peak, to the peak maximum or to the peak minimum. It has also been demonstrated that with some eluent combinations even the polar organic analytes are retained on the column and that their retention times can vary with the eluent strength. As fully ionised samples, the sodium nitrate and sodium nitrite solutions are not retained on the column. However, their retention times can vary with the concentration of the test solution and with the ionic strength and composition (percentage and organic modifier) of the mobile phase. This is because they can be excluded from part of the column material by ionic repulsion. The determination of the "true" column void volume has been the subject of much research and none of the test solutions has been universally accepted as giving the correct value. Some researchers have suggested that the void volume may even differ slightly for each analyte, as each has access to a different amount of the column, because of its size and polarity. For many practical purposes for work within one laboratory, the result is not critical and either a solution of sodium nitrate or a sample of methanol or water is usually used.

So far no international agreement has been reached on a standard method for the measurement of t_0 so that different laboratories will often use different methods (and hence different void volumes) for the same analysis. The differences in the measured values of t_0 cause corresponding changes in any calculated capacity factors making comparisons between laboratories difficult. Therefore it is important always to define the method used to measure the column void volume and the value used in any calculation.

a. *Relative retention factors*

Many of the factors that differ between assays recorded at different times, such as flow rate, column packing density and column dimensions, should have a proportional effect on the retentions of both the analyte and the column void volume marker and thus their ratio and hence the capacity factors should be unaffected. In contrast, changes in the eluent composition or temperature will only affect retained compounds and thus the calculated capacity factors will alter. However, with the variations such as might be expected between laboratories using the same method and brand of column packing material, related compounds should suffer equivalent changes. Therefore, if the retentions of analytes are expressed as relative capacity factors compared to an internal standard compound with a similar structure ($\alpha = k'_{\text{analyte}}/k'_{\text{standard}}$), the values will often be more consistent than the individual capacity factors. In many cases the internal standard peak can also be used as a reference for quantitative analyses (see later). The internal standard needs to be carefully chosen. Ideally it would be an isomer or homologue of the analyte containing the same functional groups and chromophore so as to have similar interactions and susceptibility to changes in conditions.

If the retention times are reasonably long, some laboratories use relative retention times (rather than relative capacity factors), as these values can be calculated readily by many integrators as the ratio of the two retention times, without a knowledge of the column void volume. However, the values will be more susceptible to differences in the design of the chromatograph and the extra-column void volumes, and should not be used to compare results obtained on different instruments or in different laboratories.

b. Retention indices in HPLC

An alternative method for recording the retentions of analytes in reversed-phase chromatography is as retention indices compared to a homologous series of standards [1], in a similar manner to the use of Kováts indices in GLC (Section 6.1). As with GLC there is a linear relationship between log k' and the carbon number of the members of a homologous series. However, the n-alkanes used in GLC are impractical as standards for HPLC because they would all be very strongly retained, much more than typical analytes. They also lack a chromophore and would be difficult to detect spectrophotometrically. A number of alternative series of compounds have been used [1] but most work has been based on either the alkan-2-ones [2] or aklyl aryl ketones [3]. Both series of standards are readily available and have significant ultraviolet absorption spectra.

The retention indices of analytes are largely independent of small variations in the mobile-phase composition and of the exact value used for the column void volume. It has been shown that retention indices can provide very robust retention results for drugs such as the barbiturates and local anaesthetics and have advantages in interlaboratory collaborative studies. Potentially they could provide a basis for the establishment of large databases of reproducible retention values, which could be used by different laboratories for the identification of unknown compounds.

However, because of the large differences in the selectivities of the different column materials, retention indices cannot fully compensate for the use of different brands of column materials, although they are less affected than capacity factors. It is possible to use these differences in retentions to compare the selectivities of the different column materials in a similar manner to the use of McReynolds constants in GLC [1]. The column properties are expressed as the retention indices of a series of column test compounds (toluene, nitrobenzene, *p*-cresol and 2-phenylethanol). Similar values are found for different ODS-silica columns, but there are larger differences between different bonded phases.

13.4 QUANTITATIVE ANALYSIS BY HIGH-PERFORMANCE LIQUID CHROMATOGRAPHY

Almost all HPLC analyses are primarily carried out to quantify one or more analytes and it is in this area, particularly for the determination of pharmaceuticals, that the technique has made its greatest impact. Unlike GLC, there are few limitations on the compounds that can be determined but analytes lacking chromophores are more difficult to detect and the sensitivity may therefore be limited. Most detectors used in HPLC have a wide linear range (with a few exceptions) and good stability so that calibration for quantitative analysis is usually easy. However, because all the major detectors are selective and give different responses for different compounds, each assay must be calibrated with known quantities of the analyte in a similar matrix and solvent to the sample.

The accuracy of quantitative analyses depends on the sample injection. Sample introduction using a fixed loop injection valve can be carried out with high precision (relative standard deviation 0.5%) as the volume injected is closely defined by the mechanical volume of the loop. Care must be taken when filling the loop that sufficient sample is injected to ensure that the eluent in the loop is completely replaced by the sample solution (5–10 times the loop volume) (see Section 10.2.c). If a seven-port valve is used with syringe injection then the limiting factor is the precision of the sample injections and the analyst's ability with the syringe (typically relative standard deviation 0.8%). As the reproducibility of repeated injections of a particular volume is higher than the absolute accuracy of the syringe, the same injection volume should be used for all the calibration standards and the sample solutions.

Because of the high precision of injection, particularly with automatic injectors, very frequently routine quantitative HPLC analyses are carried out without an internal standard. Instead, an external standard solution is included in each set of runs. Internal standards are more important if manual syringe injection is being used or as a check on other steps in the analytical method, such as extractions, sample transfers, or derivatisation reactions. In the last case the internal standard should also be derivatised by the reaction. If fluorescence or electrochemical detectors are being used, the internal standards should have a similar response to the analyte as both detectors can suffer from interferences. Fluorescence can be quenched by eluent components and the electrodes in electrochemical detectors often suffer from contamination or deactivation by the sample or eluent. In all determinations, internal standards serve as a useful check on the consistency of the retention times and as a confirmation of sample identity.

As noted earlier, an internal standard should be similar to the analyte, preferably a homologue or isomer, which is resolved from the analyte and would not be expected to be present as an impurity or metabolite in the sample. In reversed-phase chromatography, isomers are readily resolved, but if normal-phase chromatography is being used, compounds with a different polarity may be needed. It has been suggested that, rather than having a different standard for each assay, the most appropriate of the homologous alkyl aryl ketones or *N*-alkylacetanilides could be selected, as these standard compounds are suitable for use with a wide range of analytes having different polarities [1].

Numerous collaborative studies on the use of HPLC for quantitative determinations have confirmed that few problems are normally encountered and that HPLC is a very satisfactory method for the determination of samples ranging from pharmaceuticals to pesticides and inorganic ions.

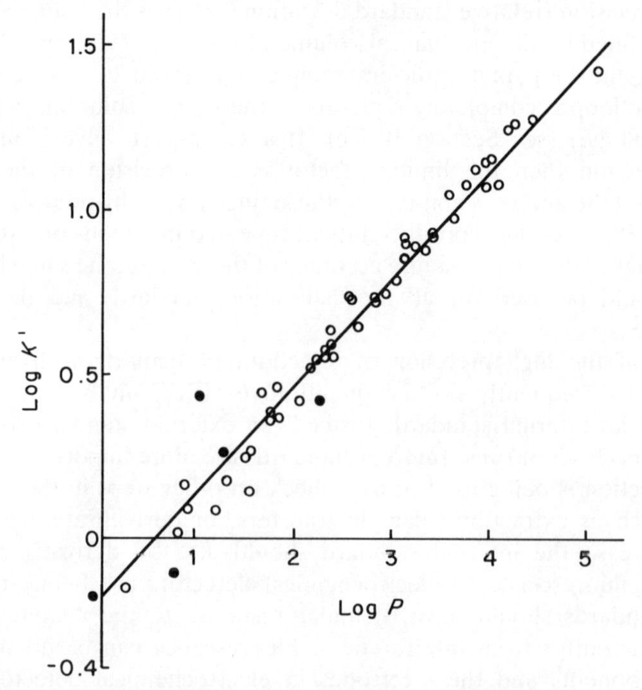

Figure 13.3 Correlation between calculated hydrophobicity (log P) of a wide range of aromatic compounds and their capacity factors on Chromosorb LC-7, 5 μm, 250 mm × 4.1 mm column. Eluent acetonitrile–water 50:50 at 1 ml min^{-1}.(Reproduced with permission from *J. High Res. Chromatogr., Chromatogr. Commun.*, 1981, **4**, 455.)

HPLC determinations are beginning to be more widely adopted as official methods and pharmacopoeia methods.

13.5 MEASUREMENT OF PHYSICAL PARAMETERS

There is considerable interest in the use of HPLC as an easy and rapid method for the estimation of partition coefficients, because for many drugs and pesticides there is a close correlation between their physiological activity and octanol/water partition coefficients (log *P* values). These partition coefficients reflect the ability of drugs to be transported around the body and to cross cell membranes, and can serve as a quick guide to possible biological activity.

For related compounds there is a close correlation between the capacity factors on reversed-phase HPLC and their partition values (Figure 13.3), but there is often a poorer correlation with compounds containing different functional groups. The capacity factor of a novel drug analyte on an ODS-silica column can be compared with those of compounds with known partition values. The partition coefficient can then be estimated by interpolation much more readily than by conventional manual liquid–liquid partition measurements, which are messy and time-consuming. In most cases, conventional methanol–water eluents have been used, but some workers claim better results can be obtained if octanol is incorporated into the mobile phase.

BIBLIOGRAPHY

Sample preparation

Many of the sample preparation methods are very similar to those used in GLC (see Chapter 6).

R.W. Frei and K. Zech (Eds), *Selective sample handling and detection in high-performance liquid chromatography. Part A*, J. Chromatogr. Library, Vol. 39A, Elsevier, Amsterdam, 1988.

Quantitative analyses

J. Asshauer and H. Ullner, "Quantitative analysis in HPLC", in H. Engelhardt (Ed.), *Practice of high performance liquid chromatography*, Springer-Verlag, Berlin, 1986, pp. 65–108.

E. Katz (Ed.), *Quantitative analysis using chromatographic techniques*, John Wiley, Chichester, 1987.

Physical parameters

T. Braumann, "Determination of hydrophobic parameters by high performance liquid chromatography: theory, experimental technique and applications in studies on quantitative structure-activity relationships", *J. Chromatogr. Chromatogr. Rev.*, 1986, **373**, 191–225.

R. Kaliszan. "Chromatography in studies of quantitative structure-activity relationships", *J. Chromatogr., Chromatogr. Rev.*, 1980, **220**, 71–83.

R. Kaliszan, " High performance liquid chromatography as a source of structural information for medicinal chemistry", *J. Chromatogr. Sci.*, 1984, **22**, 362–370.

R. Kaliszan, *Quantitative structure chromatographic retention retentionships*, John Wiley, New York, 1987.

T.L. Hafkenscheid and E. Tomlinson, "Estimation of physicochemical properties of organic solutes using HPLC retention parameters", *Adv. Chromatogr.*, 1986, **25**, 1–62.

REFERENCES

1. R. M. Smith, "Retention index scales in liquid chromatography", *Adv. Chromatogr.*, 1987, **26**, 277–320.
2. J.K. Baker and C.-Y. Ma, "Retention index scale for liquid-liquid chromatography" *J. Chromatogr.*, 1979, **169**, 107–115.
3. R.M. Smith, "Alkylarylketones as a retention index scale in liquid chromatography", *J. Chromatogr.*, 1982, **236**, 313–320.

CHAPTER 14

HIGH-PERFORMANCE LIQUID CHROMATOGRAPHY: SPECIAL TECHNIQUES

14.1 DERIVATISATION REACTIONS FOR HIGH-PERFORMANCE LIQUID CHROMATOGRAPHY

Not all compounds can be analysed easily by HPLC, either because they cannot be readily detected or because they show undesirable interactions with the column. Derivatisation can often be used to add a detectable grouping to the analyte. In most cases, the aim is to introduce a chromophore, which can be readily detected spectrophotometrically. If high sensitivity or selectivity is required fluorophores or electrochemically active groups may be introduced to give suitable products. In recent years, the separation of chiral compounds has attracted particular interest and they can be examined by a number of special techniques including derivatisation and the use of chiral stationary phases (section 14.2).

Unlike GLC, it is rarely necessary to derivatise analytes in HPLC solely to improve their separation. The principal exceptions are compounds containing a strongly ionised grouping, such as quaternary amines or sulphonic acids, as these often cause peak tailing on separation. In these cases, derivatisation to give a non-ionisable product can be used as an alternative to ion-pair chromatography, although the latter method can be regarded as a form of *in situ* derivatisation.

Derivatisation reactions can be carried out either before the sample is chromatographed (pre-column or on-column reactions) or after separation of the components on the column but before the detector (post-column reactions). The criteria for the selection of possible reactions differ in each case. In most routine laboratories, derivatisation is often regarded as the method of last resort because of the time required for sample preparation and the probability of reduced reproducibility, and almost any alternative approach will be attempted to carry out the assay directly on the sample.

Pre-column derivatisation imposes the fewest restrictions on the chemistry of the reactions and is the most commonly used technique. Many well understood and widely used characterisation reactions for organic compounds are employed. With pre-column reactions:

i. the derivatives are separated, not the original analyte compound,
ii. the reaction rate of the derivatisation is independent of the chromatographic separation. The reaction can be slow or high temperatures can be used.
iii. any excess of the reagent and any reaction by-products can be removed either before the separation in a sample purification stage or during the chromatographic separation.
iv. the eluent used for the separation is not limited by the derivatisation reaction.
v. if multiple products are formed, these will give multiple peaks on separation, so that the reaction should be quantitative for both real and standard samples.

If large numbers of samples are being prepared, the numerous manual steps required by most pre-column derivatisation reactions can be very time-consuming and boring for the operator. This can lead to errors and poor precision, as well as being expensive. These problems may be reduced by automation or by the introduction of robots to carry out many of the repetitive transferring and sample handling steps (Section 15.5).

In a few cases, the reagent can be incorporated in the mobile phase so that reaction occurs on injection. However, *on-column methods* are rarely used because the number of suitable reactions is very limited. The conversion of the analyte must be very rapid and yield the product virtually instantaneously as the sample mixes with the eluent. The choice of eluent for the separation may therefore be very restricted as it must permit the reaction to take place at room temperature but not itself react with the reagent. As the reagent is continually being passed through the column, it must not interfere with the separation. In addition, the detector must be able to distinguish between the excess reagent being eluted from the column and the derivative. Ion-pair chromatography and argentation techniques can be regarded as dynamic forms of on-column reactions, in which the analyte is converted into another form with different retention characteristics.

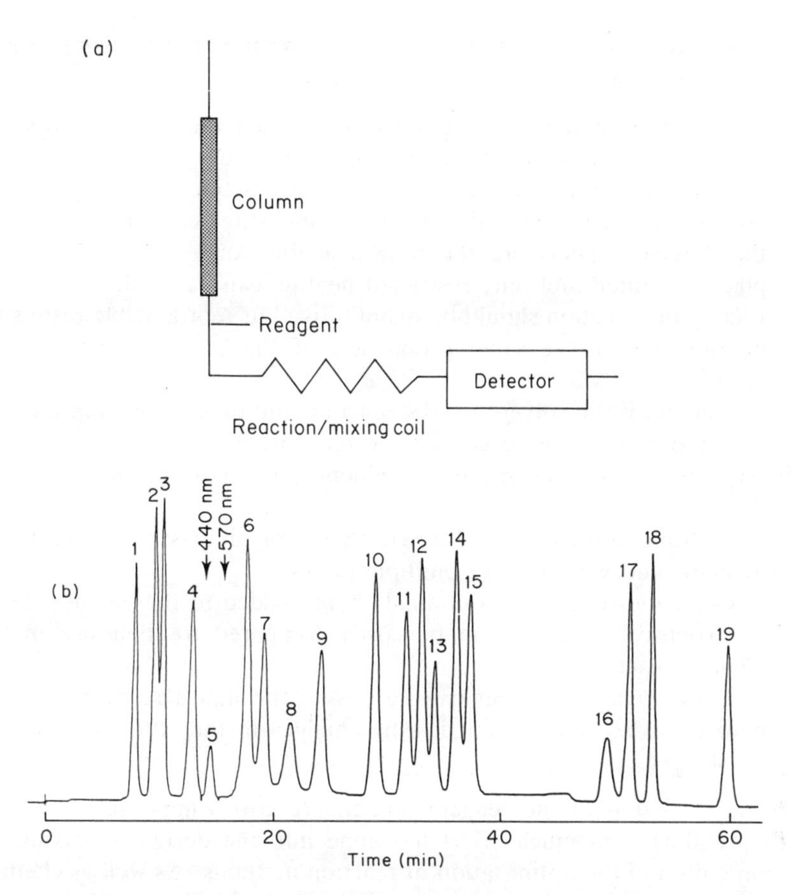

Figure 14.1 (a) Equipment for post-column derivatisation. (b) Separation of amino acids and detection with post-column derivatisation using ninhydrin. Column Micropak AA, 300 mm × 4.6 mm. Eluent: complex gradient from 0.2 M sodium citrate, pH 3.25, to 1.0 M sodium citrate, pH 7.40, at 0.3 ml min^{-1}. Spectroscopic detection at 440 nm, except for switch to 570 nm to detect the secondary amine, proline. Peaks: 1, Asp; 2, Thr; 3, Ser; 4, Glu; 5, Pro; 6, Gly; 7, Ala; 8, Cys; 9, Val; 10, Met; 11, Ile; 12, Leu; 13, Nor; 14, Tyr; 15, Phe; 16, Trp; 17, Lys; 18, His; 19, Arg. (Reproduced with permission of Varian Associates from *Amino acids, peptides and proteins.*)

Post-column derivatisation reactions have been widely used, particularly when a large number of similar samples are being examined. A separate pump is used to add a reagent to the eluent stream after the column (Figure 14.1) and the reaction occurs in a mixing coil, which may often be heated, before the product is detected in a conventional manner. The time taken to set up the procedure usually means that this approach is uneconomical if only a small number of samples are being examined. The characteristics of the post-column derivatisation technique are:

i. The separation step is independent of the derivatisation reaction and the samples are separated in their original form.
ii. The reaction must be rapid because the reaction takes place in the flowing eluent between the point at which the reagent is added and the detector. Therefore the time available for the reaction to take place is limited and only restricted heating can be used.
iii. Ideally the reaction should be quantitative but reproducible results will be obtained if the same proportion of the sample reacts in each occasion.
iv. The eluent is the solvent for the reaction and thus it must not interfere with the reaction or react with the reagent.
v. Any excess of reagent in the eluent must not interfere with the detection.
vi. If multiple products are formed they will all pass together to the detector and will not give multiple peaks.
vii. Because the reagent is continually being added to the eluent, the use of expensive reagents can be costly compared to their use in pre-column reactions.
viii. The underivatised eluent can be passed through a separate detector before the reagent is added. Any changes in the chromatogram can be studied.

The introduction of the reagent and the reactions must also not cause band spreading, and much effort has gone into the design of mixing and reaction coils and the optimisation of reaction methods. As well as chemical reagents, post-column photochemical or electrochemical methods have also been used to convert analytes into more readily detectable products.

Many different derivatisation reactions have been employed in HPLC and these have been widely reviewed (see Bibliography). In some cases the reagents have been developed specially for chromatography but most of the reactions are those that are used for the characterisation and identification of different functional groups in organic chemistry. A selection of the most widely used reactions are described below to demonstrate typical analytes and reagents.

a. *Reactions to enhance ultraviolet detection*

A major problem in HPLC is that many simple unconjugated aliphatic compounds, including alcohols, amines, carboxylic acids and carbonyl compounds, possess only weak chromophores or lack a chromophore altogether and thus cannot be easily detected spectrophotometrically. Although the refractive index detector can be used for these analytes, it has limited sensitivity. Derivatisation is therefore often used to introduce a chromophore. With the ionisable amines and acids, derivatisation often has an additional advantage that the reaction will mask the very polar functional group and the separation is improved. Many of these derivatisation reactions are based on the addition to the analyte of a substituted aromatic ring with an intense chromophore, such as a nitrophenyl group.

i. Pre-column reactions

Aliphatic alcohols can be converted into their 3,5-dinitrobenzoate esters with 3,5-dinitrobenzoyl chloride (Equation 14.1)

$$Cl{-}C({=}O){-}C_6H_3(NO_2)_2 \; + \; ROH \longrightarrow RO{-}C({=}O){-}C_6H_3(NO_2)_2 \qquad (14.1)$$

Carboxylic acids, including fatty acids (Figure 14.2 (Equation 14.2)), will readily react with phenacyl halides (or the related *p*-bromo, *p*-nitro and *p*-phenyl compounds) to give the corresponding phenacyl esters

$$BrCH_2{-}C({=}O){-}C_6H_4{-}X \; + \; RCOOH \longrightarrow RCOOCH_2{-}C({=}O){-}C_6H_4{-}X \qquad (14.2)$$

X = H, Br or NO_2

Alternatively, esters can be prepared by reaction with activated alcohols such as *O*-(*p*-nitrobenzyl)-*N,N'*-(diisopropyl)isourea (Equation 14.3)

$$(Me_2CH{-}N{=})(Me_2CH{-}NH)COCH_2{-}C_6H_4{-}NO_2 \; + \; RCOOH \longrightarrow RCOO{-}CH_2{-}C_6H_4{-}NO_2 \qquad (14.3)$$

Saturated aldehydes and ketones, which normally have only a weak chromophore, can be converted into their strongly absorbing 2,4-dinitrophenylhydrazones.

Amines will react with fluoro-2,4-dinitrobenzene (Sanger's reagent) to form weakly basic aromatic amines (Equation 14.4) or can be converted

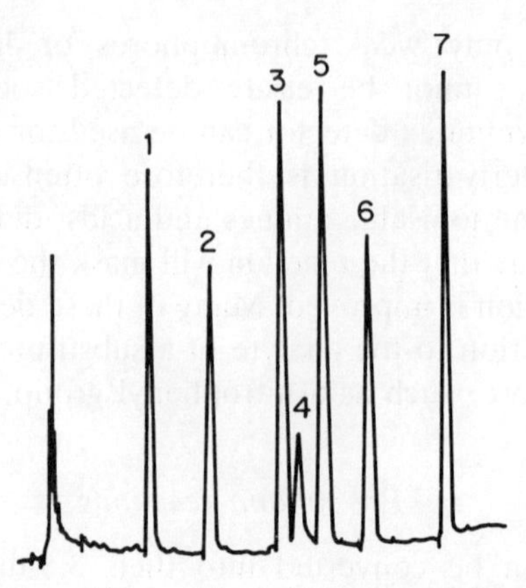

Figure 14.2 Separation of fatty acids as their *p*-bromophenacyl esters. Column LiChrosorb RP-8, 250 mm × 4.6 mm. Eluent: A, water; B, acetonitrile–THF 20:1. Programmed elution 77% B to 94% B. Spectroscopic detection at 204 nm for maximum sensitivity. Peaks: 1, lauric acid; 2, myristic acid; 3, palmitic acid; 4, oleic acid; 5, margaric acid; 6, stearic acid; 7, arachidic acid. (Reproduced with permission from *Hewlett Packard Application Note*, AN-232-17.)

into amides with 2,4-dinitrobenzenesulphonyl (or toluenesulphonyl) chloride or with activated acids such as *N*-succinimidyl-*p*-nitrophenylacetate.

$$RNH_2 + F\text{-}C_6H_3(NO_2)_2 \longrightarrow RNH\text{-}C_6H_3(NO_2)_2 \qquad (14.4)$$

ii. Post-column reactions

Probably the most widely used post-column reactions are for the detection of amino acids either by their reaction with ninhydrin to give Ruhemann's purple in the presence of cyanide as a catalyst (Equation 14.5) or by reaction with *o*-phthaldehyde (OPA) (see next section). The products of the latter reaction can be detected spectrophotometrically but more often fluorescence detection is employed.

$$RNH_2 + \text{(indane-1,2,3-trione)} \longrightarrow \text{Ruhemann's purple} \qquad (14.5)$$

Ruhemann's purple 570 nm
or 440 nm

b. *Reactions to enhance fluorescence detection*

This combination has proved very popular, as the products can be selectively and sensitively detected with little interference from non-reactive components in the sample.

i. Pre-column reactions

Carboxylic acids will react with 4-bromomethyl-7-methoxycoumarin (Equation 14.6), azido- or chloromethylanthracene to give the corresponding esters, which all show strong fluorescence.

RCOOH + $BrCH_2$–(7-methoxycoumarin) → RCOO–CH_2–(7-methoxycoumarin) (14.6)

A number of reagents have been proposed for the fluorescence detection of amines and amino acids including *o*-phthaldehyde ((Figure 14.3, Equation 14.7).

RNH_2 + *o*-phthaldehyde (CHO, CHO) → (with $HOCH_2CH_2SH$) isoindole (S–CH_2CH_2OH, N–R) (14.7)

Excitation 340 nm
Emission 455 nm

However, in some cases the isoindole product is unstable and may decompose on standing or on the column. Because amino-acids normally exist as zwitterions, they are often separated by ion-exchange chromatography.

Dansyl chloride (Equation 14.8) and fluorescamine (Equation 14.9)

RNH_2 + dansyl chloride (Me_2N–naphthalene–SO_2Cl) → Me_2N–naphthalene–SO_2–RNH (14.8)

Excitation 350-370 nm
Emission 490-530 nm

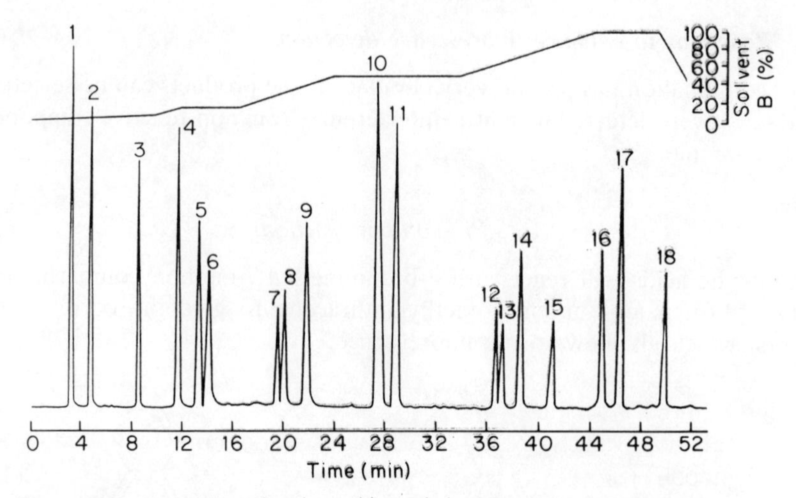

Figure 14.3 Separation of amino acids as their OPA derivatives following pre-column derivatisation. Column Ultrasphere ODS, 250 mm × 4.6 mm. Eluent: A, 0.05 M sodium acetate pH 6.8–methanol–THF 80:19:1; B, methanol–0.05 M sodium acetate pH 6.8 80:20. Programmed as shown above chromatogram. Fluorescence detection, excitation 360 nm and emission 455 nm. Peaks: 1, aspartic acid; 2, glutamic acid; 3, asparagine; 4, serine; 5, glutamine; 6, histidine; 7, glycine; 8, threonine; 9, arginine; 10, alanine; 11, tyrosine; 12, tryptophan; 13, methionine; 14, valine; 15, phenylalanine; 16, isoleucine; 17, leucine; 18, lysine monohydrochloride. (Reproduced by permission of Beckman Instruments, Inc.)

RNH_2 + [fluorescamine] → [R–N, HO, COOH product] (14.9)

Excitation 390 nm
Emission 475 nm

have also been used as reagents for amines. The latter compound reacts very rapidly and specifically with primary amines, any excess of the reagent being converted into a non-fluorescent by-product. It can also be used as a post-column reagent but is relatively expensive.

4-Chloro-7-nitrobenzo-2-oxa-1,3-diazole (NBD chloride) will react virtually specifically with secondary amines (Equation 14.10),

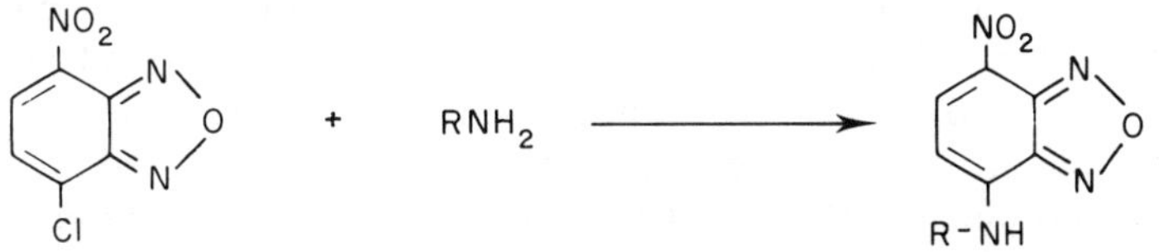

NBD-Cl + RNH_2 → NBD-NHR (14.10)

Only very slow reactions with primary amines have been reported.

ii. Post-column reactions

The reaction with OPA (see above) has also been used for the post-column derivatisation of amines and amino acids, but few other reagents are used on a routine basis for post-column fluorescence detection.

An interesting indirect method can be used to detect carbohydrates. They are oxidised by cerium(IV) salts, with the formation of fluorescent cerium(III) as a by-product, which can be readily detected. A very specific detection can also be achieved with tetrahydrocannabinol, which is converted photochemically into a fluorescent phenanthrene on irradiation of the eluent stream with photons (Equation 14.11).

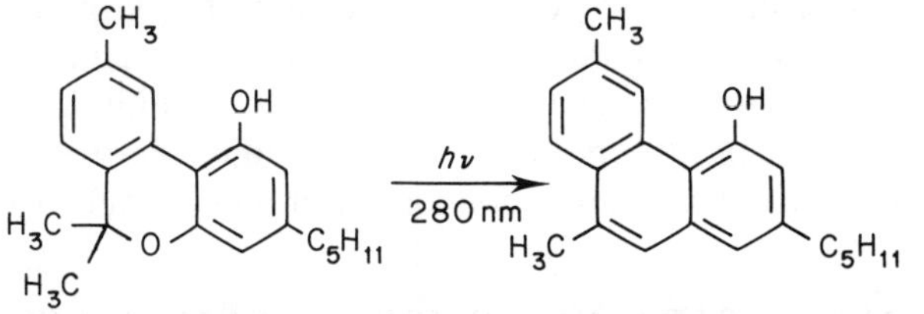

$$\xrightarrow[280\,nm]{h\nu} \qquad (14.11)$$

The product can be distinguised from naturally fluorescent compounds by repeating the assay with the irradiation lamp turned off.

c. Reactions to enhance electrochemical detection

So far, few methods have been specifically designed for this purpose. It has been proposed that reactions that yield nitroaromatic derivatives (see Equation 14.1) could be used, followed by electrochemical detection of the product in the reductive mode. Because oxidative electrochemical detection poses fewer practical problems, derivatives incorporating phenolic groups are more likely to be successful. One example is the reaction of carboxylic acids with *p*-aminophenol to give the corresponding amidophenols (Equation 14.12).

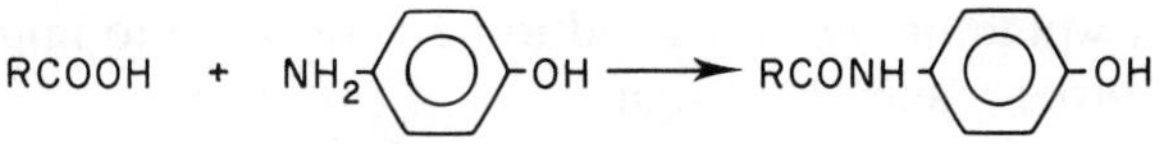

$$RCOOH + NH_2C_6H_4OH \longrightarrow RCONHC_6H_4OH \qquad (14.12)$$

Electrochemically active groups may also be formed in a two-stage coulometric electrochemical detector. In the first step a compound containing a suitable function, such as a nitrophenyl group, is reduced to the corresponding aromatic amine. This can then be detected oxidatively by a second electrochemical detector at a positive potential. The two steps provide selectivity, but the first step must give a high yield or the sensitivity may be lost.

d. Reactions to enhance separation

Because few compounds cause problems in HPLC separations, it is rarely necessary to derivatise analytes solely to improve their separation properties. However, as noted earlier, many of the reactions to enhance the detection of acids or amines also reduce their polarity and tendency to ionise, and thus improve their separation.

The formation of stable neutral chelates, using reagents such as dithiocarbamates, 8-hydroxyquinoline, or dithizone, followed by separation on either normal- or reversed-phase systems, can be employed as an alternative to ion-exchange chromatography for the separation of transition metal ions. The chelates can either be formed in a pre-column reaction or the chelating agent can be incorporated into the mobile phase to give an on-column reaction. Sometimes similar reactions, in particular the addition of pyridylazoresorcinol (PAR), can also be used in post-column derivatisation for the spectrophotometric detection of metal ions after an ion-exchange separation.

14.3 SEPARATION OF CHIRAL COMPOUNDS

Because different enantiomers of drugs and pesticides often possess different biological activities, there has been considerable interest in the application of HPLC to both their analytical and preparative separation, particularly in the pharmaceutical and agrochemicals industries. However, the silica-based columns and organic eluents normally used for HPLC are achiral and both enantiomers will interact with the separation system to the same extent, and will be unresolved.

Two main approaches have been applied to the separation of enantiomers in a similar manner to their separation by GLC. The two enantiomers can be converted into diastereoisomers, by reaction with a single enantiomer of a chiral reagent (i.e. Equation 7.13). The diastereoisomeric derivatives can then be separated on an achiral column. This approach is most attractive for analytes that would in any case need to be derivatised to improve their detection or to protect ionisable groups.

Alternatively, the enantiomers can be separated on a chiral separation system in which a chiral eluent or chiral stationary phase creates an asymmetrical environment. Chiral eluents are rarely used because few suitable asymmetrical compounds are readily available at a reasonable cost or purity to use as a mobile phase. However, more exotic eluents can be used with microbore or capillary columns because of their much lower flow rates. One approach with packed columns has been to add chiral modifiers to the mobile phase which will interact with the chiral analytes. Most interest in recent years has been in chiral stationary phases (CSP) based on either bonded silica or cellulose materials.

a. *Derivatisation for chiral separations*

If the D- and L-enantiomers of an analyte are derivatised with an optically pure asymmetrical reagent (i.e. D-enantiomer), the products will be a mixture of D,D- and D,L-diastereoisomers, whose chromatographic properties will differ. Their separation will be enhanced by conformational immobility around the chiral centres and the close proximity of the chiral centres in the derivative. A typical method is the reaction of Marfey's reagent, FDAA (1-fluoro-2,4-dinitrophenyl-5-L-alanine amide), with either amines or amino acids, which introduces both a chiral centre and a chromophore (Figure 14.4).

A closely related concept is the use of chiral ion-pair reagents or complexing reagents, which effectively result in the *in situ* formation of diastereoisomers in the eluent. One method is the introduction of a chiral copper–amino acid complex into the mobile phase, which forms outer-shell complexes with analyte amino acids.

b. *Chiral stationary phases*

In the past few years a wide range of chiral stationary phases, primarily based on four broad types, have become commercially available (Table 14.1). The first group is derived from aminopropyl-bonded silica materials, to which a chiral substituent has been attached. Many of these materials are based on work by Pirkle, who initially used an ionic linkage between the aminopropyl group and a substituted amino acid. However, these materials were sensitive to moisture and could only be used in non-aqueous eluents. Subsequently, the corresponding much stabler amide-bonded materials have become available, although even with these columns, aqueous mobile phases reduce the selectivity that can be obtained. A large number of different phases have been prepared but only selected examples are available commercially.

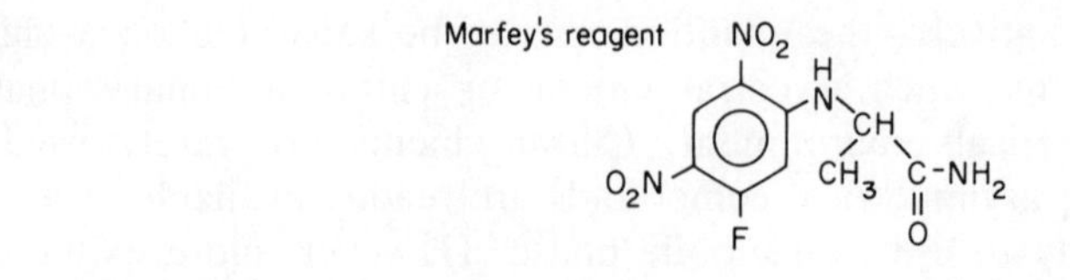

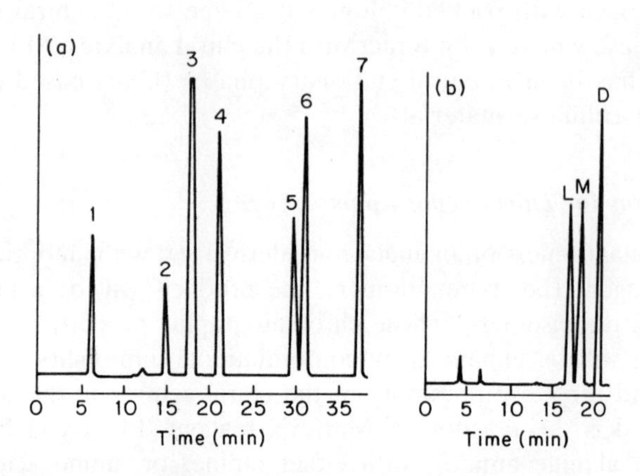

Figure 14.4 The separation of (a) D- AND L-amino acids and (b) D- and L-Dopa enantiomers as diastereoisomers after reaction with Marfey's reagent. Column Spherisorb RP-18, 100 mm × 4.6 mm. Eluent: A, 0.05 M triethylamine phosphate, pH 3.0; B, acetonitrile. Flow rate 2 ml min^{-1}. Linear gradient 10% to 40% B over 45 min. Ultraviolet detection at 340 nm. Peaks: (a) 1, L-Glu; 2, D-Glu; 3, Marfey-OH; 4, L-Met; 5, D-Met; 6, L-Phe; 7, D-Phe: (b) L, L-Dopa; M, Marfey-OH; D, D-Dopa. (Reproduced by permission of Pierce Chemical Company.)

A second group of phases has been developed by attaching the naturally asymmetrical peptides, bovine serum albumin or α_1-acid glycoproteins to a silica support. These columns appear to be generally applicable to the reversed-phase separation of a wide range of analytes (Figure 14.5) but have low sample capacity and cannot be used for preparative separations.

In the third group of column materials, the naturally chiral cyclodextrins have been attached to a silica support. These coiled carbohydrates form chiral "cavities", which can interact with chiral analytes (Figure 14.6). These materials can be used for reversed-phase separations.

The final major group of chiral stationary phases is derived from inherently chiral cellulose support materials. These are esterified with acetyl, benzoyl, or other acids to give phases with different selectivities.

Table 14.1 Chiral stationary phases for HPLC. A selection of the commercially available materials.

Material	Structure	Source[a]
Bonded amino-acid materials		
Bakerbone	Salt of (*R*)-*N*-3,5-dinitrobenzoylphenylglycine and aminopropyl-silica (non-aqueous eluents only)	B
DNBPG	Amide of (*R*)-*N*-3,5-dinitrobenzoyl-L-phenylglycine and aminopropyl-silica	B
DNBLeu	Amide of (*R*)-*N*-3,5-dinitrobenzoyl-L-leucine and aminopropyl-silica	B
D-Naphthylalanine		R
Bonded protein materials		
EnantioPac	α_1-Acid glycoprotein immobilised on DEAE (*N*,*N*-diethylaminoethyl) derivatised silica	L
Resolvosil	Bovine serum albumin immobilised on wide-pore silica	M
Bonded cyclodextrin-silica materials		
Cyclobond I	β-Cyclodextrin	A
Cyclobond II	γ-Cyclodextrin	A
Cyclobond III	α-Cyclodextrin	A
Cyclobond I acetylated	Acetylated β-cyclodextrin	A
Cellulose-based column materials		
Chiralpak O	Helical polymethacrylates	D
Chiralpak WH	Amino acid–copper ligand binding	D
Chiralpak WM	Amino acid–copper ligand binding	D
Chiracel OA	Cellulose acetate esters (–CO–Me)	D
Chiracel OB	Cellulose benzoate esters (–CO–Ph)	D
Chiracel OC	Cellulose carbamates (–O–CO–NH–Ph)	D
Chiracel OK	Cellulose ether (–CO–CH=CH–Ph)	D
ChiraPak OT(+)	Poly(triphenylmethyl methacrylate)	D
ChiraPak OP(+)	Derivative of poly(triphenylmethyl methacrylate)	D
Miscellaneous chiral materials		
LC-(*R*)-urea	Si–NH–CO–NH–CH(Me)–Ph	S

[a] Suppliers: B, Baker; R, Regis; A, Advanced Separations Technologies; L, LKB; M, Machery Nagel; D, Daicel; S, Supelco.

Most of these column materials have only been available for a limited period and comparisons so far are limited. Additional chiral stationary phases are continually being prepared and attempts are being made to devise rules or models to predict their suitability for particular analytes. However, at the present time a number of different materials may need to be examined to achieve a particular separation.

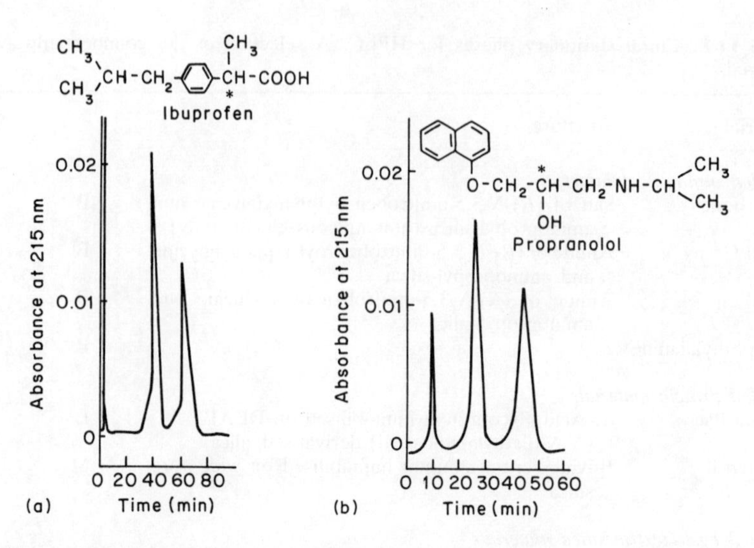

Figure 14.5 Separation of enantiomers of ibuprofen and propranolol on Enantiopac column. (a) Ibuprofen. Eluent: 8 mM sodium dihydrogen phosphate/disodium hydrogen phosphate + 0.1 M sodium chloride and 0.5% *N,N*-dimethyloctylamine, pH 6.9, 0.3 ml min^{-1} and 15°C. (b) Propranolol (as oxazolidone derivative). Eluent: 8 mM sodium dihydrogen phosphate–disodium hydrogen phosphate + 0.1 M sodium chloride and 20% 2-propanol, pH 6.9, at 0.3 ml min^{-1} and 15°C. (Reproduced with permission from Pharmacia LKB Biotechnology.)

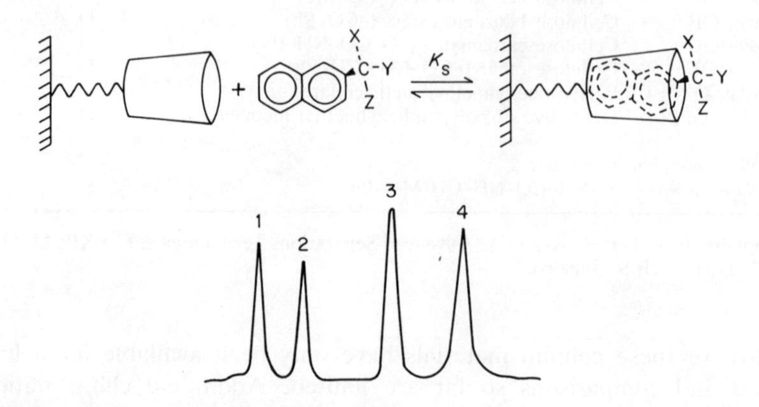

Figure 14.6 Separation of β-naphthalene derivatives of DL-alanine and DL-methionine. Column Cyclobond I (β-cyclodextrin), 100 mm × 4.6 mm. Eluent methanol–water 50:50 at 1 ml min^{-1}. Peaks: 1, L-Met; 2, D-Met; 3, L-Ala; 4, D-Ala. (Reproduced by permission of International Scientific Communications, Inc. from *International Laboratory*.)

14.4 COLUMN SWITCHING AND SAMPLE TRAPPING

A liquid chromatographic column can sometimes also be used as an extraction or trapping device to concentrate a solution that would be too dilute for direct injection. The sample can then be either eluted and separated on the same column or passed to a second column. Other column switching methods are used to increase resolution by passing part of a sample to a second column.

a. Column trapping

In order to increase the sensitivity of HPLC analyses, it is rarely possible simply to increase the injection volume without loss of resolution. Conventional liquid–liquid solvent extraction can be used to concentrate the analyte from very dilute samples before determination on HPLC. However, with multiple samples, this method is time-consuming and labour-intensive and thus expensive. An alternative technique, which can be used if the analyte is relatively non-polar in a polar matrix (or vice versa), is to use a short trapping column, which will completely retain the analyte of interest from a large volume of the matrix. Thus a reversed-phase column will trap a non-polar analyte from an aqueous matrix. Subsequent elution with a stronger eluent will then yield a concentrated sample suitable for assay by HPLC.

These short columns can be in the form of disposable cartridges containing silica, ODS-silica, or other bonded-phase materials. The material is chosen to retain the analyte but to allow as much as possible of the other components of the sample to be washed through. For the extraction of non-polar drugs from largely aqueous biological fluids such a blood or urine, an ODS column would be used and the concentrated sample recovered with methanol or acetonitrile. These cartridges can also have advantages for concentrating environmental samples in the field. However, their use still requires a number of manual operations.

Alternatively, the trapping column can replace the sample loop in an injection valve. A large volume of an aqueous sample, such as urine containing steroids, is passed through the trapping column. The analytes are trapped out and any polar components are washed to waste. When the valve is turned to the inject position, the eluent will rapidly carry the sample onto the analytical column for separation. The trapping column would be chosen to retain the sample only weakly so that the analyte would be rapidly eluted and focused onto the top of a more retentive analytical column. A typical combination would be a C_2-silica trapping column and a C_{18}-silica analytical column.

With clean samples such as polynuclear aromatic hydrocarbons in drinking water, the trapping can be carried out directly on the analytical column by first pumping the aqueous sample and then changing to the eluent to elute the analytes.

b. *Column switching, heart-cut and back-flushing methods*

Because of the viscosity of most HPLC mobile phases, the column length cannot be significantly increased to improve the resolution without unduly increasing resistance to flow and the back-pressure. However, with complex samples a single column may not fully resolve all the components of interest. In addition, if a wide range of components is present, a separation can be unduly prolonged by slowly eluted components even with eluent programming. In these cases, the effective resolution of the separation can be

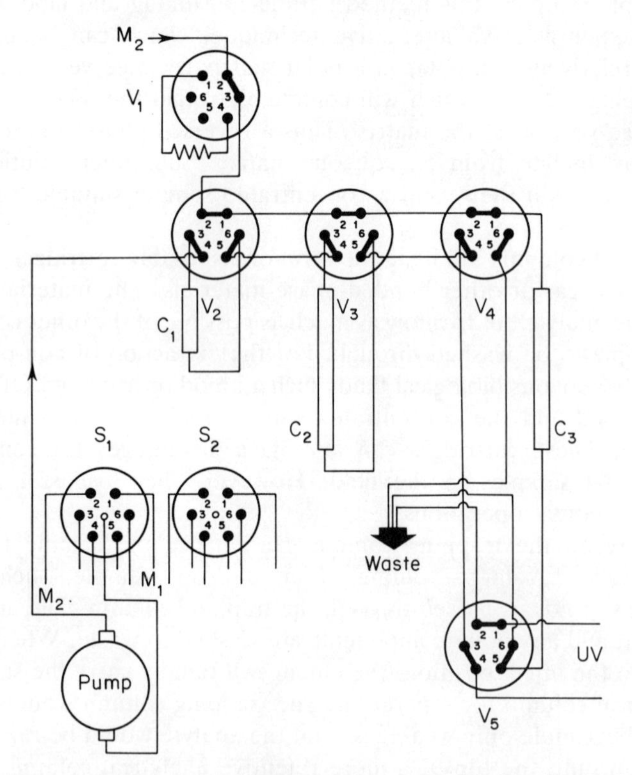

Figure 14.7 Versatile column configuration for column switching in HPLC, used for sample clean-up. C_1, pre-column; C_2, initial separation; C_3, final analytical column. (Reproduced by permission of Kontron Instruments Ltd.)

increased by heart-cutting techniques, using two linked columns. The sample is partially separated on the first column and the eluent is monitored. At an appropriate point the eluent stream containing the unresolved area of interest is switched onto a second column (Figure 14.7). Further elution on this column gives an improved separation without interference from the other components of the sample. By a careful choice of column materials and eluents, the sample can be reconcentrated as it enters the second column. Typically a C_2- or C_8 column is used as the first column, followed by a C_{18} bonded column. This will have a higher retentive power and the sample will be initially focused on the top of the column. Subsequently, the elution of this second column can be programmed. This technique is ideal for complex extracts of food samples in which many similar components are present or complex pharmaceutical samples, such as creams or lotions, in which the active ingredients may only be minor components.

Back-flushing HPLC columns, by reversing the flow of eluent, can be used to remove very highly retained samples without having to wash them through the full length of the columns. However, this method is not as popular as in GLC, because changing the direction of flow through the column can disturb the stationary-phase bed and reduce the column lifetime.

These two techniques and the earlier column trapping methods can be combined to form automated sample preparation and assay systems, with the timing of the switching and eluents being controlled by a computer. Clearly setting up a method of this type is complex and time-consuming, so it is only worth carrying out if a large number of similar samples are to be analysed.

BIBLIOGRAPHY

Derivatisation methods

Most of the general books on derivatisation for GLC listed in Chapter 7 also cover reactions for HPLC and have not been repeated here.

S. Ahuja, "Chemical derivatization for the liquid chromatography of compounds of pharmaceutical interest", *J. Chromatogr. Sci.*, 1979, **17**, 168–172.

R.W. Frei, "Assessment of the current status of reaction liquid chromatography", *J. Chromatogr.*, 1979, **165**, 75–86.

R.W. Frei and A.H.M.T. Scholten, "Reaction detectors in HPLC", *J. Chromatogr. Sci.*, 1979, **17**, 152–160.

K. Imai, T. Toyo'oka and H. Miyano, "Fluorigenic reagents for primary and secondary amines and thiols in high-performance liquid chromatography. A review", *Analyst*, 1984, **109**, 1365–1373.

T. Jupille, "UV absorption derivatization in liquid chromatography", *J. Chromatogr. Sci.*, 1979, **17**, 160–167.

P.T. Kissinger, K. Bratin, G.C. Davis and L. Pachla, "Potential utility of pre- and post-column chemical reactions with electrochemical detection in liquid chromatography", *J. Chromatogr. Sci.*, 1979, **17**, 137–146.

I.S. Krull (Ed.), *Reaction detection in liquid chromatography*, Chromatographic Science Series, Vol. 34, Marcel Dekker, New York, 1986.

I.S. Krull, K.-H. Xie, S. Colgan, U. Neue, T. Izod, R. King and B. Bidlingmeyer, "Solid phase derivatization reactions in HPLC. Polymeric reductions for carbonyl compounds", *J. Liquid Chromatogr.*, 1983, **6**, 605–626.

J.F. Lawrence, "Fluorometric derivatization in high performance liquid chromatography", *J. Chromatogr. Sci.*, 1979, **17**, 147–151.

J.F. Lawrence, "Advantages and limitations of chemical derivatization for trace analysis by liquid chromatography", *J. Chromatogr. Sci.*, 1985, **23**, 484–487.

J.F. Lawrence and R.W. Frei, *Chemical derivatization in liquid chromatography*, J. Chromatogr. Library, Vol. 7, Elsevier, Amsterdam, 1976.

M.S.F. Lei Ken Jie, "The characterization of long-chain fatty acids and derivatives by chromatography", *Adv. Chromatogr.*, 1980, **18**, 1–57.

P.J. Twitchett, P.L. Williams, and A.C. Moffat, "Photochemical detection in high-performance liquid chromatography and its application to cannabinoid analysis", *J. Chromatogr.*, 1978, **149**, 683–691.

S. van de Wal, "Post-column reaction detection systems in HPLC", *J. Liquid Chromatogr.*, 1983, **6** Suppl. 1, 37–59.

K.-H. Xie, S. Colgan and I.S. Krull, "Solid phase reactions for derivatisation in HPLC (HPLC–SPR)", *J. Liquid Chromatogr.*, 1983, **6** Suppl. 2, 125–151.

Chiral separations

S. Allenmark, *Chromatographic enantioseparation*, Ellis Horwood, Chichester, 1987.

I.S. Krull, "The liquid-chromatographic resolution of enantiomers", *Adv. Chromatogr.*, 1978, **16**, 175–210.

W.H. Pirkle and J.R. Hauske, "Broad spectrum methods for the resolution of optical isomers", *J. Org. Chem.*, 1977, **42**, 1839–1844.

W.H. Pirkle and J.R. Hauske, "Design of chiral derivatizating agents for the chromatographic resolution of optical isomers", *J. Org. Chem.*, 1977, **42**, 2436–2443.

R.W. Souter, *Chromatographic separations of stereoisomers*, CRC Press, Boca Raton, FL, 1985.

M. Zief and L.J. Crane, *Chromatographic chiral separations*, Chromatographic Science Series, Vol. 40, Marcel Dekker, New York, 1987.

Chiral stationary phases

R. Däppen, H. Arm and V.R. Meyer, "Applications and limitations of commercially available chiral stationary phases for high-performance liquid chromatography", *J. Chromatogr.*, 1986, **373**, 1–20.

W.H. Pirkle and J.M. Finn, "Chiral high-pressure liquid chromatography stationary phases. 3. General resolution of arylalkylcarbinols", *J. Org. Chem.*, 1981, **46**, 2935–2938.

W.H. Pirkle, M.H. Hyun and B. Bank, "A rational approach to the design of highly-effective chiral stationary phases", *J. Chromatogr.*, 1984, **316**, 585–604.

CHAPTER 15

DATA HANDLING AND AUTOMATION IN CHROMATOGRAPHY

15.1 COMPUTERS IN CHROMATOGRAPHY

HPLC and GLC have not been immune from the influence of the microprocessor and the information technology revolution. Computers are having a major impact in a number of areas, including instrument control, automated sample injection, data collection and integration, optimisation of separation conditions and as the heart of advanced detectors, such as the diode array, FTIR and mass spectrometers. Larger laboratories now often contain an integrated laboratory information management system (LIMS) combining data collection, data analysis and record keeping, with the facility for the direct input from chromatographs, balances and keyboards. Future developments will undoubted include a greater use of robotic systems for sample preparation steps, including extraction, dilution, addition of internal standards and derivatisation. The future development of expert systems will aid method selection and problem solving by making the accumulated knowledge of HPLC available to the individual operator.

15.2 INTEGRATION AND DATA HANDLING

The low cost of microcomputers has enabled electronic integration to be available routinely in most chromatography laboratories for both GLC and HPLC, replacing the manual measurement of retention times and peak areas. Integrators provide improved precision compared to manual methods and offer the facility to manipulate the results and directly provide ratios of peak areas to internal standards to convert peak areas into analyte concentrations by comparison with standards, and to identify peaks from their retention times.

There is, however, a tendency for the operator to uncritically accept the peak areas and retention times reported by an integrator. The same criteria should be applied as would be used if the operator were making a direct visual examination of a recorder trace. The full chromatogram should be used continually as a guide to the quality of the instrument output rather than relying on the analysed values in the final printout. These will usually neither assess peak shape nor recognise the onset of column deterioration.

It is therefore important to understand the capabilities and limitations of an integrator. Although future systems may have all the discrimination power of the human operator, it is likely that for a considerable time their basic operation will be very simple. Integrators may appear sophisticated but they operate on a series of fixed rules and instructions, using parameters set by the operator or their own default conditions. They are unable to judge if these are appropriate for the particular chromatogram being analysed but will slavishly follow their program. The integrator may be a dedicated instrument or an adapted microcomputer. Both work in essentially the same way and both can possess facilities for linking or networking to other systems for bulk storage of results and data.

a. Peak detection

The purpose of an integrator is to monitor the signal from the chromatographic detector, and to recognise the start, maximum and end of each peak. It should then use these values to determine the retention time and the area under each peak, which can be subsequently related to the amount of the analyte being eluted from the chromatograph and used to identify the analyte. The first step in this process is for the integrator to convert the continuously varying analogue output signal from the detector, at regular time intervals, into digital values using an analogue-to-digital (A/D) converter. The frequency of sampling the signal will determine the ability of the integrator accurately to measure narrow peaks, such as those found in open-tubular GLC or fast HPLC. A typical rate on a modern integrator would be 10 times a second, but older systems were much slower and are only suitable for packed column chromatographs. Before they are analysed, these values may be bunched into wider time intervals depending on the setting of the integrator. The magnitude of the detector signal is usually measured relative to an offset reference signal level at about −20 mV, chosen so that all the recorded values will be positive (Figure 15.1). Depending on the integrator, either the digitised values from the whole chromatogram (usually following bunching) or just a limited set of the most recent measured values will be stored in the memory of the integrator. The former method requires a much larger memory store but enables the chromatogram to be re-evaluated using different parameters and is within the capacity of many modern microcomputers.

As each new data point is measured, the integrator compares its value with those for the previous points. If the points show a continuing upwards trend, which rises above a preset threshold noise level and persists for a predetermined number of points, the program will recognise this as the beginning of a peak (Figure 15.2). The integrator will then go back through the earlier data points, until the baseline level is reached, and mark this as the start of the peak. Because the data points are measured at regular time intervals, the area under the peak can be determined by adding together the peak heights of all the data points from the start to the end of the peak. Therefore, as soon as the start of the peak is identified, the integrator will begin to accumulate the peak heights.

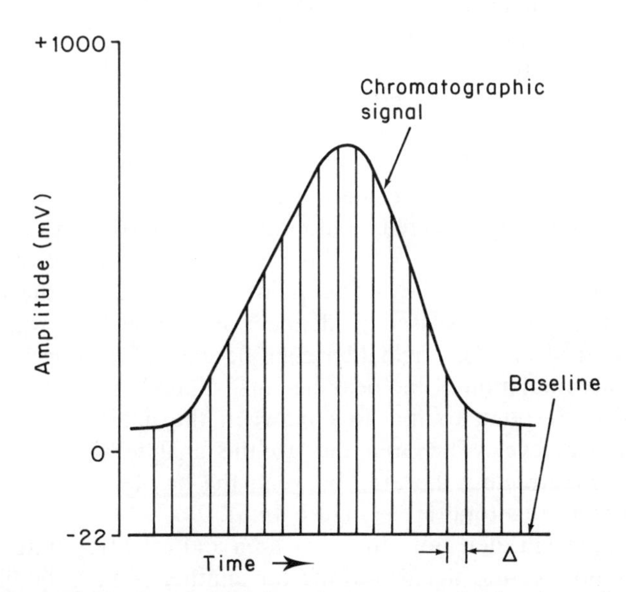

Figure 15.1 Digitisation of signal from detector at regular time intervals above an offset baseline. Δ is the width of a single A/D reading (area slice). For example, at 80 Hz, Δ = 0.0125 s. (*Hewlett-Packard 3390 manual*. Copyright © 1982 Hewlett-Packard Company. Reproduced with permission.)

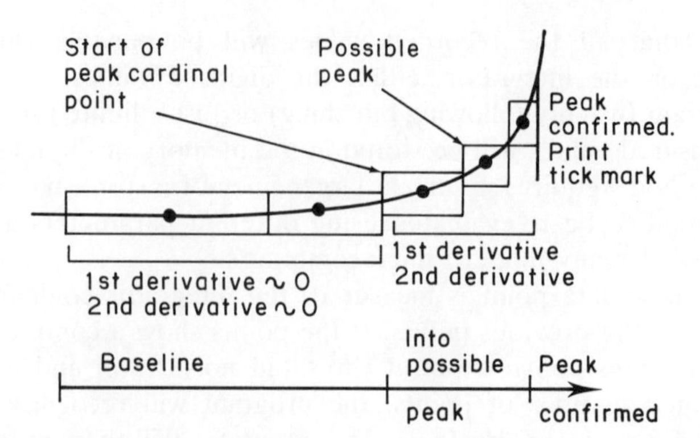

Figure 15.2 Detection of the start of a peak by the rise outside the original threshold region sustained for a specified number of data points. (*Hewlett-Packard 3390 manual*. Copyright © 1982 Hewlett-Packard Company. Reproduced with permission.)

The signal will continue to be monitored until it reaches a maximum height and will then start to decrease. After a preset number of decreasing data points have been observed, confirming that the peak maximum has been passed, the program will search back through the data to locate positively the position of maximum intensity. This value will be used to define the peak height and retention time. More sophisticated integrator programs will fit a curve to the data points near the peak maximum to locate the exact value (Figure 15.3). The retention time is usually printed on the integrator trace and included in the final report.

As the detector signal decreases after the peak maximum, the computer continues to monitor the peak heights until they fall below the threshold noise level and become steady. This point is identified as the end of the peak (Figure 15.4) and is used to define a new base line, which may be slightly different from the original baseline because of drifting of the output signal. The accumulation of the peak heights will end. The total of the peak heights will correspond to the area between the detector signal and the offset reference level. The area due to the analyte signal will then be calculated by subtracting the area between the baseline and the reference signal level from this total area (Figure 15.5).

The integrator will then store the peak area and retention time values and continue to monitor the signal, waiting for another peak to be eluted. The integration will be unaffected by slow changes in the baseline caused by column bleed on temperature programming, eluent programming or similar causes, as these will not cause a sufficiently rapid increase in the signal to

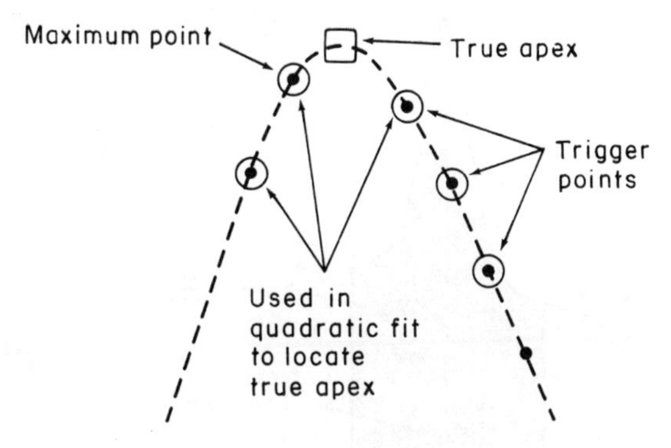

Figure 15.3 Detection of peak maximum by specified number of data points with decreasing intensity. Curve fitting over points near maximum to determine exact retention time of peak maximum and peak height. (*Hewlett-Packard 3390 manual*. Copyright © 1982 Hewlett-Packard Company. Reproduced with permission.)

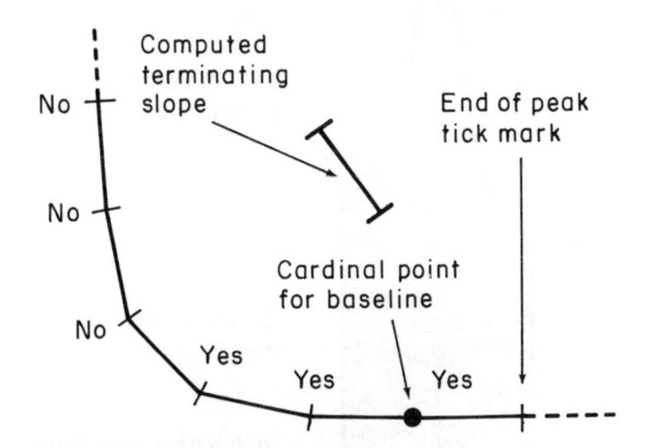

Figure 15.4 Identification of the end of peak by the decrease in slope to nearly horizontal within the threshold limits. (*Hewlett-Packard 3390 manual*. Copyright © 1982 Hewlett-Packard Company. Reproduced with permission.)

initiate the start of a peak. Instead the program will continuously reset the baseline values and the corresponding threshold limits.

At the end of the run, the integrator will print out a chromatography report typically listing peak areas, heights and retention times (Figure 15.6). This may also include post-run calculations, such as relative peak areas, areas as percentages of the total area, and relative retention times. If the system has been standardised, it may also calculate concentrations compared to external or internal standards. The capacity factors in HPLC or adjusted

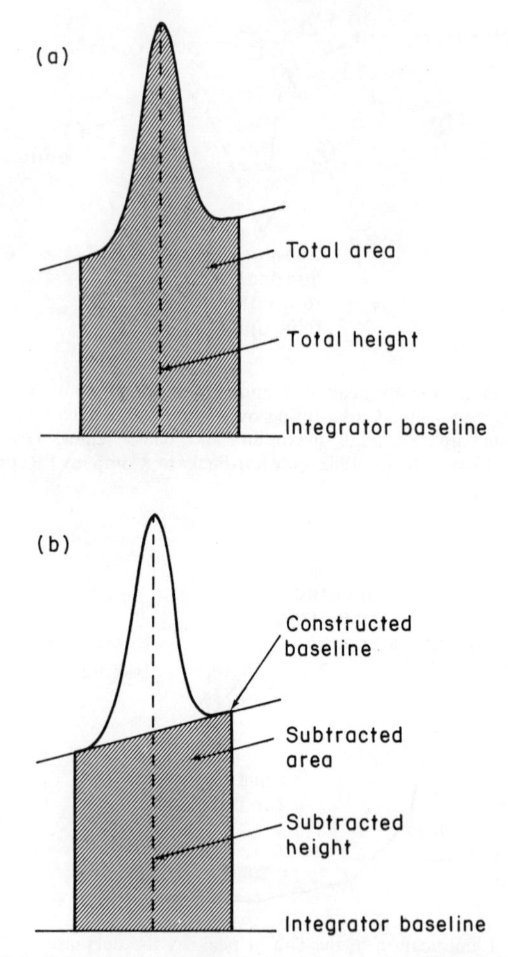

Figure 15.5 Calculation of (a) total area under curve including offset and (b) peak area after subtraction of area between offset and baseline. (*Hewlett-Packard 3390 manual*. Copyright © 1982 Hewlett-Packard Company. Reproduced with permission.)

retention times in GLC cannot be calculated, as the column void volume t_0 cannot be determined directly by the integrator.

With a chromatogram containing only well resolved individual peaks, the integrator will have few problems, but if the chromatogram contains unresolved pairs or groups of peaks, then the program has to decide the proportion of the total area under the curve that is to be assigned to each component.

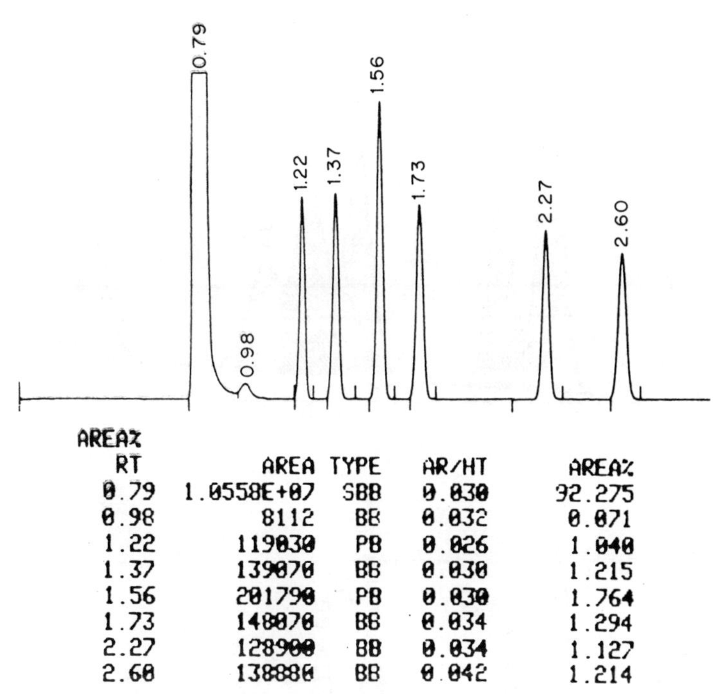

AREA%

RT	AREA	TYPE	AR/HT	AREA%
0.79	1.0558E+07	SBB	0.030	92.275
0.98	8112	BB	0.032	0.071
1.22	119030	PB	0.026	1.040
1.37	139070	BB	0.030	1.215
1.56	201790	PB	0.030	1.764
1.73	148070	BB	0.034	1.294
2.27	128900	BB	0.034	1.127
2.60	138880	BB	0.042	1.214

Figure 15.6 Typical integration report of well resolved separation giving retention times, areas and relative peak areas, showing tick marks at start and end of peaks and diagnostics on report. BB is baseline-to-baseline separation. (*Hewlett-Packard 3390 manual*. Copyright © 1982 Hewlett-Packard Company. Reproduced with permission.)

If there is a significant overlap between two peaks, the signal will go through a minimum value above the threshold level and will then begin to rise again. The integrator should recognise the minimum as a *valley* position. It will use this point to define the end of the first component and then start a new peak area calculation and search for the second peak maximum. Further unresolved components will be treated in the same way. Once the final baseline has been reached, the program will go back and arbitrarily drop a vertical line from each *valley* point to the baseline before subtracting the offset areas. The areas will then be allocated to the individual peaks (Figure 15.7a). The report of the chromatogram should indicate that the area corresponds to a valley peak either from an initial baseline or from a previous valley. It should be recognised that defining the boundary between the peaks in this way is arbitrary and will not take into account the peak shapes or tailing. The reported peak areas will therefore be approximations,

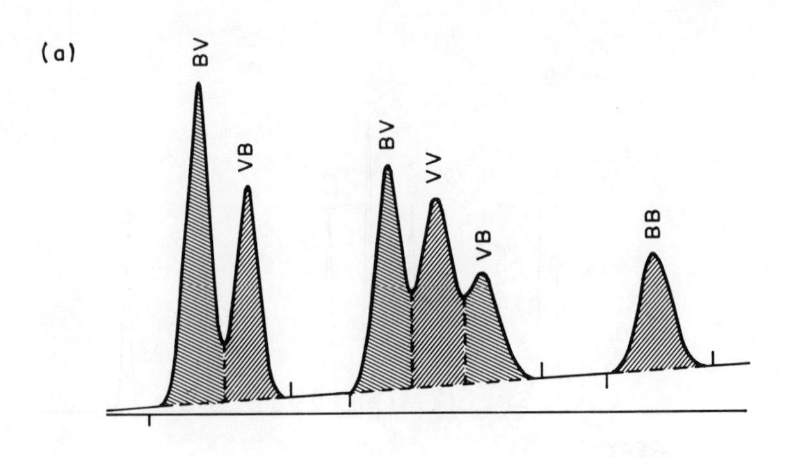

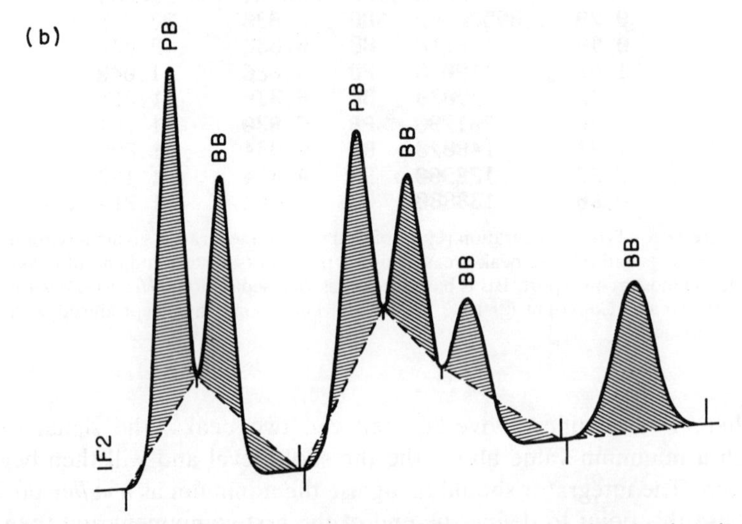

Figure 15.7 Unresolved multiple peaks. (a) Separation of peaks by dropping vertical to baseline from valley for double and multiple peaks. (b) Same separation with threshold set too high. Baseline has been erroneously assigned to each valley, underestimating areas. (*Hewlett-Packard 3390 manual*. Copyright 1982 Hewlett-Packard Company. Reproduced with permission.)

but this method is often acceptable for most studies. For accurate determinations, the chromatographic separation should be improved to give baseline resolution of all the analytes of interest, by altering the conditions or increasing the efficiency.

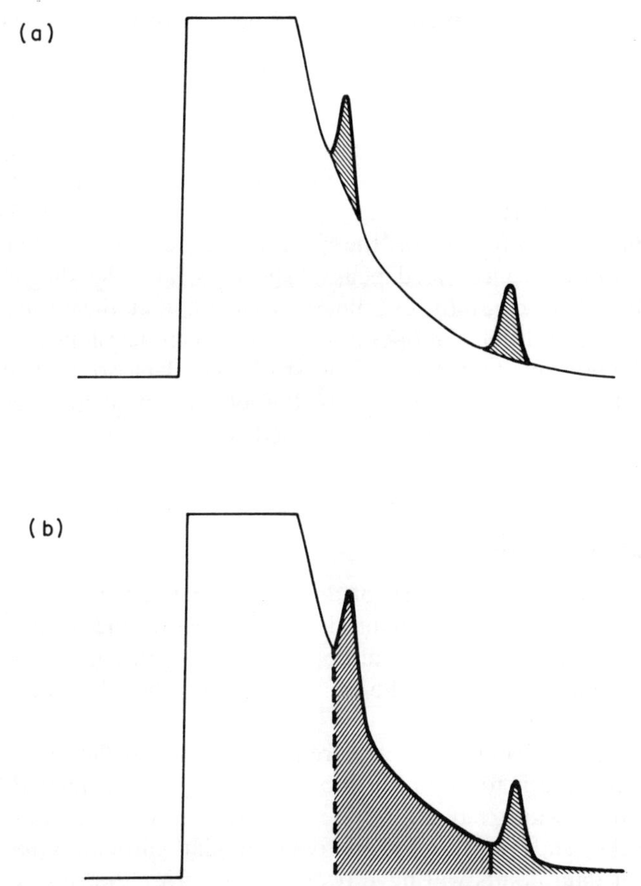

Figure 15.8 (a) Minor peak skimmed off the back of a major peak using peak skimming routine. (b) Erroneous allocation of peak areas if normal valley peak routine is followed. (*Hewlett-Packard 3390 manual*. Copyright © 1982. Hewlett-Packard Company. Reproduced with permission.)

If the signal between two nearly resolved peaks reaches a minimum within the threshold levels, then the integrator will regard each as a separate peak and will reset the baseline to the minimum point. Problems can therefore occur if the threshold range is too wide, as a misleading baseline will be drawn to the valley point, underestimating their areas (see Figure 15.7b).

Larger errors will be found if the unresolved peaks are very different in size. A common problem in GLC is the determination of an early peak being eluted on the tail of a large solvent peak. Most integrators can recognise this pattern by comparing the heights of the two peaks with the intervening valley. The program will then *skim* the small peak off the top of the major peak so that only the area above the curve of the major peak is allocated to the minor component (Figure 15.8a). This result is clearly much closer to the correct value than dropping a vertical division from the valley (Figure 15.8b) but is still an approximate value. Again the integrator report should report any peak treated in this way with an identifying code. It is sometimes difficult for the program to distinguish between these different types of unresolved peaks. The trace and the diagnostics in the report should be carefully examined to ensure that the desired approach has been used and that the operator is clear which decisions had been made by the integrator. Integrators that store the data so that they can be reanalysed can be very useful, as the effects of slightly changing the integration parameters can be assessed without needing to rerun the chromatogram.

b. Integrator parameters

The key to the successful use of an integrator is an understanding of the parameters set on the instrument and their effect on the results. Integrators come with a set of standard conditions, which may be satisfactory for simple chromatograms, but often the settings should be changed to match a particular separation.

A common problem is that the frequency at which the bunched detector signal is analysed is too rapid or too slow. This may be defined by a setting such as "peak width". If the frequency is too slow the integrator can miss sharp peaks, such as those from open-tubular chromatography, because insufficient data points will have been tested during the rise of the peak to trigger the start of the integration routine. On the other hand, an excessively rapid sampling frequency may mean that the integrator will identify sharp noise spikes as signals and therefore split peaks up into "unresolved" multiplets. In addition broader peaks may be interpreted as baseline drift because their rise of signal with time is too slow to be identified as the start of a peak. The usual indications of an incorrect setting are that the integrator will miss peaks, will identify only one end of the peak, or will be slow in recognising peak maxima. These problems can be recognised by examining the diagnostic "stop" and "start" tick marks that most integrators will place on the chromatographic trace (see Figure 15.6). Do not suppress this mode even in routine use as these indicators can often provide valuable information.

In isocratic or isothermal studies, the peaks in a chromatogram will become broader as the run progresses so that the peak width setting or frequency values may need to be updated at intervals during the run. Some integrators will do this automatically based on the measured widths of earlier peaks, otherwise the values may be redefined by the operator at predetermined points using a timed event instruction.

To avoid minor noise changes in the baseline being identified as signals and repeatedly causing data collection to be started and confusing the report, the threshold value should be set according to the noise level of the signal. This can usually be done automatically or manually. With HPLC some of this noise may come from the pump and the level should be set with the system running. Some integrators expand this region during a peak to allow for baseline drift. The final integrator report can also be instructed to ignore peaks below a certain size, but care must be taken not to exclude legitimate sample components, and the analogue recorder trace should always be compared with the integrator report.

If the integrator is being used for HPLC, negative solvent disturbance peaks, which are often present at the start of a chromatogram, can cause an erroneous negative baseline position to be established, which will result in incorrect peak areas (see Figure 11.10). In this case the integration should be disabled until the solvent peak has passed or the baseline should be redefined when the correct level has been established after the negative peak. With GLC the area of the solvent peak will dominate the separation and should be excluded from any relative peak area calculations.

15.3 DATA ANALYSIS

The results of individual separations are often used to generate calibration curves for quantitative studies or for identification purposes. Because of the small errors that inevitably occur during injection and in the preparation of standard solutions, simple graphical methods and manual curve fitting can be unreliable. Instead, use should be made of one of the numerous statistical packages available on most microcomputers to examine the data and prepare calibration curves. If the correlation is linear, a least-squares calculation can be used. As always with computers, the raw values should be examined as well as the final result to ensure that outliers or anomalous values have not been hidden by the statistics. Appropriate statistical methods can also be applied to non-linear calibrations, such as those obtained from the flame photometric detector.

The results of a separation can be stored in a data bank as part of a laboratory information management system (LIMS). This enables results from different assays to be brought together, for quality control charts to be prepared and for long- or short-term trends in results or calibration samples to be identified.

15.4 AUTOMATIC CONTROL AND INJECTION METHODS

A second major area of chromatography that is being influenced by the introduction of the microcomputer is in the control and operation of chromatographic instrumentation. With the methods presently available, any process or operation which can be monitored and adjusted electrically can be controlled by a computer.

In many GLC instruments, the oven temperature, temperature programs, carrier gas flow rates, sample injection and data collection can all be under microprocessor control. As yet most parts of a liquid chromatograph are still mechanically operated, except for flow programmers, diode array detectors and automated injectors. However, almost every repetitive operation can be carried out with greater precision under automatic rather than manual control. This aspect is particularly important for quality control assays, when virtually identical samples have to be examined over a long period of time. Automatic injection systems are available for both GLC and HPLC. These generally mimic the manual operation by taking samples from vials held in a rack and injecting them with either a syringe or loop injector onto the column. The sequence is usually controlled by the integrator and the cycle can also include programming and resetting steps. If column switching is required, then this can be incorporated into the overall automated system.

By removing the idiosyncratic actions of the individual operator, the automated instrument should work more consistently and reliably. However, checks must be built into any automated system and a standard sample included at frequent intervals. The program should be able to recognise when the results from these samples are falling outside preset limts because of instrumental drift, operational errors, or mechanical faults. Otherwise large quantities of flawed data can be collected and valuable samples may be lost. Particularly stable instrument operation is required and thermostating is needed for flow controllers in GLC and for the solvent and column in HPLC.

15.5 NEW APPLICATIONS OF COMPUTERS IN THE CHROMATOGRAPHY LABORATORY

With a general increase in the speed and memory capacity of microcomputers, coupled with lower relative costs, there are three areas where their impact is still to be fully applied in the laboratory, in addition to their general extended incorporation as controllers and operators of instruments.

a. *Data banks*

In the future it will become more common for chromatographic and other equipment to be linked together through laboratory information systems. This will enable the shared use of facilities such as memory data stores and printer/plotters, will bring together results from different instruments, and will enable the long-term storage and retrieval of results. Unlike the older networks of the 1970s, each part of the system will be able to act and operate individually but will also be able to communicate with other parts. Of particular interest to qualitative chromatography will be that coupled MS and FTIR instruments will have access to larger data banks of retention and spectral information for identification or confirmation purposes.

Almost all new instruments will be provided with standard interfaces for the transmission of control signals and for data transfer, and groups of manufacturers (and eventually hopefully all manufacturers) will establish common information protocols.

b. *Expert systems*

The increase in computing power and memory capacity will lead to the use of expert systems as sources of advice and information. At present, each chromatographer learns much of the details of instrument operation and the selection of separation conditions by experience. Each sample can pose a new problem, which has to be approached from a basis of background knowledge, methods previously reported in the literature and advice from colleagues. This is a slow process and the operator may well miss the optimum conditions.

The concept of an expert system computer program is that it gathers knowledge from a number of sources or "experts" and formulates it in a rational way, which can then be used as a source of advice and knowledge. The advantage is that, as the system builds up, it can bring together ideas from different "experts". It will be able to weigh these ideas and then to offer judgements, which would not normally be possible if only one individual and nearby colleagues could be consulted. The system will not only be able to offer advice but also to demonstrate how it reached a particular conclusion. As yet, systems of this type are still in the development stages, but within the next few years more sophisticated shells or program frameworks will become available.

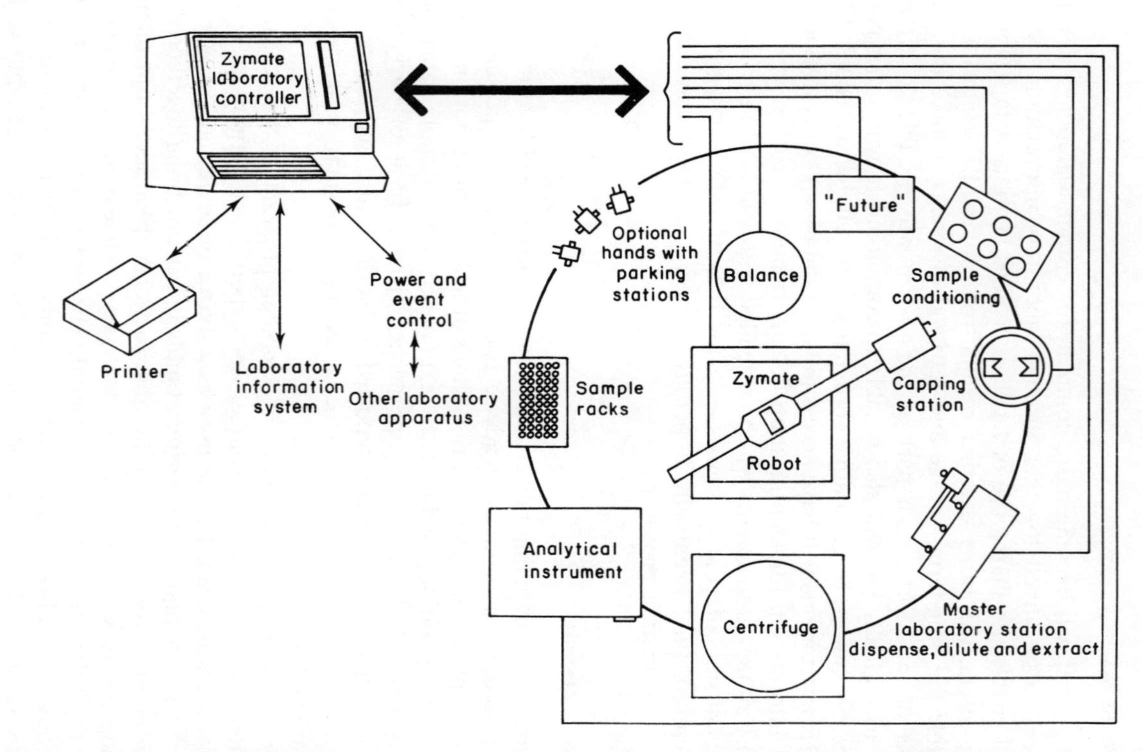

Figure 15.9 **Robot** arm system for sample preparation with different workstations. (Reproduced by permission of Zymark Corporation.)

c. *Robotics and sample preparation*

The future also holds the prospect for much greater automation of sample preparation, including derivatisation and extraction procedures. In particular, greater use will be made of robots as versatile manipulators, which can be rapidly and simply re-programmed to carry out a range of alternative pre-injection techniques, ending with the preparation of the sample rack for autosamplers or by directly presenting the finished sample to an automated injector (Figure 15.9). They can already carry out extractions and derivatisations, prepare solutions, use solid-phase columns, evaporate and centrifuge. Simpler, more dedicated, systems for derivatisation and solution chemistry are also being developed and both will be capable of being integrated directly with the computers controlling the chromatographs and with data systems.

BIBLIOGRAPHY

Data handling and interpretation

S.T. Balke, *Quantitative column liquid chromatography. A study of chemometric methods*, J. Chromatogr. Library, Vol. 29, Elsevier, Amsterdam, 1984.
R.E. Kaiser and A.J. Rackstraw, *Computer chromatography, Vol. 1*, Alfred Hüthig, Heidelberg, 1983.
C.E. Reese, "Chromatographic data acquisition and processing. Part I. Data acquisition", *J. Chromatogr. Sci.*, 1980, **18**, 201–206; "Part II. Data manipulation", *ibid.*, 249–257.

Statistical analysis

Most of these texts are general books on statistics in analytical chemistry but contain relevant sections on the preparation of calibration curves and the measurement of reproducibility.

R. Caulcutt and R. Boddy, *Statistics for analytical chemists*, Chapman and Hall, London, 1983.
K. Kafadar and K.R. Eberhardt, "Some basic statistical methods for chromatographic data", *Adv. Chromatogr.*, 1984, **24**, 1–34.
D.L. Massart and L. Kaufman, *The interpretation of analytical chemical data by the use of cluster analysis*, John Wiley, New York, 1983.
D.L. Massart, B.G.M. Vandeginste, S.N. Deming, Y. Michotte and L. Kaufman, *Chemometrics. A textbook*, Elsevier, Amsterdam, 1987.
J.C. Miller and J.N. Miller, *Statistics for analytical chemistry*, 2nd ed. Ellis Horwood, Chichester, 1988.
M.A. Sharaf, D.L. Illman and B.R. Kowalski, *Chemometrics*, John Wiley, Chichester, 1987.

Computers in the laboratory

In this rapidly growing field, textbooks are rapidly outdated, particularly those describing specific systems. *Analytical Chemistry* runs a useful updating series "A/C Interface" in the A pages edited by R.E. Dessy, and *Journal of Chemical Education* has a series on basic laboratory experiments using computers and interfacing.

P. Kucera and P. Riggio, *Applications of computers in chromatography*, J. Chromatogr. Library, Vol., Elsevier, Amsterdam, in press.

Expert systems

R.E. Dessy (Ed.), "Expert systems. Part I", *Anal. Chem.*, 1984, **56**, 1200A–1212A; "Part II", *ibid.*, 1312A–1332A.

Robotics

R. Dessy (Ed.), "Robots in the laboratory. Part I", *Anal. Chem.*, 1983, **55**, 1100A–1114A; "Part II", *ibid.*, 1232A–1242A.

CHAPTER 16

FUTURE DEVELOPMENTS IN CHROMATOGRAPHY

16.1 ADVANCES IN EXISTING CHROMATOGRAPHIC METHODS

The immediate future development of chromatography will depend on ideas that can now be recognised in the experimental and development stages and on trends elsewhere in analytical chemistry which have yet to have an impact on separation methods.

a. *Gas–liquid chromatography*

A greater understanding of the techniques for the preparation of bonded-phase open-tubular columns should lead to an improvement in the reproducibility of retention times and hopefully a smaller number of similar but slightly different stationary phases. Both steps will enable results obtained in different laboratories to be compared more easily and should lead to the widespread establishment and use of data banks of retention indices for the identification of unknown analytes. New stationary phases are likely to be developed for chiral separations with more general selectivity and resolving power.

New microprocessor-based mass spectrometric detectors are already on the market using concepts such as Fourier transform MS and ion-trap spectroscopy. These should lead to simpler and hence cheaper dedicated GC–MS systems with comprehensive data banks of spectra for rapid sample identification.

b. *High-performance liquid chromatography*

New concepts for HPLC detectors are frequently being devised but few appear to offer any advantage over existing detectors. The main thrust is likely to be in perfecting interfaces for LC–MS and overcoming some of the problems with the present systems. The electrochemical detector is likely to become more widely applied, because of its high sensitivity, as long as the problems of stability and reproducibility can be overcome.

The major changes are likely to be in the area of column materials as manufacturers try to produce more homogeneous and more reproducible products derived from synthetic polymers. Work will also concentrate on surface-treated silicas to overcome the inherent disadvantages of the present silica-based column materials. However, the large accumulation of methods based on current ODS-silicas means that any new stationary phase will have to offer a significant advantage to overcome the costs associated with re-evaluating and revalidating a revised method with a new column material.

Completely new column materials are also being investigated and a number of groups have been investigating porous graphitic carbon stationary-phase materials, which appears to offer some interesting mechanical and chromatographic properties (Figure 16.1). As in GLC, more selective and specific chiral column materials will be developed and additional work with chiral mobile phases using small-bore columns is likely. This may be aided by further work with open-tubular columns for liquid chromatography, although in order to obtain reasonable efficiencies the internal diameter of the column needs to be less than 20 μm. This causes inherent problems of very limited sample capacity and detection sensitivity, and thus may be restricted to a few readily detectable compounds that are electrochemically active or contain fluorophores.

Probably the major development will be an explosive increase in the use of instrumental HPLC methods for the separation of biopolymers, including peptides, enzymes and nucleotides. These will make use of reversed-phase and in particular hydrophobic interaction separations on wide-pore columns, with their promise of the retention of biological activity.

There is a continuing interest in true liquid liquid separation systems and there has been considerable recent interest in counter-current chromatography in which the separation is carried out using two immiscible liquid phases (see Bibliography) and commercial instruments are now available. The main advantage of this and droplet counter-current chromatography appears to be high sample capacity and mild separation conditions, free from interactions with a stationary phase. However, as yet relatively poor efficiencies and long separation times have inhibited their widespread use.

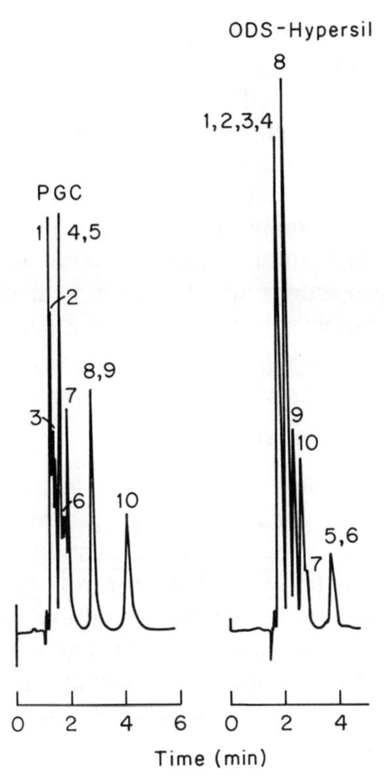

Figure 16.1 Separation of monofunctional derivatives of benzene on porous graphitic carbon (PGC) stationary phase compared to ODS-Hypersil. Both columns 100 mm × 5 mm. Eluents: for PGC column, methanol–water 90:10, and for ODS-Hypersil column, methanol–water 70:30, at 1 ml min^{-1}. Ultraviolet detection at 254 nm. Peaks: 1, benzene; 2, aniline; 3, phenol; 4, benzyl alcohol; 5, toluene; 6, chlorobenzene; 7, anisole; 8, acetophenone; 9, nitrobenzene; 10, methyl benzoate. (Reproduced by permission of Elsevier Science Publishers from *J. Chromatogr.*, 1986, **352**, 20.)

Preparative and process-scale liquid chromatography is likely to become a routine chemical engineering method capable of separations on the kilogram to tonne scale for both fine chemicals and biological products. These separations will include chiral methods and the purification of synthetic intermediates.

The related analytical technique of flow injection analysis often uses equipment common to HPLC. In this case the sample is injected into a flowing carrier stream containing a reagent. After passing through a mixing coil the analyte is detected by using either a spectrophotometric or electrochemical detector. The principal difference from HPLC is that there is no column or separation step and the selectivity is achieved solely by the chemistry of the reaction or the specificity of the detector. The sample throughput can be very high with up to 600 samples an hour. In the chromatography laboratory this technique could potentially be used for pre-column derivatisation with automatic injection of the product from the flowing stream onto the column. This would avoid the sample handling and transferring steps, which often cause problems and are time-consuming. Some of the selective reactions and detection methods used in flow injection analysis may also be applicable to post-column derivatisation techniques in HPLC.

c. *Thin-layer chromatography*

In the last few years greater reproducibility in the production of the thin layers of stationary phase in HPTLC has lead to an increased interest in TLC methods. These advances and the availability of better densitometers have raised the acceptability of TLC for routine quantitative analyses, particularly for limit tests and screening of multiple samples. The automation of sample application to give more uniform sample sizes and even spots has also had an impact. The commercial availability of new elution methods, particularly overpressure TLC, offer higher retention reproducibility and more versatility.

15.2 NEW SEPARATION METHODS

a. *Supercritical fluid chromatography*

Although it was first demonstrated in 1962 that a supercritical fluid could be used as the mobile phase, this mode of separation has only recently begun to attract widespread interest in the analytical laboratory. This has been prompted by the availability in the last few years of commercial instruments for supercritical fluid chromatography (SFC) and demonstrations of its practical application.

A supercritical fluid is produced when a compound is above its critical pressure and temperature (Figure 16.2). When a gas is raised above its critical temperature and then the pressure is increased, it will not undergo a phase change and condense to form a liquid. Instead, it will steadily become more dense and above the critical pressure will form a single supercritical fluid phase. The properties of a supercritical fluid lie intermediate

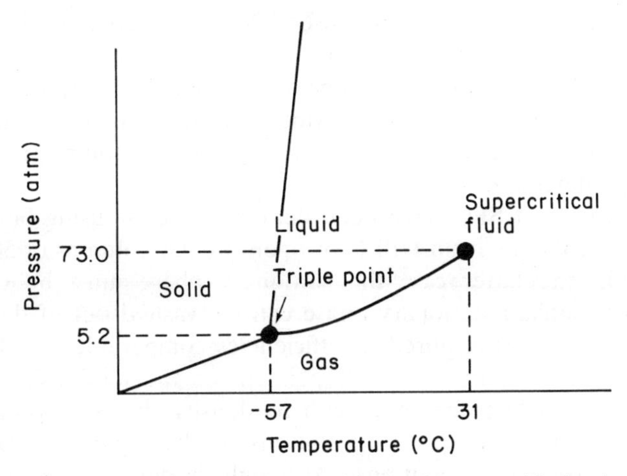

Figure 16.2 Phase diagram for carbon dioxide showing supercritical region above critical point at 31°C and 73.0 atm. (Reproduced by permission of Suprex Corp.)

Table 16.1 Critical pressures and temperatures of compounds used as mobile phases in supercritical fluid chromatography.

Compound	Critical temperature (°C)	Critical pressure (atm)
Xenon	16.5	57.6
Carbon dioxide	31.1	72.9
Nitrous oxide	36.4	71.5
Sulphur hexafluoride	45.5	37.1
Ammonia	132.4	111.3
Pentane	196.5	33.3
Hexane	234.5	29.3

between those of a liquid and a gas and vary considerably with the applied pressure. Generally they have viscosities and diffusion coefficients lower than liquids but densities and solubilising properties similar to liquids. The higher the density the more liquid-like are the properties and at lower densities the more gas-like. These properties are potentially very useful for a chromatographic mobile phase as mass transfer problems should be considerably reduced compared to conventional liquids. A limited range of compounds can be easily converted into supercritical fluids at reasonable temperatures and pressures and most have been examined for their chromatographic properties (Table 16.1). However, many of these compounds are unsuitable for routine use. The lower alkanes, hexane and pentane, would offer a considerable fire hazard and ammonia is particularly corrosive to column materials. Almost all analytical applications have therefore concentrated on carbon dioxide because of its low critical values and low toxicity, with some interest in the expensive xenon particularly with coupled FTIR detection.

The applications of SFC have been demonstrated by using both packed reversed-phase columns from HPLC or open-tubular columns of 50–100 μm from GLC. In the latter case the stationary phase must be chemically bonded, as crosslinked stationary phase can be washed out of the column, and the narrow bore is required, if efficiencies comparable to GLC are to be achieved. The retention is very dependent on the solubilising power of the mobile phase, which increases with its density. Pressure programming can therefore be used to speed up elutions or the separation can be run under constant-pressure conditions. Although carbon dioxide is a strong eluent, it is very non-polar. The separation can be modified and the peak shapes of polar analytes improved by adding small amounts of methanol or glycol ethers to the carbon dioxide, but their effect appears to be mainly on the stationary phase rather than increasing the polarity of the mobile phase. Not unexpectedly there have been practical problems in designing pumps for SFC. Syringe pumps can be used for open-tubular columns or modified HPLC pumps for packed columns. In most cases the pump head needs to be cooled so that the carbon dioxide is being pumped into the column as a liquid. The columns are usually held in modified GLC ovens to ensure constant temperature control. Temperature programming is not used, as increasing the temperature slows down the elution.

One of the attractions of SFC is that a wide range of detectors can be used including the universal FID, which is fitted to most commercial systems, and the selective TID and flame photometric detectors, all of which do not respond to the carbon dioxide mobile phase. Some studies have also used ultraviolet and fluorescence spectrophotometric detectors with high-pressure flow cells and these are particularly suitable for the analysis of aromatic hydrocarbons. One area of great interest is the coupling of SFC to mass spectrometry and this has been the subject of considerable research.

The separation of complex mixtures of hydrocarbons with efficiencies equivalent to capillary GLC has been obtained with carbon dioxide as the eluent (see Figure 2.14), but the technique is not restricted to non-polar compounds and it has also been applied to drugs, flavour compounds and high-molecular-weight natural products (Figure 16.3). One of the advantages of SFC is that the temperatures required are much lower than for GLC and thermal decomposition is minimised.

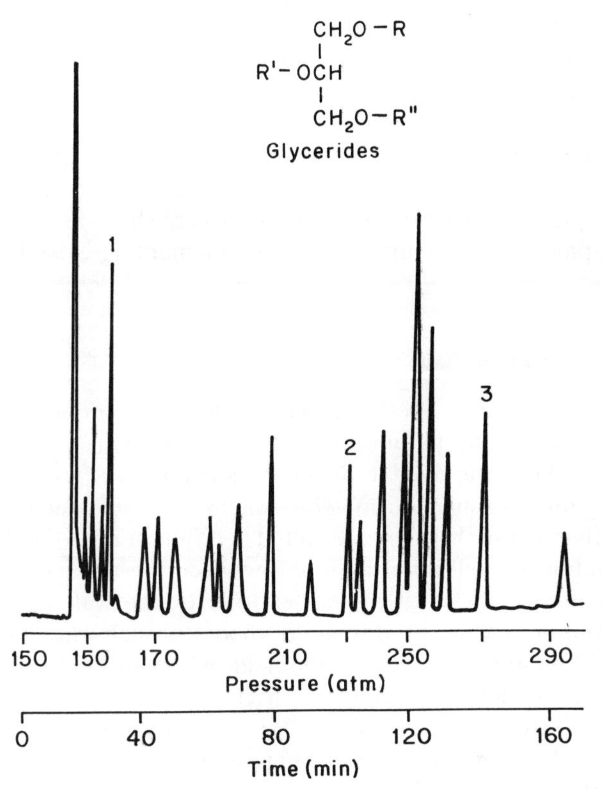

Figure 16.3 Separation of mixture of mono-, di- and triglycerides using SFC. Column DB-5, 19.5 m × 200 μm. Film thickness 0.25 μm. Mobile phase carbon dioxide at 90°C programmed from 150 to 300 atm at 1 atm min^{-1} after 20 min isobaric period. Flame ionisation detector. Peaks: 1, monoglyceride (R = CO $(CH_2)_{12}CH_3$, R′,R″ = H); 2, diglyceride (R,R″ = $CO(CH_2)_7CH{:}CH(CH_2)_7CH_3$, R′ = H); 3, triglyceride (R,R′,R″ = $CO(CH_2)_9CH{:}CH(CH_2)_7CH_3$). (Reproduced by permission of Dr Alfred Hüthig Verlag from *JHRC, CC*, **9**, 12 (1986).)

There is still considerable debate about the future role of SFC compared to HPLC and GLC. It appears to offer advantages for higher-molecular-weight or thermally labile compounds. Unlike HPLC, it can use a universal detector and can readily be coupled to mass spectrometry. It is claimed that the separation conditions can be rapidly and simply altered with a minimum of equilibration required. Because the selectivity differs from HPLC or GLC, it may be able to provide an alternative technique for the confirmation of the purity of drug samples and the demonstration of peak homogeneity. It holds potential for preparative separations as it will be very easy to remove the mobile phase to leave the pure sample. Supercritical fluids are already used in many areas of the food industry as powerful extraction solvents and there is interest in the direct linking of extraction (SFE) and separation (SFC) systems for the determination of flavour components, pesticide residues and similar areas.

It seems likely that the next few years will see a rapid expansion of work in this field and, if it can be shown to have the ability to analyse otherwise difficult samples, it will become firmly established as a third major chromatographic technique linking and complementing existing HPLC and GLC methods.

b. *Field flow fractionation*

A family of separation methods related to chromatography has been developed based on the parabolic flow rate of a liquid moving through a narrow tube. A field is applied to the flow system, such as gravity, centrifugal force, temperature, or magnetism, which aligns different analytes at different positions in the tube so that they are eluted at different rates. With centrifugal force an equilibrium will be established between the applied field trying to force the analyte particles or molecules towards the wall of the tube and Brownian motion trying to distribute them through the liquid. Larger molecules will be more affected by the field, which will tend to concentrate them in the more slowly moving stream close to the walls of the column. They will therefore travel more slowly through the column and a separation can be achieved on the basis of size (Figure 16.4). These methods are only just becoming commercially available and appear to have particular application for the separation of fine powders such a paint pigments and

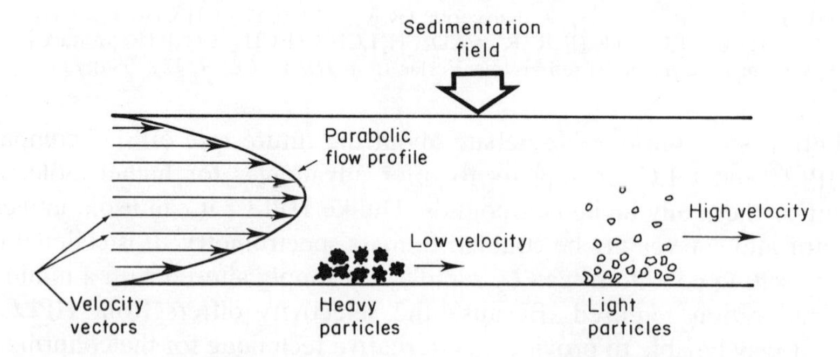

Figure 16.4 Principle of sedimentation field flow fractionation. Heavier particles are concentrated in slower-moving region near wall of tube by centrifugal sedimentation field and elute more slowly than lighter particles. (Reproduced by permission of Labmate Ltd.)

latex particles. As the method involves a liquid mobile phase and no stationary phase, these methods may have particular advantages for preparative separations and are viewed with interest in biotechnology for the separation of macromolecules such as proteins and peptides.

c. *Affinity chromatography*

In recent years affinity chromatography has been developed as a very specific method for the separation of biologically interesting molecules. The stationary phase is coated with a grouping that specifically interacts with a particular analyte of interest. When a mixture is run through the column, only this analyte compound is retained and the remainder of the sample is washed to waste. By altering the eluent, the analyte can then be stripped off the column free from contamination. Although it can be used as a clean-up method before an analytical separation, the technique is usually so specific that no further analysis is required. Most of the recent work has gone into methods for the preparation of more specific column materials for a wider range of analytes. Almost all the applications have been in the areas of biochemistry and biotechnology.

BIBLIOGRAPHY

High-performance liquid chromatography

J.H. Knox, B. Kaur and G.R. Millward, "Structure and performance of porous graphitic carbon in liquid chromatography", *J. Chromatogr.*, 1986, **352**, 3–25.
K.K. Unger, "Porous carbon packings for liquid chromatography", *Anal. Chem.*, 1983, **55**, 361A–375A.

Flow injection analysis

D. Betteridge, "Flow injection analysis", *Anal. Chem.*, 1978, **50**, 832A–846A.
J. Růžička, "Flow injection analysis. From test tube to integrated microconduits", *Anal. Chem.*, 1983, **55**, 1040A–1053A.
K.K. Stewart, "Flow injection analysis. New tools for old assays. New approach to analytical measurements", *Anal. Chem.*, 1983, **55**, 931A–940A.
M. Valcárcel and M.D. Luque de Castro, *Flow-injection analysis. Principles and applications*, Ellis Horwood, Chichester, 1987.

Countercurrent chromatography

K. Hostettmann, "Droplet counter-current chromatography", *Adv. Chromatogr.*, 1983, **21**, 165–186.

K. Hostettmann, M. Hostettmann-Kaldas, and K. Nakanishi, "Droplet countercurrent chromatography for the preparative isolation of various glycosides", *J. Chromatogr.*, 1979, **170**, 355–361.

Y. Ito and W.D. Conway, "Development of countercurrent chromatography", *Anal. Chem.*, 1984, **56**, 534A–554A.

N.B. Mandava and Y. Ito, *Countercurrent Chromatography Theory and Practice*, Chromatographic Science Series, Vol. 44, Marcel Dekker, New York, 1988.

Y. Ogihara, O. Inoue, H. Otsuka, K. Kawai, T. Tanimura and S. Shibata, "Droplet counter-current chromatography for the separation of plant products", *J. Chromatogr.*, 1976, **128**, 218–223.

Supercritical fluid chromatography

J.C. Fjeldsted and M.L. Lee, "Capillary supercritical fluid chromatography", *Anal. Chem.*, 1984, **56**, 619A–628A.

D.R. Gere, "Supercritical fluid chromatography", *Science*, 1983, **222**, 253–259.

M. Novotný, "Biochemical and environmental applications of microcolumn and supercritical fluid chromatography", *J. Pharm. Biomed. Anal.*, 1984, **2**, 207–221.

M. Novotný, "New detector strategies through supercritical fluid chromatography", *J. High Res. Chromatogr. Chromatogr., Commun.*, 1986, **9**, 137–144.

M. Novotný, S.R. Springston, P.A. Peaden, J.C. Fjeldsted, and M.L. Lee, "Capillary supercritical fluid chromatography", *Anal. Chem.*, 1981, **53**, 407A–414A.

P.J. Schoenmakers and F.C.C.J.G. Verhoeven, "Supercritical-fluid chromatography — prospects and problems", *Trends Anal. Chem.*, 1987, **6**, 10–17.

R.M. Smith (Ed.), *Supercritical fluid chromatography*, RSC Chromatography Monographs, Vol. 1, Royal Society of Chemistry, London, 1988.

T.G. Squires and M.E. Paulaitis (Eds.), *Supercritical fluids: chemical engineering principles and applications*, ACS Symposium Series, Vol. 329, American Chemical Society, Washington, DC, 1987.

U. van Wasen, I. Swaid and G.M. Schneider, "Physicochemical principles and applications of supercritical fluid chromatography (SFC)", *Angew. Chem., Int. Ed.*, 1980, **19**, 575–587.

C.M. White and R.K. Houck, "Supercritical-fluid chromatography and some of its applications: a review", *J. High Res. Chromatogr., Chromatogr. Commun.*, 1986, **9**, 4–17.

Affinity chromatography

P. Mohr and K. Pomerening, *Affinity chromatography. Practical and theoretical aspects*, Chromatographic Science Series, Vol. 33, Marcel Dekker, New York, 1985.

H. Schott, *Affinity chromatography*, Chromatographic Science Series, Vol. 27, Marcel Dekker, New York, 1984.

W.H. Scouten, *Affinity chromatography*, John Wiley, New York, 1981.

J. Turková, *Affinity chromatography*, J. Chromatogr. Library, Vol. 12, Elsevier, Amsterdam, 1978.

R.R. Walters, "Affinity chromatography", *Anal. Chem.*, 1985, **57**, 1099A–1114A.

Field flow fractionation

J. Janča, *Field-flow fractionation. Analysis of Macromolecules and particles*, Chromatographic Science Series, Vol. 39, Marcel Dekker, New York, 1987.

J. Janča, K. Klepárník, V. Jahnová, and J. Chmelík, "Progress in field-flow fractionation: theory, methodology, and applications", *J. Liquid Chromatogr.*, 1984, **7** Suppl. 1, 1–39.

W.W. Yau and J.J. Kirkland, "Nonequilibrium effects in sedimentation field flow fractionation", *Anal. Chem.*, 1984, **56**, 1461–1466.

APPENDIX 1

SOURCES OF CHROMATOGRAPHIC METHODS

A1.1 LITERATURE OF CHROMATOGRAPHY

The chromatography literature can be divided into three main areas. Primary research publications are reports of original research work, published mainly in learned journals. In the secondary review literature the authors, who are usually themselves experts in the field, bring together and survey research work from a number of original papers on a limited topic. Finally there are books and monographs that have greater depth and a wider coverage but are inevitably less up-to-date.

The most important source of information about previous work is the original literature. However, it can often be very difficult or time-consuming to find the information required as a guide to a particular assay or method. Each paper will primarily describe a single application or technique and they may often not compare different methods or suggest the most suitable conditions or technique for a particular assay. Any search for information directly through the primary literature can be daunting as there are now over 150 000 papers on chromatography.

Abstracting and indexing journals such as *Chemical Abstracts* can often be a useful starting point, particularly if a method is required for a specific compound, such as a drug or pesticide. The search can be aided by using one of the many on-line computer search services, which can look for either the formula, the chemical name, or the *Chemical Abstracts* Registry number of the compound of interest. Restricting the study to a particular technique may also reduce the number of responses to a manageable level, i.e.

hexachlorobenzene and GLC. General topics such as “drug analysis” or “pesticide residues” are too broad and will produce many thousands of responses.

In order to keep track of current developments in chromatography in the research literature, the chromatographer can be aided by specialised abstracting services such as *CA Selects: Gas chromatography*, or *CA Selects: High performance liquid chromatography* (both prepared by *Chemical Abstracts* Service), or *Chromatography Abstracts* (formerly *Gas and Liquid Chromatography Abstracts*, which are prepared by The Chromatographic Society). More general analytical abstracting services such as the full issues of *Chemical Abstracts* or *Analytical Abstracts* can also be used. In addition the *Journal of Chromatography* contains frequent bibliographies of papers and index words.

a. Primary literature

Papers on chromatographic methods and their applications appear in many journals in areas ranging from analytical chemistry and physical chemistry to pharmaceutical chemistry and agriculture.

The primary specialised chromatography journals include:

Journal of Chromatography including its *Biochemical Applications* and *Chromatographic Review* sections, which are numbered in the main sequence. This journal is probably the single most important primary chromatography publication and publishes virtually weekly issues.
Chromatographia
Journal of Analytical and Applied Pyrolysis
Journal High Resolution Chromatography and Chromatographic Communications.
Journal of Liquid Chromatography
Journal of Planar Chromatography

General analytical chemistry journals of importance, which should be scanned by most chromatographers include:

Analytical Letters
Analytica Chimica Acta
Analytical Chemistry (often includes review articles in the A pages)
Analytical Proceedings
Analyst
Talanta
Zeitschrift für Analytische Chemie

Each application area for chromatography will have more specialised journals, which may often carry papers with particularly strong coverage of methods and applications. Some examples include:

Biomedical Chromatography
Journal of Pharmaceutical and Biomedical Analysis
Journal of Agriculture and Food Chemistry
Journal of Forensic Chemistry

b. Review journals in chromatography

The secondary literature of chromatography is very strong and comprehensive specialised reviews of 30–60 pages on all aspects of chromatography can be found in:

Advances in Chromatography
High Performance Liquid Chromatography — Advances and Perspectives
Progress in Chromatography
Trends in Analytical Chemistry

Trade journals have useful notes on methods and on new equipment and can often be obtained on free subscription. Most of those listed also include regular review articles and current updates as well as contributions with practical advice on instrumentation and operating techniques. Typical examples include:

Chromatography Forum
European Chromatography News
International Laboratory (and *American Laboratory*)
LC.GC Magazine. Magazine of gas and liquid chromatography

Instrument and equipment manufacturers often distribute house journals or newsletters, which include useful examples of applications as well as details of new equipment:

Chrompack News
Chromatogram (Beckman)
Chromatography Newsletter (Perkin Elmer)
Chromatography Review (Spectra Physics)
Peak (Hewlett Packard)
Retention Times (Tracor)

The Supelco Reporter
Topics (SGE)
Waters Column

and many others. The catalogues of equipment manufacturers and chromatographic supply houses are an invaluable source of information on column materials, stationary phases, solvent properties and methods, and a wide collection can be a useful reference library for any chromatographer.

c. *Books on chromatography*

For general information or for in-depth coverage of a particular area, the many books and monographs on chromatography are probably the first resource that should be examined. Because these usually take some time to write and publish, they will not reflect the most recent advances. However, the applications that are described will usually have stood the test of time, unlike research papers that often omit the problems or limitations of a method as they report only the experience of the one laboratory. Many books on the methodology and techniques of chromatography have been included in the bibliographies at the ends of the individual chapters. The listings in the following sections will therefore concentrate on the more accessible books and compilations, which are primarily aimed at the applications and uses of gas or liquid chromatography. Although extensive, the list is not comprehensive and most books published before 1980 and most non-English language publications have been omitted. Particularly in HPLC, older books will often describe column materials and methods that are now outdated.

A1.2 GENERAL BOOKS ON CHROMATOGRAPHY APPLICATIONS

a. *General applications*

G. Zweig and J. Sherma (Eds.), *CRC handbook of chromatography*, CRC Press, Boca Raton, FL. (A number of volumes on specific topics.)

b. *Pharmaceutical, clinical and biomedical applications*

A.P. De Leenheer, W.E. Lambert and M.G.M. De Ruyter (Eds.), *Modern chromatographic analysis of the vitamins*, Chromatographic Science Series, Vol. 30, Marcel Dekker, New York, 1985.

A. Kuksis (Ed.), *Chromatography of lipids in biomedical research and clinical diagnosis*, J. Chromatogr. Library, Vol. 37, Elsevier, Amsterdam, 1987.

T. Mills, W.N. Price, P.T. Price and J.C. Roberson, *Instrumental data for drug analysis, Vols. 1 and 2*, Elsevier, New York, 1982; *Vol. 2*, 1985.

J.W. Munson (Ed.), *Pharmaceutical analysis. Modern methods. Parts A and B*, Marcel Dekker, New York, 1982, 1984.

E. Reid and I.D. Wilson (Eds.), *Drug determination in therapeutic and forensic contexts*, Plenum, New York, 1984.

K. Tsuji and W. Morozowich (Eds.), *GLC and HPLC determination of therapeutic agents*, Chromatographic Science Series, Vol. 9, Marcel Dekker, New York, 1978.

R. Verpoorte and A. Baerheim Svendsen, *Chromatography of alkaloids*. Part B: *gas–liquid chromatography and high performance liquid chromatography,* J. Chromatogr. Library, Vol. 23B, Elsevier, Amsterdam, 1984.

G.H. Wagman and M.J. Weinstein, *Chromatography of antibiotics,* J. Chromatogr. Library, Vol. 26, Elsevier, Amsterdam, 1984.

D.M. Wieland, M.C. Tobes and M.J. Mangner (Eds.), *Analytical and chromatographic techniques in radiopharmaceutical chemistry*, Springer Verlag, New York, 1985.

c. Food, agricultural and environmental applications

G. Charalambous, *Analysis of food and beverages*, Academic Press, New York, 1984.

J. Gilbert (Ed.), *Analysis of food contaminants*, Elsevier, London, 1984.

R.L. Grob (Ed.), *Chromatographic analysis of the environment*, Marcel Dekker, New York, 1982.

R.L. Grob and M.A. Kaiser, *Environmental problem solving using gas and liquid chromatography*, J. Chromatogr. Library, Vol. 21, Elsevier, Amsterdam, 1982.

d. Miscellaneous applications

K.H. Altgelt and T.H. Gouw (Eds.), *Chromatography in petroleum analysis*, Chromatographic Science Series, Vol. 11, Marcel Dekker, New York, 1979.

J.C. Macdonald (Ed.), *Inorganic chromatographic analysis*, John Wiley, New York, 1985.

G. Schwedt, *Chromatographic methods in inorganic analysis,* Alfred Hüthig, Heidelberg, 1981.

A1.3 APPLICATIONS OF GAS–LIQUID CHROMATOGRAPHY

a. General applications

W.G. Jennings (Ed.), *Applications of glass capillary gas chromatography*, Chromatographic Science Series, Vol. 15, Marcel Dekker, New York, 1981.

R.J. Laub and R.L. Pecsok, *Physicochemical applications of gas chromatography*, John Wiley, New York, 1978.

b. Pharmaceutical, clinical and biomedical applications

H. Jaeger (Ed.), *Glass capillary chromatography in clinical medicine and pharmacology*, Marcel Dekker, New York, 1985.

H. Jaeger (Ed.), *Capillary gas chromatography–mass spectrometry in medicine and pharmacology*, Alfred Hüthig, Heidelberg, 1987.

K. Pfleger, H. Maurer and A. Weber, *Mass spectral and GC data of drugs, poisons and their metabolites*. Part I. *Introduction tables, GC data*, VCH Verlag, Weinheim, 1985.

c. Food, agricultural and environmental applications

W. Bertsch, *Analysis of volatiles in the environment*, Alfred Hüthig, Heidelberg, 1986.

V. Formáček and K.-H. Kubeczka, *Essential oils analysis by capillary gas chromatography and carbon-13 NMR spectroscopy*, Wiley-Heyden, Chichester, 1982.

W. Jennings and T. Shibamato, *Qualitative analysis of flavor and fragrance volatiles by glass capillary gas chromatography*, Academic Press, New York, 1980.

G. Odham, L. Larsson and P.-A. Mårdh (Eds.), *Gas chromatography/mass spectrometry applications in microbiology*, Plenum, New York, 1984.

F.I. Onuska and F.W. Karasek, *Open tubular column gas chromatography in environmental sciences*, Plenum, New York, 1984.

P. Sandra and C. Bicchi (Eds.), *Capillary gas chromatography in essential oil analysis*, Alfred Hüthig, Heidelberg, 1987.

P. Schreier, *Chromatographic studies of biogenesis of plant volatiles*, Alfred Hüthig, Heidelberg, 1984.

d. *Miscellaneous applications*

V.G. Berezkin, V.R. Alishoyev and I.B. Nemirovskaya, *Gas chromatography of polymers,* J. Chromatogr. Library, Vol.10, Elsevier, Amsterdam, 1977 (reprinted 1983).

J.R. Conder and C.L. Young, *Physicochemical measurement by gas chromatography*, John Wiley, New York, 1979.

C.J. Cowper and A.J. DeRose, *The analysis of gases by chromatography*, Pergamon Series in Analytical Chemistry, Vol. 7, Pergamon, Oxford, 1983.

T.R. Crompton, *Gas chromatography of organometallic compounds*, Plenum, New York, 1982.

S.A. Leibman and E.J. Levy, *Pyrolysis and GC in polymer analysis*, Chromatographic Science Series, Vol. 29, Marcel Dekker, New York, 1985.

T. Paryjczak, *Gas chromatography in adsorption and catalysis,* Ellis Horwood, Chichester, 1987

A1.4. APPLICATIONS OF HIGH-PERFORMANCE LIQUID CHROMATOGRAPHY

a. *General applications*

H. Colin, A.M. Krstulovic, J. Escoffier and G. Guiochon, *A guide to the HPLC literature*. Vol. 1, (*1966–1979*), Vol 2, (*1980–1981*), Vol. 3, *1982*. John Wiley, New York, 1984–85.

A.M. Krstulovic and P.R. Brown, *Reversed phase high-performance liquid chromatography: theory, practice and biochemical applications*, Wiley Interscience, New York, 1982.

A. Pryde and M.T. Gilbert, *Applications of high performance liquid chromatography,* Chapman and Hall, London, 1979

b. *Pharmaceutical, clinical and biomedical applications*

P.R. Brown (Ed.), *HPLC in nucleic acid research*, Chromatographic Science Series, Vol. 28, Marcel Dekker, New York, 1984.

W.W. Christie, *High-performance liquid chromatography and lipids,* Pergamon, Oxford, 1987.

D.B. Drucker, *Microbiological applications of high-performance liquid chromatography,* Cambridge University Press, Cambridge, 1987.

L. Fishbein, *Chromatography of environmental hazards,* Vol. 4, *Drugs of abuse*, Elsevier, Amsterdam, 1982.

W.S. Hancock and J.T. Sparrow, *HPLC analysis of biological compounds. A laboratory guide,* Chromatographic Science Series, Vol. 26, Marcel Dekker, New York, 1984.

G.L. Hawk, *Biological/biomedical applications of liquid chromatography.* Parts I–IV, Chromatographic Science Series, Vols. 10, 12, 18 and 20, Marcel Dekker, New York, 1979, 1982.

A. Henschen, K.-P. Hupe, F. Lottspeich and W. Voelter (Eds.), *High performance liquid chromatography in biochemistry*, VCH Verlag, Weinheim, 1985

P.M. Kabra and L.J. Marton, *Clinical liquid chromatography.* Vols. 1 and 2, CRC Press, Boca Raton, FL, 1984.

M.P. Kautsky (Ed.), *Steroid analysis by HPLC. Recent applications*, Chromatographic Science Series, Vol. 16, Marcel Dekker, New York, 1981.

A.M. Krstulovic (Ed.), *Quantitative analysis of catecholamines and related compounds*, John Wiley, Chichester, 1986.

I.S. Lurie and J.D. Wittwer (Eds.), *High-performance liquid chromatography in forensic chemistry*, Chromatographic Science Series, Vol. 24, Marcel Dekker, New York, 1983.

E.F. Rossomando, *High performance liquid chromatography in enzymatic analysis*, John Wiley, New York, 1987.

I.W. Wainer (Ed.), *Liquid chromatography in pharmaceutical development: an introduction*, Aster, Springfield, 1985.

S.H.Y. Wong (Ed.), *Therapeutic drug monitoring and toxicology by liquid chromatography*, Chromatographic Science Series, Vol. 32, Marcel Dekker, New York, 1985.

c. Food, agricultural and environmental applications

G. Charalambous (Ed.), *Liquid chromatographic analysis of food and beverages*, Academic Press, New York, 1979.

J.F. Lawrence (Ed.), *Liquid chromatography in environmental analysis*, Humana Press, New York, 1984.

R. Macrae (Ed.), *HPLC in food analysis*, Academic Press, London, 1982.

d. Applications to the separation of polymers

B.G. Belinkii and L.Z. Vilenchik, *Modern liquid chromatography of macromolecules,* J. Chromatogr. Library, Vol. 25, Elsevier, Amsterdam, 1983.

G. Glöckner, *Polymer characterisation by liquid chromatography*, J. Chromatogr. Library, Vol. 34, Elsevier, Amsterdam, 1986.

J. Janča (Ed.), *Steric exclusion liquid chromatography of polymers*, Chromatographic Science Series, Vol. 25, Marcel Dekker, New York, 1984.

O. Mikeš (Ed.), *High performance liquid chromatography of biopolymers and their fragments*, J. Chromatogr. Library, Elsevier, Amsterdam, in preparation.

A1.5 APPLICATIONS OF THIN-LAYER CHROMATOGRAPHY

A. Baerheim-Svendsen and R. Verpoorte, *Chromatography of alkaloids. Part A. Thin layer chromatography*, J. Chromatogr. Library, Vol. 23A, Elsevier, Amsterdam, 1983.

G. Musumarra, G. Scarlata, G. Romano, S. Clementi and S. Wold, "Application of principal components analysis to the TLC data for 596 basic and neutral drugs in four eluent systems", *J. Chromatogr. Sci.*, 1984, **22**, 538–547.

A.H. Stead, R. Gill, T. Wright, J.P. Gibbs and A.C. Moffat, "Standardised thin-layer chromatographic systems for the identification of drugs and poisons. A review", *Analyst*, 1982, **107**, 1106–1168 (reprinted 1983 as a monograph, Royal Society of Chemistry, London).

J.C. Touchstone (Ed.), *Advances in thin-layer chromatography: clinical and environmental applications*, John Wiley, New York, 1982.

L.R. Treiber (Ed.), *Quantitative thin-layer chromatography and its industrial applications*, Chromatographic Science Series, Vol. 36, Marcel Dekker, New York, 1986.

H. Wagner, S. Bladt and E.M. Zgainski, *Plant drug analysis. A thin layer chromatography atlas*, Springer Verlag, Berlin, 1984.

APPENDIX 2

PRACTICAL PROBLEMS IN CHROMATOGRAPHY

A2.1 PROBLEMS IN CHROMATOGRAPHY

Unless the results of a chromatographic separation are reliable, they are valueless to the analyst. It is therefore important to be able to recognise and rectify any problems in the operation of the equipment or in the separation and detection of the analytes.

There are two main areas that cause problems. Firstly, there are problems encountered during the development of a method, and secondly, problems that appear during operation, after the method has been established and verified. Although the former are important, they are primarily in the field of method development, the selection of column materials and sample preparation, and have been discussed earlier. It is the latter operational problems that will primarily be discussed in this Appendix.

The operator needs to be able to verify the operation of the equipment, the separation and identification of the analytes and any quantitative conclusions. There are four main aspects of this operation, problem prevention, problem recognition, problem location and problem rectification. These aspects are particularly important for any laboratory carrying out analyses as part of a study for submission to governmental regulatory bodies. These will require that the validity and integrity of the data can be demonstrable to comply with good laboratory practice (GLP) regulations.

a. *Problem prevention*

An important step in problem prevention is that the operator must have an understanding of the aims and operation of the analysis and is not just blindly following instructions. This would include a knowledge of the analytical method and results and of the instrumental limitations of the equipment. The operator should be aware of the need for good working practices, including cleanliness, an organised routine, adherence to method protocols, sample clean-up and the use of the correct materials such as clean gases in GLC or the correct grade of solvents for HPLC.

This approach should be coupled with routine preventative maintenance of the chromatograph either within the laboratory or by the manufacturer's engineers, which can play an important role in ensuring the trouble-free operation of systems. Records should be kept of both routine servicing and repairs. After any servicing or changes to the system, the operation of the equipment should be reverified, using test samples, and the results should be retained as part of the laboratory record.

b. *Problem recognition*

Before a problem can be rectified, it must be recognised. Problems usually manifest themselves by their effect on the analytical results, appearing as irreproducible values, abnormal responses, or poor peak shapes. For this reason the analyst should regularly examine the raw chromatographic data and not just rely on the tabulated output of an integrator or data system.

It is therefore essential that every analysis in routine use has a criterion of reproducibility. This is usually based on a test sample mixture, whose retentions and peak areas should match standard values or fall between predefined limits. The test mixture for a particular assay will usually contain both the analyte and an internal standard, so that both absolute and relative effects can be identified. General test mixtures may also be used in both GLC and HPLC as an overall guide to column efficiency and retention.

Most abnormal responses and poor peak shapes are sorted out during method development, but if they appear during routine operation, they usually indicate a major failure in the system. Changes in the column are the most probable cause (see later), whereas problems with detectors usually cause total loss of the signal. Problems with pumps or gas flows will cause gradual or erratic changes in retention times.

c. *Problem location and rectification*

Usually the identification of the cause of the problem is the major step towards its solution. The principal problem areas are sample preparation, injection technique, the equipment, data interpretation and operator errors. Many of these problems are general to both HPLC and GLC, although equipment problems may require a different approach in the two techniques.

i. Sample preparation

Once a method has been established, most of the problems that can occur with the sample are operator errors. These include the use of the wrong sample, incorrect work-up of the sample, the use of the wrong solvent to prepare the sample, or the omission of an internal standard. Derivatisation reactions may fail because of changes in the reaction conditions or reagent. This can be detected by using an internal standard of similar structure to the analyte, which should also be derivatised. Changes in the sample matrix may alter the separation or work-up and may require revalidation of the method.

ii. Injection technique

In GLC, manual and to a lesser extent automatic injections are a common source of problems. The speed of needle insertion and rate of depression of the plunger add to irreproducibility. These variations are particularly serious if the sample is viscous, as different amounts of sample will be left in the syringe. If the sample solvent is too volatile, it may boil off as the needle is inserted through the septum, causing the sample to enter the cooler zone just below the septum, delaying transfer to the column. Injections onto capillary columns pose particular problems, which were discussed earlier (section 3.2).

Syringes are a major problem area in GLC (and to a lesser extent in HPLC) because such small volumes of liquid are being injected. It can be difficult to ensure that the nominal volume does not include any small air bubbles, particularly with the smaller 1–5 μl syringes, which hold all the sample in the needle. For accurate quantitative work, it is therefore essential to use an internal standard in the sample and to base any results on the ratio of the peak areas of the analyte and standard. Any significant deviations from the expected absolute peak areas of the standard should also be noted as these may indicate incomplete injection or loss of sample.

Syringes should have the correct needle length so that the sample is delivered into the centre of the heated injection zone. Damaged syringes with bent needles or cracked barrels should be discarded or repaired. The former will damage the septa and the latter may leak and deliver unreliable volumes. The insertion of the syringe through the septum can be a source

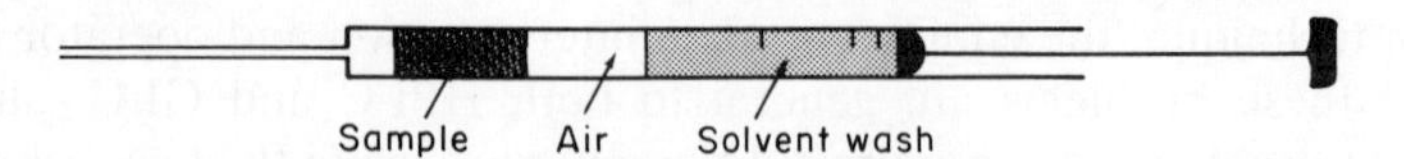

Figure A.1 Syringe loaded to use solvent flush injection method in GLC.

of poor reproducibility and a consistent smooth routine technique should be cultivated. The needle should be fully inserted into the septum, the sample plunger positively depressed and the syringe withdrawn in a regular smooth operation. The precision of GLC injections can often be improved by using a solvent wash in which a small volume of pure solvent is first drawn into the syringe followed by the sample solution and then an air gap (Figure A.1). The air gap ensures that as the needle is pushed through the septum there will be no boil-off of solvent (and sample) from the tip of the needle. Particular care must be taken to clean syringes between samples by washing with a suitable solvent. Carry-over in the syringe can result in residual traces of concentrated samples being injected into the chromatograph, causing ghosting in subsequent traces run at a higher sensitivity.

If the retention times in GLC are inconsistent and sample peak sizes vary erratically, a common cause is that the septum is leaking either continuously or during injection. A simple test is to wipe spit across the top of the injection port; if it bubbles then gas is escaping and the septum should be replaced. BE CAREFUL: the top of the injector is often very hot. Septa have a fairly limited lifetime (typically 20–100 injections) and in some cases routine replacement after a specified number of samples or on changing the column may be desirable. Lifetimes with automatic injection are usually better as the needle pierces the septum in the same place on each injection. Before removing a septum from the injection port, make sure that the carrier gas flow through the column has ceased or pressure remaining in the column may expel the packing material out of the column.

In HPLC the use of a valve injector removes most of the potential problem areas. However, care must be taken with a fixed loop injector that an ample excess of sample is used to wash all the eluent completely out of the loop, typically 5–10 times the loop volume. Alternatively, if a seven-port valve is being used, no more than 50% of the loop volume should be injected so that there is no danger of sample escaping from the loop. After the sample has been loaded into an HPLC valve in the *LOAD* position, the syringe needle should be left in the inlet port of the valve as it is rotated to the *INJECT* position to inject the sample. Premature removal may draw back some of the sample from the loop or allow the sample to run out of the loop. Domed tipped syringe needles should be used and these should have the correct external diameter to fit snugly into the seal on the injection port. In order to prevent sample carry-over, the valve ports, particularly the injection port, should be cleaned at intervals by flushing with solvent or mobile phase.

In both GLC and HPLC a constant injection volume of solutions of different concentrations should be used (rather than injections of different volumes) for preparing calibration curves, as the precision of injection is better than the absolute accuracy of the syringe.

A2.2 EQUIPMENT PROBLEMS IN GAS–LIQUID CHROMATOGRAPHY

GLC instruments have been evolving over the last 30 years and are now generally very reliable, but problems can still occur in routine operation with the carrier gas, column oven, columns and detectors.

a. Carrier gas

Bottled carrier gases are rarely a problem as long as a grade intended for GLC is used, which should be free of oxygen and water vapour. Oxygen-free nitrogen (OFN) is readily available. If higher purity is required for sensitive and high-temperature work, it can be passed through charcoal and molecular sieve absorption tubes. Particular care must be taken if an electron capture detector (ECD) is being used and, as well as using absorption tubes, metal tubing must be used for all the carrier gas connections. The widely used nylon tubing is porous and air can permeate through the walls and contaminate the carrier gas. If the laboratory has a fixed gas line system, this may need to be cleaned before an electron capture detector can be used, to remove any soldering flux or solvent vapours. If hydrogen is used as the carrier gas, for safety there should be an automatic leak tester in the oven.

Most carrier gas problems appear as incorrect retention times caused because the flow has been turned off or down at the cylinder or regulator, or the wrong flow settings have been used. The usual test is to check the flow rate at the exit of the column with a bubble flow meter using a stopwatch. The values should match those specified in the method protocol. If there is no hydrogen or air flow the carrier can also be measured at the jet of a flame ionisation detector. When the column is changed, the connections should be tested for leaks by painting the joints either with a dilute soap solution or for open-tubular columns, with methanol.

b. Column oven

The oven is probably the most robust part of the GLC. Most problems result from operator errors in the setting of the temperature or the programme. If a temperature programme is being repeated, sufficient time must be allowed at the end of the cycle for the oven to re-establish a stable initial temperature. The oven may need to be cooled for some time at a lower temperature than the initial temperature, particularly if the start of the programme is below 50°C, otherwise residual heat in the oven walls and detector block may raise the temperature above the nominal initial setting. New microprocessor-controlled ovens often have comprehensive feedback systems and will indicate when the oven temperature is ready for injection.

Total failure of the oven is usually caused by a fuse failure either because of electrical overload or because the temperature of the thermal cut-out fuse has been exceeded. This may indicate that the column has been overheated and the column may have been damaged. If an ECD with a tritium source (now fairly rare) is being used, the thermal limit must be set to 200°C.

c. Columns

Unless the maximum recommended temperature (MRT) has been grossly exceeded, columns only age fairly slowly so that any dramatic change in the separation usually has an alternative cause. However, if a column is operated for some time near its MRT or if oxygen contaminates the carrier gas, the liquid phase may be degraded, which can result in changes in selectivity and retention times or in peak tailing. The degraded column cannot be repaired and should be repacked with fresh column material.

During method development, tailing peaks may indicate that the solvent is unsuitable, usually because there is too large a polarity difference between the liquid phase and the analyte (such as alcohols on a non-polar dimethylsilicone). Open-tubular columns are usually more versatile and suffer less problems. A leading peak (Figure A.2) indicates that the column is overloaded and the injection should be repeated with a smaller sample or using a larger split ratio. Samples that contain involatile components can leave rubbish at the top of the column. This can build up and eventually interact with the sample, causing tailing or loss of selected components. The top of the column can be repacked or, with open-tubular columns, the first few coils can be broken off. The newer bonded stationary phases for open-tubular columns can be washed by passing a solvent through the column.

The performance of columns and detectors should be checked at regular intervals using a simple test mixture, and the retentions and efficiency

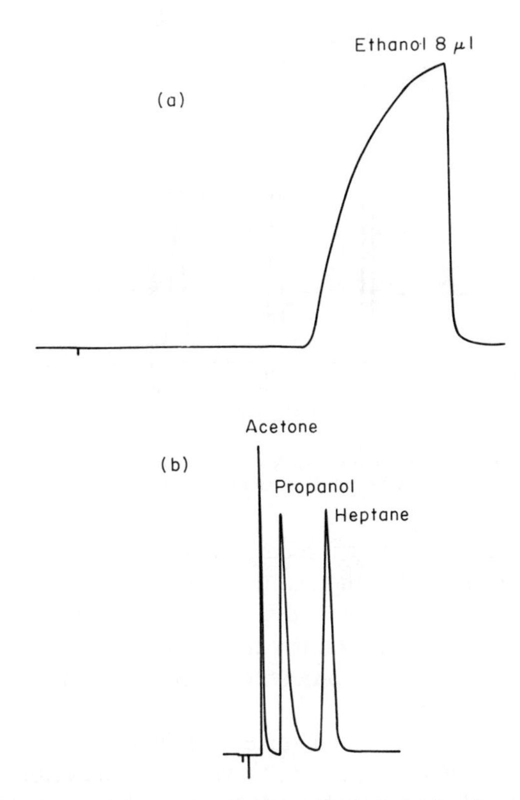

Figure A.2 Examples of distorted peak shapes in GLC. (a) Leading peak caused by column overload (10% polypropyleneglycol; 40°C; 50 × 10^4). (b) Peak tailing of propanol compared to heptane caused by interaction with the support material or stationary phase because of polarity difference between the analyte and liquid phase (10% Apiezon L; 45°C).

determined and recorded (Figure A.3). The more complex Grob test mixture is mainly designed to measure stationary-phase characteristics (see Figure 4.12). General column care is based on monitoring the usage of the column, including any conditioning, and ensuring that the columns are permanently labelled with the stationary phase, support material and an identifying number, which can be related to the preparation and test chromatograms.

d. Detectors

Flame ionisation detectors (FID) are very robust and cause few problems in operation. A total lack of response even to the solvent suggests that the

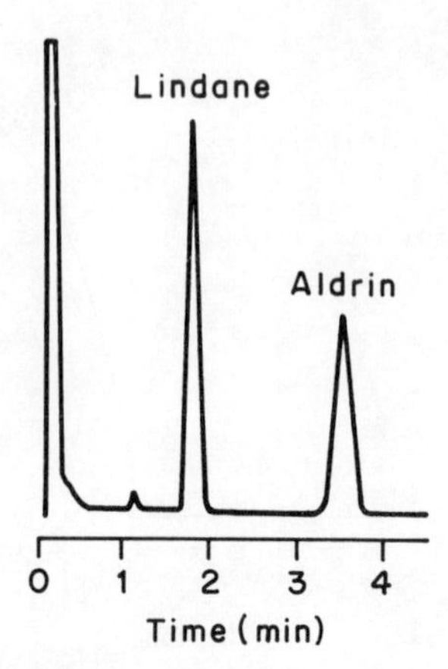

Figure A.3 Simple test mixture for GLC to confirm electron capture detector operation and column efficiency. (Courtesy of Hewlett-Packard Company.)

flame is not lit. This can be checked by holding a piece of cold glass, such as a microscope slide, above the exit from the FID. Normally a condensation film will rapidly form on the glass. If the flame is unstable, giving an unexpectedly noisy baseline, or is extinguished by the solvent peak, then the hydrogen or air supply is probably restricted or incorrectly set and the flow rates should be checked.

A noisy baseline with irregular spikes may indicate that the instrument is being used at the limit of its sensitivity or the detector jet has become contaminated by bleed from the column and should be cleaned (see the instrument manual). As always, comparison of the recorder trace with previous test samples should indicate if the problem is worsening.

Regular spikes on the baseline indicate electrical crosstalk from another instrument or equipment in the laboratory. Switching small transformers on or off or nearby lift motors are often a source. Improvements in instrument design mean that noise from these problems is usually only seen on older instruments or recorders. Changes in the baseline position on changing the attenuation of the amplifier indicate that the electrical zero of the recorder and the detector have not been set correctly, and the user should refer to the instrument instruction manual for the set-up procedure.

Electron capture detectors (ECD) bring their own special problems. Flat-topped peaks and non-linear calibrations can be caused by overloading the detector. For most electron capturing analytes, the ECD linear range will be close to the limit of detection of an FID. Contamination of ECDs is a major problem and special care must be taken to exclude oxygen from the carrier gas (see earlier). The detector may be cleaned by operating at high temperature but should not be dismantled to expose the radioactive source. The manufacturer's instructions should be carefully followed. The detector should be checked at intervals for any leakage of activity according to the relevant regulations for sealed sources. Halogenated solvents, such as chloroform or dichloromethane, must be kept away from the instrument as their diffusion to the detector through the air in the laboratory can reduce the standing current. Chlorinated solvents must never be used in the preparation of the samples. Similarly the ECD cannot be used with columns that contain electron capturing groups, such as trifluoromethylsilicone phases, as any column bleed will contaminate the detector.

Very slow changes in the baseline with any detector may be caused by changes in column bleeding on programming, the elution of a very highly retained sample or the slow decomposition of residues in the injection port.

A2.3 EQUIPMENT PROBLEMS IN HIGH-PERFORMANCE LIQUID CHROMATOGRAPHY

The equipment used in HPLC contains more moving mechanical parts than in GLC and these are subjected to greater stresses. The liquid mobile phase is also more aggressive than the gases used in GLC. HPLC is therefore inherently more likely to suffer degradation and loss of performance, so that the operator must monitor the separations more closely. Operating faults can include problems with the sample, injector valve, pumps and most commonly with the column and eluent. Except for the electrochemical detector, whose electrodes can become contaminated and need polishing, the detectors used in HPLC are generally very stable and free from major operating problems. Normally the only problem with spectrophotometric detectors is total lamp failure, which is first noticed as a very noisy background and no signals, or aging of deuterium lamps, prior to failure, when the background noise level at short wavelengths (<230 nm) is markedly worse than at higher wavelengths.

a. Pumps and flow problems

Mechanical wear on HPLC pumps usually shows up as changes in retention times. The most common faults are leaks from the piston seals of reciprocating pumps and sticking check valves. The piston seals on most pumps can be easily replaced once the problem is recognised. Valve problems are often more difficult to identify. Small amounts of dirt and debris from the pump or samples can prevent the balls in the check valves from seating properly. If the inlet valve is not closing, then the eluent will be pumped back into the reservoir. If a small air bubble is introduced into the inlet tube, it will move back and forth as the eluent is sucked into the pump head and expelled back to the reservoir. The effects can be misleading as, if the column is disconnected, the pump may appear to work, but this is only because the resistance to flow is less through the open exit. If the outlet check valve is not working, the eluent will leak back from the column, little eluent will be taken up from the reservoir and an air bubble in the tubing will remain nearly static or move slowly towards the pump. Air bubbles in the pump head from a degassing eluent may also cause irregular flows, which can show up as "pump noise" on recorder traces from ultraviolet or refractive index detectors.

These valve problems can sometimes be rectified by increasing the pumping rate and running the eluent to waste in order to wash dirt particles or air out of the system. If problems with a check valve persist, then consult the manufacturer's manual. Often the valves can be removed and cleaned.

Degassing may be worsened if the eluent cannot flow freely through the solvent inlet frit in the reservoir because it has become blocked with dust or deposits. This will cause the pump head to be starved of eluent. The filter should be cleaned with acid or replaced. Any more drastic failure of a pump can usually only be repaired by a service engineer.

b. Samples and solvent

The solvent used to prepare the samples in HPLC can have a significant effect on the efficiency of the separation and the peak shape (Chapter 13). Generally the preferred solvent has the same composition as the mobile phase. Any problems should be identified during the validation of the method, and the solvent selected for the sample should form part of the method protocol, as any changes may alter the efficiency. Care should also be taken that neither the analyte nor other components of the sample will precipitate on injection into the mobile phase, as they may block the column or connecting tubing.

c. Injector valve

Poor peak area reproducibility may result from leaks in a rotary injection valve. If eluent is leaking from the inlet port or exit port of the valve, then the surface of the rotor seal has become scored. The eluent is seeping, under the pressure of the column, from one port to another. The seal should be replaced, which takes only a few minutes. Leaks out of the side of the valve usually indicate that the column pressure is too high and the eluent is being forced past the faces of the valve. Valves can normally be tightened to hold a higher pressure but the system should be checked for the cause of the abnormally high pressure, as it may indicate a blockage elsewhere in the system.

d. Column and eluent

This is the main problem area in HPLC and causes changes in both the efficiency and selectivity of the separation. The performance should be tested at regular intervals by using either a specific separation test mixture or an analyte–internal standard test mixture.

The composition of the eluent has a major effect on the retention times of samples on the column. The method should carefully define whether the volume or weight of each component should be used in preparing the eluent (see Table 13.1). Particular attention should be paid to the adjustment of the pH of a buffer. This should be carried out before the organic components of the mobile phase are added, because pH electrodes give erroneous values in non-aqueous solvents. The purity of the components of the eluent is important, because if they give a high background signal this may reduce the overall sensitivity. A high background signal will also increase the susceptibility of the detector to any fluctuations in the eluent flow rate. Particular care is needed with the electrochemical detector, because of its high sensitivity, and high-grade buffer components should be used to reduce the background signal. Special precautions may be needed to remove and exclude oxygen from the eluent and sample if the electrochemical detector is being used in the reductive mode.

The solvent may need to be degassed if programmed elution is being used, otherwise bubbles can form in the detector and produce spikes on the recorder trace (Section 10.4). This problem can sometimes also occur with isocratic elution, but can often be overcome by creating a slight back-pressure in the detector cell. Isocratic eluents can often be recycled back to the eluent reservoir, as any sample components will be greatly diluted and have no effect on the separation. However, care must be taken with mobile phases containing tetrahydrofuran, as it may evaporate and alter the composition. With electrochemical detection, the eluent appears to pick up electrochemically active impurities during its passage through the column and detector, and it should not normally be recycled.

Programmed separations are particularly sensitive to the quality of the eluent because impurities will be trapped and concentrated on the column when the eluent is weak at the start of the run. They will then be eluted as sharp peaks as the eluent strength increases (see Figure 10.3). Methanol and THF cause few problems, but acetonitrile is difficult to purify and should be tested in a blank separation. Although distilled water is normally sufficiently pure for HPLC, it should be stored in glass rather than plastic containers. It can extract sufficient plasticisers (phthalate esters) from the plastic to produce significant peaks. If a buffer has been used in an HPLC separation, care must be taken that this is washed out of the system before a neat organic phase is introduced, or the buffer salts may be precipitated and will be almost impossible to remove from blocked injector valves or pumps.

It is often forgotten that HPLC separations are temperature dependent and many separations are carried out at ambient temperature, but even a 10°C change can cause large changes in retentions and for reproducible results the columns and eluents should be thermostated.

The principal cause of separation degradation is a deterioration of the column. This appears initially as reduced efficiency (Figure A.4) or split peaks (Figure A.5). The peak shape may also degrade and tailing may occur. Sometimes the column pressure will rise markedly, causing leaks. Usually these problems are a consequence of the dissolution of part of the column material, leaving channels in the column or a void at the top of the column, which acts as a large dead volume. In some cases, rubbish from the eluent or sample forms an impervious bed across the top of the column

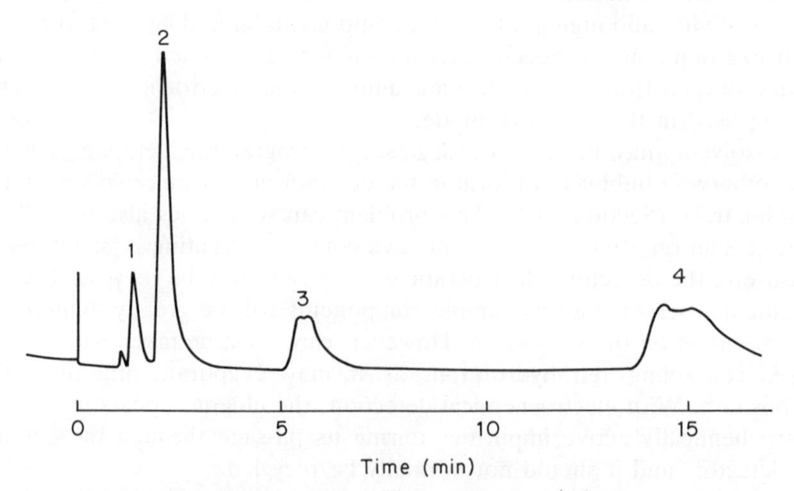

Figure A.4 Chromatogram of HPLC test mixture (see Figure 10.13) on an old column, which is starting to deteriorate. Peaks: 1, benzamide; 2, acetophenone; 3; benzophenone; 4, biphenyl. Peaks are broadened and last peak is splitting. Column ODS-Hypersil, 100 mm × 5 mm. Eluent: methanol/water 60:40 at 1 ml min^{-1}. Ultraviolet detection at 254 nm.

packing material. Often these problems can be remedied by removing the first 3–6 mm of the column packing and repacking the top by hand with a thick slurry of new packing material or small glass beads. In serious cases, the column will have to be replaced. Using a radial or axial compression column can sometimes help as the external force on the column material should take up any losses as they occur. The protocol for a separation should indicate the lowest permitted efficiency before column replacement.

Preventative maintenance and careful sample preparation can often considerably extend the life of the column and its performance. Any suspended material in the sample should be removed by filtering through a millipore filter before injection. For alkyl-bonded silica columns the pH of the eluent should be within the range pH 2.5–8.0 to avoid attack on the silica. A sacrificial pre-column, packed with silica, placed before the injection valve, will saturate the eluent, reducing silica dissolution, and a guard column between the injection valve and the analytical column will trap insoluble materials. These should be checked at regular intervals and replaced before they are exhausted.

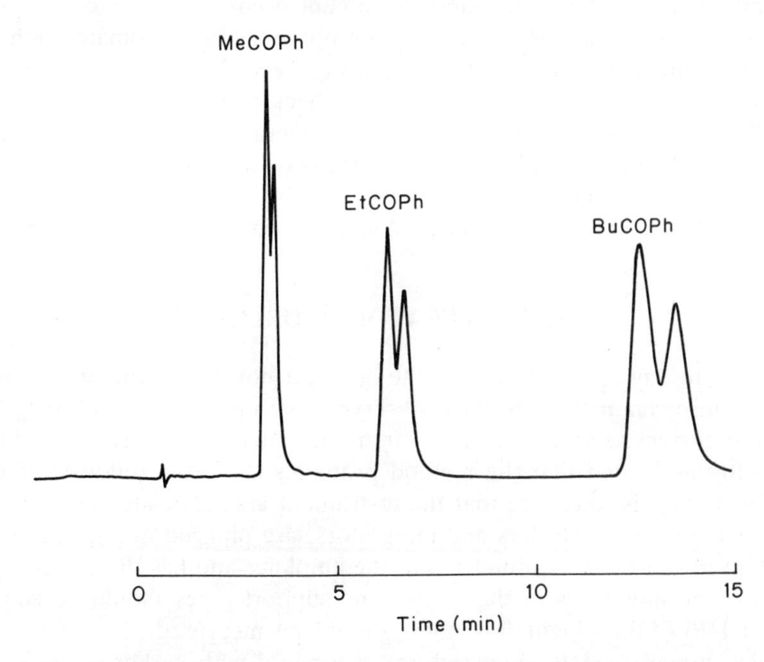

Figure A.5 Peak splitting in HPLC due to channelling in the column. This can be recognised because all the peaks show the same distortion. Column ODS-Techsil, 100 mm × 5 mm. Eluent: methanol–water 60:40. Spectrophotometric detection at 254 nm.

With normal-phase separations it is extremely important to control the moisture level of the eluent or sample solvent, as even a minor change or a small volume of moist solvent can dramatically degrade the column performance. This will also alter the selectivity of the column, particularly if very non-polar eluents are being used.

A2.4 RECORDS AND DATA INTERPRETATION

Both normal laboratory practice and GLP regulations emphasise the need for accurate and comprehensive record keeping. All chromatograms should be labelled as they are recorded with all the details of the separation and the sample. This is also an essential first step in quality control, so that changes in results over a period of time can be recognised. The information should include date, time, operator, sample and code number or logbook reference, injection size and solvent, column (support, liquid phase and percentage), conditions, including any temperature or eluent programme and mobile-phase flow rates, and instrument used.

It is important that the operator does not treat the chromatograph as a black box and simply accepts the output of a recorder or integrator without assessing its validity. The operator should check that the peaks are on-scale, that the resolution is acceptable, that the integrator has correctly identified the start and end of each peak and has not been misled by noise or baseline disturbances, and that any decisions, such as the assignment of the divisions between unresolved peaks or peak skimming, have been correctly made.

A2.5 OPERATOR PROBLEMS

Most day-to-day errors are due to the operator not the instrument. So when a problem or anomalous result is observed, the operator should first check that the correct sample, correct column and eluent, and correct conditions are being used, and that the method protocols are being followed exactly. The next stage is to ensure that the instrument and associated components, such as detectors, recorders and integrators, are plugged in and turned on, and that the recorder is connected to the amplifier and has the correct span. In GLC the flow rates of the carrier and support gases should be checked and in HPLC the eluent flow rates should be measured. The system test samples should then be analysed and compared with earlier records.

If problems are found, first consult the appropriate manuals, which should be stored in a known fixed location in the laboratory. Discuss the results with colleagues or supervisors, who may have seen the problem before.

Operator checklist

1. Is everything turned on and connected?
2. Are settings/conditions as in instructions? Test.
3. Is sample injection correct, septum leaking, or syringe blocked?
4. Try test mixture and reference samples. Do they match file copy?

Much of the successful operation of any instrument depends on the experience of the operator. Inevitably, in these brief discussions of chromatographic problems, some aspects have been omitted and the operator will learn many aspects and problems the hard way. Problems and operating tips are continually being discussed in the literature and at meetings, and useful comments and guides can be found in the books and references given in the bibliography.

BIBLIOGRAPHY

Instrument manuals for individual components — "As a last resort, read the instructions!"

General chromatography problems

Journal of Chromatographic Science runs a regular column of questions and replies on practical aspects of chromatography.

LC.GC Magazine. Magazine of gas and liquid chromatography runs a comprehensive series of articles on trouble shooting in HPLC and GLC covering all aspects, many with particular reference to problems with columns.

J.Q. Walker, M.T. Jackson and J.B. Maynard, *Chromatographic systems. Maintenance and troubleshooting*, Academic Press, New York, 1972.

Gas–liquid chromatography

F.L. Bayer, "An overview of chromatographic instrumentation. Problems and solutions", *J. Chromatogr. Sci.*, 1982, **20**, 393–401

Cleaning flame ionisation detectors — When and how, Bulletin 783C, Supelco, Bellafonte, PA, 1980.

H.M. McNair and E.J. Bonelli, *Basic gas chromatography*, Varian, Palo Alto, CA, 1969, Ch. VIII.

Troubleshooting guide. How to locate gas chromatography problems and solve them yourself, Bulletin 792, Supelco, Bellafonte, PA, 1983.

High performance liquid chromatography

P.A. Bristow, *Liquid chromatography in practice*, HETP, Macclesfield, 1976.

J.W. Dolan and L.R. Snyder, *Troubleshooting HPLC systems*, Video-course, Elsevier, Amsterdam, 1987.

HPLC troubleshooting guide. How to identify, isolate, and correct many HPLC problems, Bulletin 826, Supelco, Bellafonte, PA, 1986.

J. Kirschbaum, S. Perlman and R.B. Poet, "Anomalies in HPLC", *J. Chromatogr. Sci.*, 1983, **20**, 336–340.

J. Kirschbaum, S. Perlman, J. Joseph and J. Adamovics, "Ensuring accuracy of HPLC assays", *J. Chromatogr. Sci.*, 1984, **22**, 27–30.

J. Kirschbaum, S. Perlman, J. Adamovics and J. Joseph, "Anomalies in HPLC, III", *J. Chromatogr. Sci.*, 1985, **23**, 493–498.

F.M. Rabel, "Use and maintenance of microparticle high performance liquid chromatography columns", *J. Chromatogr. Sci.*, 1980, **18**, 394–408.

D.J. Runser, *Maintaining and trouble shooting HPLC systems: a user's guide*, John Wiley, London, 1981.

Standard practice for testing fixed-wavelength photometric detectors used in liquid chromatography, ASTM E 685–79, American Society for Testing and Materials, Philadelphia, 1979.

APPENDIX 3

TERMS AND DEFINITIONS

t_0	Column void volume (IUPAC and ASTM t_M)
t_R	Retention time from injection
t'_R	Adjusted retention time ($t_R - t_0$)
k'	Capacity factor ASTM k; IUPAC D_m)
K	Distribution constant (IUPAC K_D)
K_{IP}	Formation constant of ion-pair complex
h	Height equivalent to a theoretical plate (HETP) (often H)
H	Height equivalent to an effective theoretical plate (HEETP)
h_r	Reduced plate height (h often used)
V_S	Volume of stationary phase in column
V_M	Volume of mobile phase in column
α	Phase ratio V_s/V_M
α	Separation ratio, relative capacity factors or relative adjusted retention times (k'_2/k'_1)
w_h	Width of peak at half-height (2.35 σ)
w_b	Width of peak at base (4 σ)
n	Number of theoretical plates. Column efficiency (often N)
N	Number of efficiency theoretical plates. Effective column efficiency (N_{eff})
b/a	Tailing factor
R_s	Resolution of two peaks
u	Average linear velocity of mobile phase (L/t_0)
d_P	Particle size
d_c	Column internal diameter (open-tubular)
d_f	Film thickness of the stationary phase
L	Column length
A	Peak area

D_M Diffusion rate in the mobile phase
D_S Diffusion rate in the stationary phase

For full list and discussion on anomalies and differences see Chapter 2, references 4–6.

SUBJECT INDEX